MATHEMATICAL AND NUMERICAL MODELING IN POROUS MEDIA: APPLICATIONS IN GEOSCIENCES

Multiphysics Modeling

Series Editors

Jochen Bundschuh
University of Southern Queensland (USQ), Toowoomba, Australia
Royal Institute of Technology (KTH), Stockholm, Sweden

Mario César Suárez Arriaga
Department of Applied Mathematics and Earth Sciences,
School of Physics and Mathematical Sciences, Michoacán University UMSNH,
Morelia, Michoacán, Mexico

ISSN: 1877-0274

Volume 6

Mathematical and Numerical Modeling in Porous Media: Applications in Geosciences

Editors

Martín A. Díaz Viera
Instituto Mexicano del Petróleo (IMP), México

Pratap N. Sahay
Centro de Investigación Científica y de Educación Superior de Ensenada (CICESE), México

Manuel Coronado
Instituto Mexicano del Petróleo (IMP), México

Arturo Ortiz Tapia
Instituto Mexicano del Petróleo (IMP), México

CRC Press
Taylor & Francis Group
Boca Raton London New York

CRC Press is an imprint of the
Taylor & Francis Group, an **informa** business

A BALKEMA BOOK

CRC Press
Taylor & Francis Group
6000 Broken Sound Parkway NW, Suite 300
Boca Raton, FL 33487-2742

First issued in paperback 2018

CRC Press/Balkema is an imprint of the Taylor & Francis Group, an informa business

© 2005 by Taylor and Francis Group, LLC

Typeset by MPS Limited, Chennai, India

No claim to original U.S. Government works

ISBN-13: 978-0-415-66537-7 (hbk)
ISBN-13: 978-1-138-07639-6 (pbk)

Published by: CRC Press/Balkema
                P.O. Box 447, 2300 AK Leiden, The Netherlands
                e-mail: Pub.NL@taylorandfrancis.com
                www.crcpress.com – www.taylorandfrancis.com

Library of Congress Cataloging-in-Publication Data

Mathematical and numerical modeling in porous media : applications in geosciences /
Martín A. Diaz Viera ... [et al.].
    p. cm. – (Multiphysics modeling, ISSN 1877-0274 ; v. 6)
    Includes bibliographical references and index.
    ISBN 978-0-415-66537-7 (hardback) – ISBN 978-0-203-11388-2 (ebook)
        1. Geophysics–Mathematical models.   2. Porous materials–Mathematical models.
I. Diaz Viera, Martín A.
    QC809.M37M38   2012
    550.1′51–dc23

                                                           2012000068

Visit the Taylor & Francis Web site at
http://www.taylorandfrancis.com

and the CRC Press Web site at
http://www.crcpress.com

About the book series

Numerical modeling is the process of obtaining approximate solutions to problems of scientific and/or engineering interest. The book series addresses novel mathematical and numerical techniques with an interdisciplinary emphasis that cuts across all fields of science, engineering and technology. It focuses on breakthrough research in a richly varied range of applications in physical, chemical, biological, geoscientific, medical and other fields in response to the explosively growing interest in numerical modeling in general and its expansion to ever more sophisticated physics. The goal of this series is to bridge the knowledge gap among engineers, scientists, and software developers trained in a variety of disciplines and to improve knowledge transfer among these groups involved in research, development and/or education.

This book series offers a unique collection of worked problems in different fields of engineering and applied mathematics and science, with a welcome emphasis on coupling techniques. The book series satisfies the need for up-to-date information on numerical modeling. Faster computers and newly developed or improved numerical methods such as boundary element and meshless methods or genetic codes have made numerical modeling the most efficient state-of-the-art tool for integrating scientific and technological knowledge in the description of phenomena and processes in engineered and natural systems. In general, these challenging problems are fundamentally coupled processes that involve dynamically evolving fluid flow, mass transport, heat transfer, deformation of solids, and chemical and biological reactions.

This series provides an understanding of complicated coupled phenomena and processes, its forecasting, and approaches in problem solving for a diverse group of applications, including natural resources exploration and exploitation (e.g. water resources and geothermal and petroleum reservoirs), natural disaster risk reduction (earthquakes, volcanic eruptions, tsunamis), evaluation and mitigation of human induced phenomena (climate change), and optimization of engineering systems (e.g. construction design, manufacturing processes).

Jochen Bundschuh
Mario César Suárez Arriaga
(Series Editors)

Editorial board of the book series

Table of Contents

Section 3: Statistical and stochastic characterization

Section 4: Waves

Preface

This book is the result of a selected collection of outstanding contributions presented at two recent international workshops, one on porous media in Geosciences and the other on advances in numerical methods. The first of them was the 8th North American Workshop on Applications of the Physics of Porous Media, 2009 (PORO2009) held in Ensenada, Baja California in Mexico, with participants from diverse countries such as Mexico, Canada, United States, Australia and Spain. The subjects treated covered among other topics: basic theory on porous media processes, poroelasticity, fluid transport, waves, geomechanics, geostatistics and oil recovery. The second workshop was the V International Congress on Numerical Methods 2010, held at Guanajuato Mexico, in which scientists from Mexico, Brazil, Spain, United States and Argentina participated. The topics presented include, supercomputing, parallel computing, optimization and inverse problems, fracture mechanics, no-mesh methods, non-lineal dynamic computing, and diverse applications on oil reservoirs, geohydrology, air quality modeling and road construction.

This volume is divided in four broad sections that corresponds to the state-of-the-art in porous media research presented in those workshops, (i) fundamental concepts, (ii) analytical and numerical flow and transport modeling, (iii) statistical and stochastic characterization methods, and (iv) wave propagation. Section 1 concerns with basic concepts in permeability, up-scaling techniques and thermoporoelasticity. Section 2 deals with analytical and numerical models for flow and transport in porous media, enhanced oil recovery methods (microbial growth and in-situ combustion), tracer tests and reservoir advection-diffusion simulation. Section 3 presents some developments on geostatistical characterization and stochastic simulation of oil fields. Finally, Section 4 is about slow shear waves, porosity waves and saturation waves in porous media.

The book provides to readers a general overview of some of the current main research areas in porous media and numerical methods, but it also offers deeper details to those interested specifically in some of the topics. By this reasoning, the book may prove very useful to a broad audience at student level as well as to experts interested in knowing recent advances on mathematical and numerical modeling in porous media with applications in Geosciences.

We hope that this book become an advent point for encouraging students to get involved into the study of porous media, and also an inspiration for researchers.

The Editors

Acknowledgements

This book is the result of selected contributions presented mostly at two forums: the 8th North American Workshop on Applications of the Physics of Porous Media, 2009 and the Special Session on Mathematical and Numerical Modeling of Oil Recovery Methods in the V International Congress of Numerical Methods, 2010. Many governmental and private institutions have supported the author's participation in the mentioned scientific events, as well as the development of the research works appearing in this book. In particular, we would like to thank the publication funding support of two institutions: Consejo Nacional de Ciencia y Tecnología (CONACYT, México) through the Red de Modelos Matemáticos y Computacionales and Centro de Investigación Científica y de Educación Superior de Ensenada (CICESE, México). We also thank the editorial board members of the book series Multiphysics Modeling for their editorial help.

About the editors

The editors of this book are respected scientists specialized in diverse porous media applications on Geosciences. They have published multiple works on this topic, and some of them have a vast experience organizing technical meetings and workshops. This volume is a collection of various relevant papers presented at some of these recent events.

Martín A. Díaz-Viera, Instituto Mexicano del Petróleo, Mexico. He is an Engineer in Applied Mathematics graduated from *Moscow Power Engineering Institute*, Russia. He received the Red Diploma and a First Grade Diploma for the best thesis of a foreigner. He received his Masters and Doctorate of Science (in the Earth System Modeling option) in the Graduate Program in Earth Sciences at the *Institute of Geophysics of UNAM*, the latter with honors. He has published more than 25 national and international publications and has taught over 20 graduate level courses on Geostatistics, Mathematical Modeling and Numerical Methods. He is currently a researcher at the *Instituto Mexicano del Petróleo* conducting research in the areas of Geostatistics and Stochastic Models for Reservoir Characterization and Mathematical and Computational Modeling of Oil Recovery Processes. He is a member of several scientific societies and associations such as the *Society of Petroleum Engineers*, the *International Association for Mathematical Geosciences* and the *Mexican Society in Numerical Methods in Engineering and Applied Sciences*. E-mail:mdiazv@imp.mx.

Pratap N. Sahay, Centro de Investigación Científica y de Educación Superior de Ensenada, México. He received his Ph.D. in geophysics from the *University of Alberta* in 1986. After a postdoctoral fellowship at the *Theoretical Physics Institute of the University of Alberta*, he joined the *Centro de Investigación Científica y Educación Superior de Ensenada* (CICESE) in 1987 where he is a professor of seismology and leads the Porous Media Laboratory performing theoretical and applied research in petroleum seismology. E-mail: pratap@cicese.mx.

Manuel Coronado, Instituto Mexicano del Petróleo, Mexico. He is a physicist graduated at the *National University of Mexico* (1972–1978). He obtained a M.Sc. in Physics in the same institution (1978–1980). He received his Ph.D. degree from the *Technical University of Munich* and the *Max-Planck Institute for Plasma Physics* in Germany (1980–1984), working on fusion plasma transport in magnetically confined toroidal systems. He worked at the *Instituto de Ciencias Nucleares* of the *National University of Mexico* for ten years on the area of plasma physics and fusion energy (1984–1994). He spent a sabbatical leave in the *University of Wisconsin* at Madison developing models for plasma rotation and transport (1991–1992). Later he joined an emerging Mexican private research center for polymer and paint research, *Centro de Investigación en Polímeros*, (1994–2001) working among other areas in polymer rheology. He is currently a research scientist in the *Instituto Mexicano del Petróleo* since 2001. His present research area is tracer transport applied to dynamic characterization of oil reservoirs. Dr. Coronado is the author of multiple papers on plasma physics and on tracer tests in underground porous formations. He is member of the *Mexican Academy of Sciences*. E-mail: mcoronad@imp.mx.

Arturo Ortiz Tapia, Instituto Mexicano del Petróleo, México. He received his engineering degree at the *Universidad de Celaya* (México, 1989–1994), where he graduated with the title of Agricultural Engineer with specialty in biotechnology, with the thesis *"Development of a more efficient tissue culture for the cloning of Brassica oleracea"*, graduating with the highest honors, in 1994. Then he pursued applied mathematics and theoretical physics studies at *Cambridge University*, United Kingdom, from 1994–1995, and then obtaining an M. Phil. in theoretical chemistry at the Department of Chemistry of the same university (1995–1996) with the thesis *"Chaos in Inert Gas Clusters"*. He then continued to finish his doctorate at the Faculty of Electrical Engineering of the *Czech Technical University* (Czech Republic, 1997–2001), graduating with the dissertation *"Self-Organization in Z-Pinch Plasmas"*. Back in Mexico he was hired by the *Instituto Mexicano del Petróleo*, where he has been doing research in porous media since 2001. His main research interests are the Systematic Mathematical Modeling of Continuous Media, applied numerical methods, and applied theory of numbers. Other mathematical interests include the geometrical properties of helices and spirals, and their relationship with recursivity, and spatial and temporal patterns of differential equations in arbitrary dimensions. E-mail: aortizt@imp.mx.

Contributors

Andrés Eduardo Moctezuma Berthier graduated in 1986, as Petroleum Engineer from the *Universidad Nacional Autónoma de México* (UNAM), with the thesis *"Oil and Condensate Estabilization"*. In 1998 finishes his M. Engineering also at UNAM with the thesis "Determination of Relative Permeabilities in Gas-Condensated Reservoirs". Later he finished his doctorate in Petroleum Engineering at the *University Pierre et Marie Curie*, in Paris, France, with the dissertation *"Desplacements inmiscibles dans de carbonates vacuolaires"*. Since 1989 he has been part of the research staff of the *Instituto Mexicano del Petróleo*, and since 2008 is coordinating its Enhanced Hydrocarbon Recovery program. E-mail: amoctezu@imp.mx.

Angelica G. Vital-Ocampo is a Chemical Engineer graduated from the *Autonomous University of Morelos State*, México. She obtained both her M.S. and Ph.D. degrees in Mechanical Engineering, from the *National Center for Research and Technology Development*, in 2002 and 2011, respectively. She is pursuing a postdoctoral fellowship at *Instituto Mexicano del Petróleo* since March 2011. Her research area is Mathematical and Numerical Modeling of Multiphase Flow in Porous Media. E-mail: avital@imp.mx.

Arturo Erdely is a B.Sc in Actuarial Science, M.Sc and Ph.D in Mathematics (Probability and Statistics) from *Universidad Nacional Autónoma de México*. He is currently full-time professor at *Universidad Nacional Autónoma de México*, and his main research interest is in Copula Theory and its applications to dependence modeling. E-mail: aerdely@apolo.acatlan.unam.mx.

Daniel M. Tartakovsky received his MSc in Fluid Mechanics/Applied Mathematics from *Kazan State University* (Russia) in 1991; and PhD in Hydrology from the *University of Arizona*, Tucson (USA) in 1996. He is currently a Professor of Fluid Mechanics in Department of Mechanical and Aerospace Engineering at *University of California*, San Diego. His areas of research include uncertainty quantification, probabilistic risk assessment, stochastic partial differential equations, hybrid numerical algorithms, subsurface flow and contaminant transport, multiphase flow, well hydraulics, surface water/groundwater interaction, inverse modeling, subsurface imaging, decisions under uncertainty. E-mail: dmt@ucsd.edu.

Gustavo Murillo-Muñetón is a Petroleum Geologist currently working for the *Instituto Mexicano del Petroleo*. He got his bachelor degree from the *Instituto Politecnico Nacional* in Mexico. He holds a Master Degree from the *University of Southern California* in igneous and metamorphic petrology. He obtained his Ph.D. from the *Texas A&M University* in the area of sedimentology and stratigraphy. His research interests are stratigraphy, sedimentology and diagenesis of sedimentary systems mainly carbonate rocks. E-mail: gmurill@imp.mx.

Ilenia Battiato received her first MSc in Environmental Engineering from *Politecnico di Milano*, Italy in 2005; her second MSc in Engineering Physics from *University of California*, San Diego in 2008; and a PhD in Engineering Science with specialization in computational sciences from *University of California*, San Diego in 2010. She is currently Assistant Professor of Thermofluid Sciences in the Department of Mechanical Engineering at *Clemson University*. Her research involves analytical and numerical modeling of transport processes in crowded environments at a variety of scales with applications to environmental flows, nanotechnology, biological systems and granular matter. Current studies include fluidization threshold of wet granulates, hybrid numerical algorithms for flow and transport in porous media, multi-scale models of carbon nanotube forests. E-mail: ibattiat@gmail.com.

Javier Méndez-Venegas is a B.Sc. and M.Sc. in Statistics from *Universidad Autónoma Chapingo* in 2005 and Colegio de Postgraduados in 2008, respectively. Presently he is working on his Ph.D. thesis in Mathematical Modeling of Earth Systems at the *Universidad Nacional Autónoma de México* since 2008. His present research areas are Mathematical Modeling, Geostatistics and Stochastic Modeling. E-mail: lemendez84@yahoo.com.mx.

Jetzabeth Ramírez Sabag earned her Doctor and Master degrees from the *Universidad Nacional Autónoma de México*, UNAM, both in Petroleum Engineering. She has 20 years of experience in modeling tracer flow in porous media, and for more than ten years is devoted to research in the design, operation and interpretation of tracers tests and its industrial applications in Mexican reservoirs. She has taught Advanced Reservoir Engineering as a Professor in the Department of Petroleum Engineering in the UNAM over the past 20 years. She currently works in the research program for Hydrocarbon Recovery of the *Instituto Mexicano del Petróleo*. E-mail: jrsabag@imp.mx.

Luis G. Velasquillo-Martínez received his Ph.D. from the *Institut de Physique du Globe* de Paris, France. For several years his research interests included the neotectonic and seismic hazard of Isthmus of Tehuantepec in Southern of Mexico. He joined the *Instituto Mexicano del Petróleo*, Mexico, in 2000 as a petroleum geophysicst and has worked several research projects in petroleum exploration and reservoir studies. His current areas of interest are the characterization of full field fracture distribution in carbonate rocks and the implications for the static reservoir characterization. E-mail: lgvelas@imp.mx.

Mario César Suárez Arriaga studied Physics in the School of Sciences, *National University of Mexico* (UNAM), Mathematics and Mechanics at the *University of Toulouse* and at the Institute of Theoretical and Applied Mechanics, *University of Paris VI*, France. He received his PhD on geothermal systems from the School of Engineering, UNAM (2000). His main areas of scientific research are the numerical modeling of complex natural systems, continuum mechanics and geothermal energy. He worked 19 years in geothermal reservoir engineering at the *Comisión Federal de Electricidad* in Mexico, and started the first practical evaluations of submarine geothermal Mexican reservoirs to produce electricity. Presently, he works as a research-professor of Applied Mathematics and Mechanics at the School of Sciences, *Michoacan University*, Mexico. E-mail: mcsa50@gmail.com.

Norman Udey received a B.Sc. in Honours Physics from the *University of Alberta*, a Ph.D. in Theoretical Physics, *University of Alberta*, and a Special Certificate in Computing Science (Specialization in Software Design), *University of Alberta*. He is currently working at *Wavefront Reservoir Technologies Ltd.* as a research scientist. His primary areas of interest are the theory and computer simulation of the thermodynamics and dynamical wave processes of porous media. His secondary, supporting interests are software engineering, computer graphics, data modeling and visualization, numerical analysis and algorithms, and reservoir simulation. E-mail: normanu@onthewavefront.com.

Octavio Cazarez-Candia is a Mechanical Engineer graduated from the Technological Institute of Zacatepec, Morelos, México, in 1992. He obtained both his M.S. and Ph.D. degrees in Mechanical Engineering from the *National Center for Research and Technology Development*, in 1995 and 2001, respectively. He works at the *Instituto Mexicano del Petróleo* since 2003. His research area is on mathematical modeling of multiphase flow through porous media, oil wells and pipes. Dr. Cazarez is the author of multiple papers on two and three phase flow in wells and pipes, and on in-situ combustion in porous media. E-mail: ocazarez@imp.mx.

Oscar Valdiviezo-Mijangos is a Physicist who obtained his M. Eng. and PhD in Mathematical Modeling of Earth Systems (Effective Properties of Rocks) at the *Universidad Nacional Autónoma de México*. After a postdoctoral research fellowship in the area of tracer technology at the Instituto Mexicano del Petróleo, he is currently a research scientist there. His research areas are mathematical modeling, transport in porous media, inverse problem in tracer flow, and non linear optimization. E-mail: ovaldivi@imp.mx.

Ricardo Casar González is a Geological Engineer from the *Universidad Nacional Autónoma de México*, where he obtained a Master Degree in Operation Research and a Doctor degree in Exploration Engineering. He also received a Specialization Diploma in Geostatistics at the Center of Geostatistics of the *Paris School of Mines*, France. Currently he works at the *Instituto Mexicano del Petróleo*. His research areas are Geostatistics Applied to Earth Sciences and Stochastic Modeling for Reservoir Characterization. E-mail: rcasar@imp.mx.

T.J.T. Spanos received his B.A. in Mathematics and Physics from the *University of Lethbridge*. He subsequently completed a M.Sc. at the *University of Alberta* on black hole dynamics in General Relativity. His Ph.D. was in Geophysics at the University of Alberta were he began working on geophysical fluid dynamics. This led to an *Alberta Oil Sands Technology and Research Authority* (AOSTRA) post doctoral fellowship in the Theoretical Physics Institute at the *University of Alberta* where he began working on the physics of porous media. Dr Spanos was then awarded an AOSTRA chair in physics in the Geophysics group at the *University of Alberta* and subsequently Professor of Physics. Dr. Spanos is currently Professor Emeritus at the *University of Alberta* and Research Physicist at Wavefront Technology Solutions a company that he co-founded. E-mail: tim@phys.ualberta.ca.

Tobias M. Müller received a diploma (German MSc equivalent) from the *Universität Karlsruhe* (Germany) and a PhD from the *Freie Universität Berlin* (Germany), both in Geophysics. Since 2008 he is a research team leader at the Earth Science & Resource Engineering division of CSIRO (*Commonwealth Scientific and Industrial Research Organization*) in Perth, Australia. His research interests include seismic wave propagation in random media, poroelasticity and the physics of rocks. E-mail: tobias.mueller@csiro.au.

Victor Henández-Maldonado is an Engineer in Chemical Metallurgy from the *Universidad Nacional Autónoma de México*. He obtained his MSc in Industrial Mathematics and Computing from *Instituto Mexicano del Petróleo*. He is currently a PhD student in Mathematical Modeling for reservoir characterization at the *Instituto Mexicano del Petróleo*. His research areas are mathematical modeling, geostatistics and stochastic modeling for reservoir characterization. E-mail: vmhernann@yahoo.com.mx.

Vinicio Suro Pérez is a Geophysical Engineer from the *Universidad Nacional Autónoma de México*. He obtained a PhD in Geostatistics from the *Stanford University* (USA). For many years he has worked for Petroleos Mexicanos (Pemex), the Mexico's National Oil Company, in the Exploration and Production División, helding important positions as Reserve Manager, Sub Director of Planning and Sub Director of the Pemex South Region. Since recently he is director of the Instituto Mexicano del Petróleo. He has a broad experience in Petrophysics, Reservoir Engineering, Reservoir Characterization and Geostatistics. E-mail: vinicio.suro@pemex.com.

Section 1:
Fundamental concepts

CHAPTER 1

Relative permeability

T.J.T. Spanos

1.1 INTRODUCTION

Relative permeability is a concept proposed in the 1930's and 1940's to describe multiphase flow using equations constructed in analogy with Darcy's equation for single phase flow. Although this description was treated with some skepticism at first it eventually became the standard description of multiphase flow in the engineering literature. This occurred because there were no credible alternatives at the time. However it is now possible to rigorously construct the equations of multiphase flow and examine the role of relative permeability in some detail. The concept of relative permeability, or as referred to by Rose (2000) "extensions of Darcy's Law" has been discussed by numerous authors Muscat (1937, 1949), Hubbert (1940 1956), Buckley and Leverett (1942), Rose (1949), Yuster (1951), Slattery (1970), Bear (1972), Scheidegger (1974), Whittaker (1980), de la Cruz and Spanos (1983), Dullien (1992), Bentsen (1994), Spanos (2002). As well a rather outrageous concept has also crept in to the debate, that relative permeability is associated with Onsager's relations. The papers that discuss this have mistakenly associated the thermodynamics of molecular mixtures with porous media. In the case of porous media the phases are mixed at various scales. As a result Onsager's relations place very strong restrictions on the dynamic interactions of the various phases such that the laws of physics are obeyed at all scales. These relations are straightforward to construct and have nothing to do with permeability. In the present paper it is shown that relative permeability is not a parameter at all but rather a function of the dynamic variables and that relative permeability may be determined by both generalizing the equations of motion and introducing an additional degree of freedom into the motions. The associated additional dynamic variable, saturation, introduces an additional equation of motion associated with the large scale pressure difference between the fluid phases. This additional large scale equation of motion allows for a resolution of both the Buckley-Leverett paradox and the closure problem encountered in many engineering analyses using volume averaging to construct large scale flow equations and relative permeability. The construction of Onsager's relation in the case of thermostatic compression is presented in the Appendix as an illustration of the constraints they impose on the dynamical motions of porous media and the associated parameters. In the present discussion the following scales will be considered:

- Microscale (molecular scale)
  Physical interactions described in terms of molecular dynamics. Mixing of the phases is described by molecular diffusion.

- Macroscale (pore scale)
  Physical processes described in terms of continuum equations:
  + Fluid motions – Navier Stokes equation.
  + Solid motions – Elasticity Equation, plus frictional sliding of grains etc.
  + Miscible fluid interactions – Convection diffusion equation, chemical potential, mass fractions of the component fluid phases. The component phases for simple motions are described by linear equations, which interact according to prescribed boundary conditions or chemical mixing.

- Megascale (laboratory scale)
  The component fluid and solid phases cannot be described independently. At this scale we are looking at averaged behaviour of the component phases and their interactions. The phases at this scale are described by a nonlinear classical field theory which accounts for both molecular scale interactions of the phases and the mascoscale interactions.

- Mesoscale (intermediate scale)
  At this scale one is between scales and as a result structure is introduced that cannot be described by a continuum theory. For example acoustic wave propagation would need to be described in terms of scattering theory.

- Gigascale (crustal scale)
  The scale of crustal processes in the Earth.

In the following analysis it is seen that the concept of relative permeability requires some generalization from the form currently in use by engineers. As well the role of megascopic capillary pressure in frontal displacements is described. But first the origins and limitations of Darcy's equation are presented as a starting point. Then the physics of multiphase flow is presented in conjunction with a detailed discussion of relative permeability.

1.2   DARCY'S EQUATION

Steady state single phase flow in porous media was quantified by Darcy's Law in 1856

$$\frac{\mu}{K}\mathbf{q} = -\nabla p + \rho g \hat{\mathbf{z}} \tag{1.1}$$

This relation is based on Darcy's experiment which measures the volume of fluid that flows through a homogeneous porous medium of constant cross sectional area, out an outlet in a fixed period of time and at a constant pressure gradient.

Consider a uniformly packed porous medium of length h and cross sectional area A in a cylindrical jacket. Place a reservoir at each end with an inflow at one end and an out flow at the other. This flow may be established by connecting these reservoirs to two large reservoirs at fixed but different heights. This difference in height is referred to as the head and maintains a constant pressure (the constant force per unit area exerted on the surface of the porous medium, including both solid and fluid, by the fluid in the reservoir bounding the porous medium) at the two ends of the porous medium. One may monitor this pressure at the endpoints or any points in between by attaching tubes to the porous medium and observing the height of the fluid level in the tubes.

In Darcy's experiment it is observed that the volumetric flow rate Q is proportional to the cross sectional area of the porous medium, A, times the change in height in the tubes, $h_2 - h_1$ divided by the length of the porous medium h

$$Q = const \frac{A(h_2 - h_1)}{h} \tag{1.2}$$

By varying the fluid viscosity one observes that this relationship becomes:

$$Q = \frac{K'}{\mu} \frac{A(h_2 - h_1)}{h} \tag{1.3}$$

where μ is the fluid viscosity and the constant K' is a property of the particular porous medium being studied. Here K' is independent of μ, A, h, $(h_2 - h_1)$ and Q. This makes K a physical parameter meaning that like viscosity it may be determined by an infinite number of different

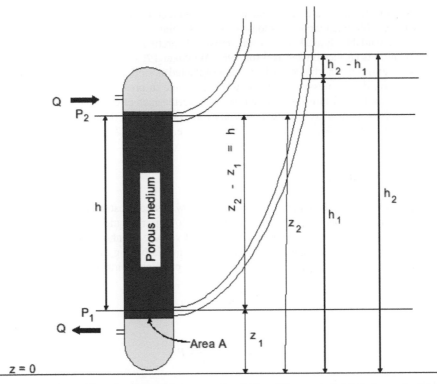

Figure 1.1 This figure illustrates Darcy's experiment in which a difference in height between the inlet and outlet reservoirs, causes a pressure drop, and a volumetric flow rate Q through the porous medium. Here the porous medium is homogeneous with a length h and cross sectional area A. This experiment determines the permeability of the porous medium which is independent of fluid viscosity, the length and cross sectional area of the porous medium and the difference in height between the two reservoirs.

experiments that always yield the same value for K'. On the other hand Q, A, h and $(h_2 - h_1)$ are quantities that may be varied in the experiment and are therefore called variables. Darcy's Law results from this experiment and is generally stated as the result that the flow rate (Darcy velocity) $q = Q/A$ is linearly related to the pressure gradient

$$-\nabla p = \rho \mathbf{g} \frac{(h_2 - h_1)}{h} \tag{1.4}$$

Yielding the relation called Darcy's equation (which may also include body forces such as gravity)

$$\frac{\mu}{K} \mathbf{q} = -\nabla p \tag{1.5}$$

where $K = K'/\rho g$ is call the permeability of the porous medium and K is independent of ρ and g (i.e. K' is linearly proportional to ρg).

In order to obtain a theoretical understanding of this relationship one must start with the well understood macroscopic equations that describe fluid flow at the pore scale and obtain a mega-scopic description at the scale of thousands of pores. Megascopic equations governing fluid flow and deformations of the porous matrix reflect the properties and behaviour of the constituent materials and their interaction. Hence in theoretical attempts to construct these equations it is important to maintain a strong connection with the well established equations governing the behaviour of the macrocroscopically segregated components, which interact at the numerous interfaces in accordance with suitable boundary conditions. From this point of view some averaging scheme is, therefore, unavoidable. Equation (1.5) was derived theoretically, using volume averaging, by a number of authors (Anderson and Jackson 1967, Whitaker 1967, Slattery 1967, Newman 1977). In these derivations the inertial terms have been neglected as well as the dynamic interaction with the solid matrix due to pressure diffusion when a change in pressure across the porous matrix occurs. However these details must be included to describe such effects as the evolution to Darcy flow or fluid flow associated seismic or porosity wave propagation. That discussion is outside of the scope of this chapter but may be found elsewhere (e.g. Spanos 2002).

1.3   HETEROGENEITY

Now that steady state (independent of time) Darcy flow has been established in a homogeneous (uniform) isotropic (independent of the direction of flow) porous medium the next question to ask is what if these restrictions are relaxed. As a first venture to inhomogeneous porous media one may now set up Darcy experiments, with porous media having different permeabilities, in parallel and series. First consider five porous media in parallel as shown in Figure 1.2.

For each of the porous media

$$\frac{\mu}{K_i}\mathbf{q}_i = -\nabla p \tag{1.6}$$

for $i = 1$ to 5. Thus:

$$\mathbf{q}_T = -\frac{K_T}{\mu}\nabla P \tag{1.7}$$

where:

$$\mathbf{q}_T = \sum_{i=1}^{5} \mathbf{q}_i \tag{1.8}$$

and:

$$K_T = \sum_{i=1}^{5} K_i \tag{1.9}$$

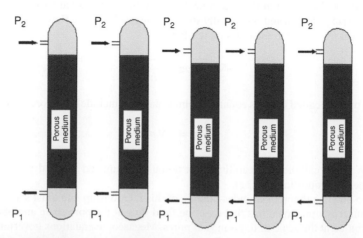

Figure 1.2   This figure shows five porous media hooked up in parallel. Each porous media is connected to the same inlet reservoir and the same outlet reservoir. Thus each reservoir has an identical head and thus the same pressure drop between the inlet and outlet.

Here Equation (1.6) is Darcy's equation. Now consider five porous media in series as shown in Figure 1.3.

For each porous media

$$\frac{\mu}{K_i}\mathbf{q} = -\nabla p_i \tag{1.10}$$

for $i = 1$ to 5. Thus:

$$\frac{\mu}{K_T}\mathbf{q} = \nabla P_T \tag{1.11}$$

where:

$$P_T = P_6 - P_1 \tag{1.12}$$

and:

$$\frac{1}{K_T} = \sum_{i=1}^{5} \frac{1}{K_i} \tag{1.13}$$

So in both of these examples Darcy's equation describes the fluid motions and the permeability of the overall system may be calculated.

Now consider a porous medium constructed by combining different homogenous pieces of various sizes and shapes. We may construct the equations of motion for each homogeneous piece. Then going up in scale we may construct the equations of motion for the composite medium. This demonstrates that the form of the equations of motion depends on the properties and structure of the heterogeneities. Here first restrict this discussion to single phase steady flow. Then the megascopic incompressible flow of a fluid through a homogeneous porous medium is given by

$$\frac{\mu}{K}\mathbf{q} = -\nabla p \tag{1.14}$$

$$\nabla \cdot \mathbf{q} = 0 \tag{1.15}$$

These equations describe flow at a scale large compared to the pore scale and small in comparison to structural inhomogeneities. Now assume that these homogeneous structures occur within volume elements Ω. Now applying averaging to these equations up to a gigascale larger than the homogeneous structures one obtains

$$0 = \frac{1}{\Omega}\int_{\Omega}\left(\frac{\mu}{K}\mathbf{q} + \nabla p\right)d\Omega \tag{1.16}$$

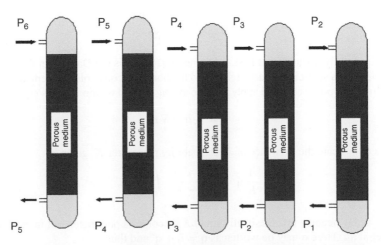

Figure 1.3 This figure shows five porous media hooked up in series. Thus the outlet pressure of the first porous medium is the inlet pressure for the second and so on. So each porous medium causes a pressure drop in succession with the same constant volumetric flow rate in each.

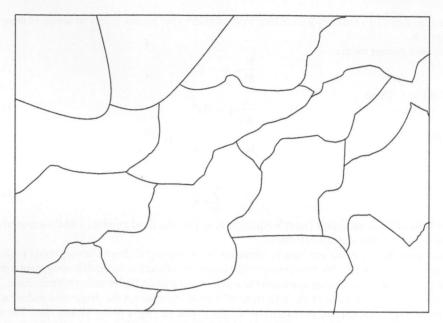

Figure 1.4   From a porous medium composed of homogeneous porous media at a smaller scale the equations of motion depend on the size and structure of the homogeneous porous media. The form of the equations of motion is not Darcy's equation and permeability not a meaningful physical parameter.

and thus:

$$0 = \frac{\mu}{\Omega} \int_{\Omega} \frac{\mathbf{q}}{K} d\Omega + \nabla \frac{1}{\Omega} \int_{\Omega} p \, d\Omega \qquad (1.17)$$

Now assume there are a number of homogeneous pieces $\beta = 1, \ldots, n$ within Ω. Then

$$0 = \frac{\mu}{\Omega} \sum_{\beta} \int_{\Omega_\beta} \frac{\mathbf{q}}{K} d\Omega + \nabla \bar{P}_o \qquad (1.18)$$

where:

$$\bar{P} = \frac{1}{\Omega} \int_{\Omega} p \, d\Omega \qquad (1.19)$$

This result is slightly disturbing at first sight since Eq. (1.18) is not Darcian in form. However in the case where Ω is sufficiently large relative to the homogeneous volume elements Ω_β, then we may start from the pore scale and apply volume averaging and obtain Darcy's equation

$$0 = \frac{\mu}{K} \bar{\mathbf{q}} + \nabla \bar{P} \qquad (1.20)$$

This seems to indicate the existence of a permeability parameter K' given by

$$\frac{1}{K'} \bar{\mathbf{q}} = \frac{1}{\Omega} \sum_{\beta} \frac{1}{K_\beta} \int_{\Omega_\beta} \mathbf{q} \, d\Omega \qquad (1.21)$$

In the limit as Ω_β become infinitesimally small K' becomes a parameter, but in general K' is not a parameter at all. Here $\mathbf{q}$ may be written as $\mathbf{q} = \bar{\mathbf{q}} + \mathbf{q}'$ and thus

$$\frac{1}{K'} \bar{\mathbf{q}} = \frac{1}{\Omega} \sum_{\beta} \frac{1}{K_\beta} \int_{\Omega_\beta} (\bar{\mathbf{q}} + \mathbf{q}') \, d\Omega \qquad (1.22)$$

or:

$$\frac{1}{K'}\bar{\mathbf{q}} = \frac{1}{\Omega}\bar{\mathbf{q}}\sum_\beta \frac{\Omega_\beta}{K_\beta} + \sum_\beta \frac{1}{K_\beta}\frac{1}{\Omega}\int_{\Omega_\beta}\mathbf{q}'d\Omega \tag{1.23}$$

This relation clearly demonstrates that, in general, permeability cannot be defined as a parameter for a megascopically heterogeneous porous medium.

All naturally occurring porous media are heterogeneous at the pore scale. Laboratory samples may be constructed from glass beads of uniform size in a specific packing. This would appear to be the only example of a porous medium homogeneous at this scale. But random heterogeneities at this scale are averaged out when going to the megascale provided no additional structure is introduced between these scales. This is the scale at which permeability is defined. From the previous argument it is seen that if new structure is now introduced then permeability is no longer defined as a parameter at larger scales.

Often one hears engineers and geologist refer to permeability as being scale dependent or a reservoir having two types of permeability (say fracture and medium). From the previous analysis we see that simply means that Darcy's equation is not the equation of motion for flow in such media and permeability is not a physical parameter. Another common error is to describe permeability as being porosity dependent, but one may simply take a hexagonal packing of identical glass beads. One may then choose different sizes of glass beads for each different porous media, and one may vary permeability by many orders of magnitude but keeping the porosity exactly the same. This simple experiment illustrates that these two parameters may be varied independently. It has however been observed that for a given rock type, naturally occurring porous media have a correlation between permeability and porosity. This is the origin of numerous empirical permeability versus porosity relations.

1.4   LUBRICATION THEORY

Lubrication theory gives some additional insight into the equations describing porous media because the equations are of the same form. Following this construction the distinct differences between the two theories will be pointed out. First consider two parallel plates separated by 1 mm or less and assume steady flow between the plates.

The equation of motion for this flow between the plates is given by the Navier-Stokes equation for steady flow

$$\mu\frac{\partial^2 v}{\partial z^2} = \frac{\partial P}{\partial x} \tag{1.24}$$

Here there is no motion parallel to z so $\partial P/\partial z$ is equal to zero. For steady flow $\partial P/\partial x$ is a constant. Integrating Equation (1.24) yields

$$v = A + Bz + \frac{1}{2\mu}\frac{\partial P}{\partial x}z^2 \tag{1.25}$$

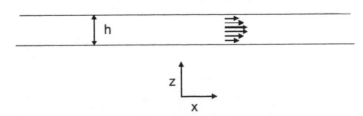

Figure 1.5   Laminar flow occurs between two parallel plates. When the average flow between the plates is considered an equation identical in form to Darcy's equation is obtained.

The constants A and B are determined by the no-slip condition that the velocity must vanish at the plates $z = 0$ and $z = b$. Here $v = 0$ at $z = 0$ requires $A = 0$ and $v = 0$ at $z = b$ requires

$$B = -\frac{1}{2\mu}\frac{\partial P}{\partial x}b \tag{1.26}$$

and therefore

$$v = -\frac{1}{2\mu}\frac{\partial P}{\partial x}z(b - z) \tag{1.27}$$

We may now reduce the problem by one degree of freedom by considering the x component velocity in one dimension

$$\bar{v} = \frac{1}{b}\int_0^b v\,dz \tag{1.28}$$

Therefore

$$\frac{12\mu}{b^2}\bar{v} = -\frac{\partial P}{\partial x} \tag{1.29}$$

and now note that this of the same form as Darcy's equation with

$$K = \frac{b^2}{12} \tag{1.30}$$

Lubrication theory differs from flow in porous media in that the equations of lubrication theory are obtained by removing a dimension by averaging. Darcy's equation describes the interaction of the fluid with the solid in three dimensions. Lubrication theory, however, does give some insight into how permeability is a property of the geometry of the medium.

Another problem that can be described using lubrication theory is flow in capillary tubes. Here the equation of motion for steady flow given by Equation (1.24) may be written in cylindrical coordinates subject to the conditions

$$v_z = v_z(r)\,, \quad v_r = v_\phi = 0 \tag{1.31}$$

as:

$$\frac{\mu}{r}\frac{d}{dr}\left(r\frac{dv_z(r)}{dr}\right) = \frac{dP}{dz} \tag{1.32}$$

Now observe that the left hand side of this equation is a function of r only and the right hand side is a function of z only. Therefore both sides must be constant. Now rewriting Equation (1.32) in the form

$$\frac{d}{dr}\left(r\frac{dv_z(r)}{dr}\right) = \frac{dP}{dz}\frac{r}{\mu} \tag{1.33}$$

and integrating yields

$$\frac{dv_z(r)}{dr} = \frac{A}{r} + \frac{dP}{dz}\frac{r}{2\mu} \tag{1.34}$$

Integrating again yields

$$v_z(r) = B + A\ln(r) + \frac{dP}{dz}\frac{r^2}{4\mu} \tag{1.35}$$

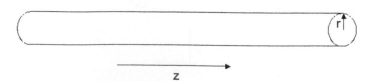

$$z$$

Figure 1.6   Laminar flow in a capillary tube is considered. Again when only the average volumetric flow rate is considered an equation identical in form to Darcy's equation is obtained.

Here A and B are determined from the no slip condition at the boundary of the tube and the condition that $v_z(r)$ is finite at the centre. Therefore $A = 0$ and $B = -(dP/dz)(b^2/4\mu)$ where b is the radius of the tube. Therefore

$$v_z(r) = -\frac{1}{4\mu}\frac{dP}{dz}(b^2 - r^2) \tag{1.36}$$

The mean velocity is

$$\bar{v} = \frac{\int_0^b 2\pi r v_z(r)\,dr}{Area} \tag{1.37}$$

which yields

$$\bar{v} = -\frac{b^2}{8\mu}\frac{dP}{dz} \tag{1.38}$$

So once again the mean velocity is linearly proportional to the pressure gradient and permeability is defined by the geometry of the tube.

1.5   MULTIPHASE FLOW IN POROUS MEDIA

For the purposes of this discussion of a porous medium, the pore scale is assumed to be a scale at which the fluids and solid may be described independently and they interact with each other at their boundaries.

At the pore scale the independent thermodynamics and thermomechanics of the component phases has been firmly established. Each has a unique energy potential, an independent equation of motion and each has an entropy defined by its molecular scale properties. However at the pore scale the phases interact with each other macroscopically. It is the nature of these interactions that must be preserved and described in a larger scale thermomechanical or thermodynamic description (i.e. the laws of physics must hold at all scales and it is this constraint that yields a unique description at the megascale).

In order to get additional understanding of the problem that must be addressed in going up in scale consider the following thought experiment. Consider cubic membranes the size of the small squares shown in Figure 1.7 containing a fluid.

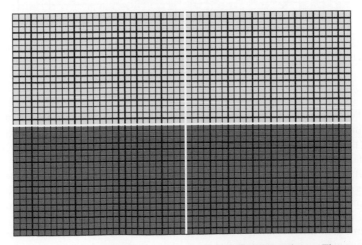

Figure 1.7   Fluid is contained in membranes the size of the small squares shown. The upper membranes contain water and the lower ones glycerin. The smallest scale at which the motions may be described is the scale of the four large squares.

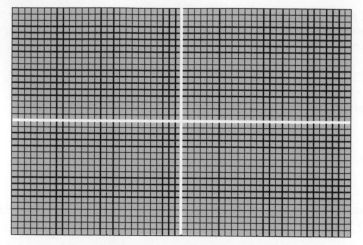

Figure 1.8   When the membranes dissolve the fluids mix at the molecular scale. The mixing is described by the chemical potential and the mass fractions of the fluids.

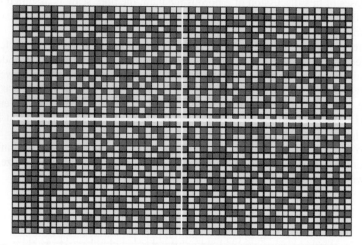

Figure 1.9   Here the membranes are moved around randomly so that approximately the same number of membrane containing each liquid are in each of the large squares.

Assume the membranes in the upper half with the light blue colour contain water and the membranes in the lower half with the darker blue colour contain glycerin. In the first experiment allow all of the membranes to dissolve and wait until the fluids are completely mixed at the molecular scale as shown in Figure 1.8.

Here it is observed that the fluids mix with each other at the molecular scale. The thermodynamic description is given by the chemical potential and the mass fractions of the fluids. Here the mixing occurs at the molecular scale and the energy involved is the chemical energy. This process involves only molecular scale thermodynamics.

Now assume that the membranes all remain intact and the fluids are mixed by moving the membranes around in a semi-random fashion as shown in Figure 1.9.

Now, before analyzing this experiment, perform another one in which the membranes are move around in orderly fashion as shown in Figure 1.10.

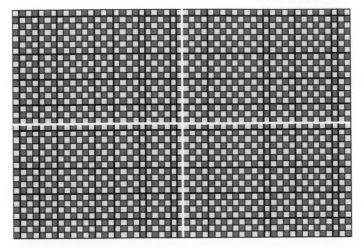

Figure 1.10   Here the membranes are moved around systematically so that they alternate along each line and column in each of the large squares.

In the last two experiments mixing occurs at what we could refer to as the pore scale and the results are observed at a much larger scale.

The mixing that occurs in the last two experiments has absolutely nothing to do with molecular scale thermodynamics. The mixing occurs though mechanical energy at the scale of the membranes and the fluids can be described independently by their volume fractions. Furthermore their volume fractions are mixed differently in Figures 1.9 and 1.10. If the thermodynamics is now considered at the scale of the larger cubes, shown by the white lines, then the mixing is described by the volume fractions and an order potential. If these processes were now generalized to consider the dynamic mixing of the phases then saturation becomes a dynamic variable. For a porous medium in which fluids are being mixed at the pore scale this is the role of the following equation

$$P_1 - P_2 = \beta_2 \frac{\partial s}{\partial t} \tag{1.39}$$

This equation makes a statement which is obvious: if the saturation is changing in a volume element then there must be a pressure difference between the average pressure in the displacing phase and the average pressure in the displaced phase even in the absence of surface tension. This statement is obvious because there is a pressure gradient in the direction of flow. The implications of this additional equation of motion is that the additional degree of freedom seen in the above thought experiments must now be accounted for (pore scale mixing of fluids). This introduces dispersion into the equations of motion. This means that the additional degree of freedom introduced by making saturation a dynamic variable has introduced an additional physical process. So we may now have the mixing of fluids at the molecular scale (molecular diffusion), the mixing of the fluids at the pore scale (dispersion) or the mixing of the fluids at the megascale (viscous fingering).

Equation (1.39) is often omitted in derivations arising from volume averaging. That is because its origins come from the mechanical energy description given above at the pore scale. Thus it is new information introduced at the pores scale and cannot be obtained from the volume averaging of macroscopic equation whose origins come from the molecular scale. This is the origin of the so-called closure problem encountered by many engineers.

Including surface tension alters the values of some of the parameters in the equations of motion. It also alters the form of the dynamic pressure Equation (1.39). The dynamic pressure equation now takes the form (c.f. Spanos 2002)

$$\frac{\partial(P_1 - P_2)}{\partial t} = -\beta_1 \frac{\partial s}{\partial t} + \beta_2 \frac{\partial^2 s}{\partial t^2} \qquad (1.40)$$

The equations of motion for steady state multiphase flow in porous media are given by Equation (1.40) and the equations of motion for the two fluids

$$Q_{11}\mathbf{q}_1 - Q_{12}\mathbf{q}_2 = -(\nabla p_1 - \rho_1 \mathbf{g}) \qquad (1.41)$$

$$-Q_{21}\mathbf{q}_1 + Q_{22}\mathbf{q}_2 = -(\nabla p_2 - \rho_2 \mathbf{g}) \qquad (1.42)$$

The equations of multiphase flow in porous media were first constructed by de la Cruz and Spanos (1983). The equations of flow in porous media used in Petroleum engineering literature and in the ground water Hydrology literature contain numerous errors in physics. A summary of the errors made in constructing the equations in those fields will be given here.

Now the discussion of multiphase flow in porous media will begin by considering miscible flow in porous media. The equations of motion for the flow of miscible fluid phases has generally been modeled by the convection diffusion equation. This equation is based on the mixing of fluid phases at the molecular scale and completely misses the dynamic processes that occur in porous media. The problem however is much greater than the convection (or advection) diffusion equation not being the correct equation to describe the flow of miscible fluid flow in porous media. The system of equations being used is self-contradictory and also is not complete. Degrees of freedom, which can be observed in simple thought experiments, are not accounted for because there aren't enough equations. The origin of this error is that the thermodynamics being used does not apply to porous media at the scale it is being applied. A detailed description of the miscible flow problem in porous media is given in Udey and Spanos (1993) and in Spanos (2002). There it is shown that the correct equations of motion for modeling miscible flow in porous media are: a Fokker-Planck equation which unlike the convection diffusion equation allows for a natural description of longitudinal and lateral dispersion as well as a dynamic pressure equation which says that if one phase is displacing another there is a pressure difference between them. Without this dynamic pressure equation the equations of motion are not complete. This can be easily demonstrated with a simple thought experiment. Assume that one fluid is displacing another in a porous medium. Further assume that although the phases are miscible a negligible amount of molecular diffusion occurs during the course of the experiment. Since the fluids remain distinct we may consider them in terms of their volume fractions rather than their mass fractions (i.e. saturation instead of concentration). Here if one phase is displacing another then the amount of the displacing phase within a volume element must increase in time and the amount of displaced phase must be reduced. This requires a change in saturation in time within the volume element. The average pressure of the displaced fluid must be less than that of the displacing fluid since the pressure gradient is in the direction of flow. So if there is a saturation gradient then the saturation in the volume element must be changing in time and the average pressure in the two phases must be different. This may be expressed as

$$P_1 - P_2 = \beta_2 \frac{\partial s}{\partial t} \qquad (1.43)$$

A simple experiment to illustrate the dispersional motion predicted by this theory can be performed by constructing a homogenous porous medium of about 1 Darcy permeability and then displacing clear water with blue dyed water with about a meter head. During the course of the experiment little diffusion will occur however it can be observed that the saturation contours will initially separate in an unstable fashion and then reach a terminal velocity at which point they will propagate as waves. The behaviour, of this very simple experiment, cannot be described by the convection diffusion models used by hydrologists. A theoretical description of this process is given by Udey (2009).

Including surface tension alters the values of some of the parameters in the equations of motion. It also alters the form of the dynamic pressure equation (Eq. 1.43). The dynamic pressure equation now takes the form (c.f. Spanos 2002)

$$\frac{\partial(P_1 - P_2)}{\partial t} = -\beta_1 \frac{\partial s}{\partial t} + \beta_2 \frac{\partial^2 s}{\partial t^2} \tag{1.44}$$

The equations of motion for steady state multiphase flow in porous media are given by Equation (1.44) and the equations of motion for the two fluids

$$Q_{11}\mathbf{q}_1 - Q_{12}\mathbf{q}_2 = (\nabla p_1 - \rho_1 \mathbf{g}) \tag{1.45}$$

$$-Q_{21}\mathbf{q}_1 + Q_{22}\mathbf{q}_2 = (\nabla p_2 - \rho_2 \mathbf{g}) \tag{1.46}$$

Now compare this description of multiphase flow with relative permeability

$$\frac{\mu_1}{KK_{r1}}\mathbf{q}_1 = (\nabla p_1 - \rho_1 \mathbf{g}) \tag{1.47}$$

$$\frac{\mu_2}{KK_{r2}}\mathbf{q}_2 = (\nabla p_2 - \rho_2 \mathbf{g}) \tag{1.48}$$

and in this description capillary pressure is imposed as a zeroth order constraint. Here K_{r1} and K_{r2} are not physical parameters but rather are functions of the parameters and variables given in Equations (1.44), (1.45) and (1.46). For a quantity like viscosity say to be a physical parameter it must be able to be measured by any flow that the Navier-Stokes equation describes. Thus it may be measured by an infinite number of different experiments and one must always obtain the same result. When K_{r1} and K_{r2} are assumed to be parameters they do not account for forces that exist between the fluid phases when such effects as dispersion are occurring. So for example they cannot simulate dispersion. This can be seen by observing that Equations (1.47) and (1.48) each have only one flow velocity, meaning there is no relative motion of the fluids. If there were relative motion then there would be a force associated with that relative motion as is seen in Equations (1.45) and (1.46). The model Equations (1.47) and (1.48) are associated with is based on molecular scale thermodynamics. Furthermore with Equations (1.47) and (1.48) as the sole equations of motion i.e. without Equation (1.44) the equations are stating that saturation is uniform and no displacement is taking place (i.e. $(\partial s_1/\partial t) = 0$ and $(\partial^2 s_1/\partial t^2) = 0$). In the zero surface tension limit this leads to what is called the Buckley-Leverett paradox. Here since the equations are making an unphysical statement they yield an unphysical solution when a displacement takes place. Here Figures 1.11 and 1.12 represent oil and water relative permeability curves measured under steady state conditions (constant saturation and steady flow).

But K_{ro} and K_{rw} can be calculated from Equations (1.44), (1.45) and (1.46) and are found to depend on Q_{11}, Q_{12}, Q_{22}, Q_{21}, β_1, β_2, q_1 and q_2. Now using Equations (1.47) and (1.48) the fractional flow of water for a water displacement process is given by Figure 1.13.

Here the inflection point in this graph causes unphysical behaviour in the flow. This may be seen by plotting the position as a function of saturation, as shown in Figure 1.14.

Here multiple values of saturation are predicted at the same position in space. When the velocity of saturation contours is plotted (see Figure 1.15) it is observed that low values of saturation are moving slower than some higher values of saturation, an unphysical result.

The origin of this problem may be observed by plotting the velocity of the saturation contours as a function of saturation where it is observed that some higher water saturation contours move faster than lower saturation contours and thus would pass them a completely unphysical result.

The origin and resolution of this paradox is completely transparent when studied in terms of Equations (1.44), (1.45) and (1.46). When this is done it is observed that the relative permeabilities are functions of q_1, q_2, Q_{11}, Q_{12}, Q_{22}, Q_{21}, β_1 and β_2. As a result the values of the relative permeabilities for frontal displacement are different than they are for constant saturation steady state flow. When the new values for frontal displacement are calculated they yield Figures 1.16 and 1.17.

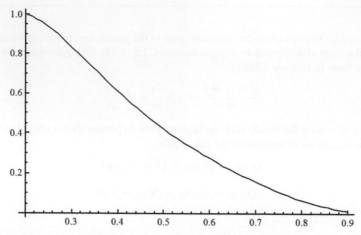

Figure 1.11 The relative permeability to oil, K_{ro}, versus saturation as measured in constant saturation steady state experiments.

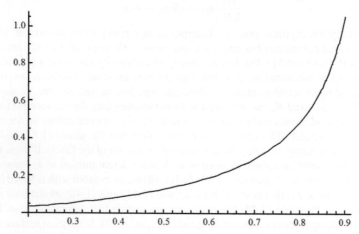

Figure 1.12 The relative permeability to water, K_{rw}, versus saturation as measured in constant saturation steady state experiments.

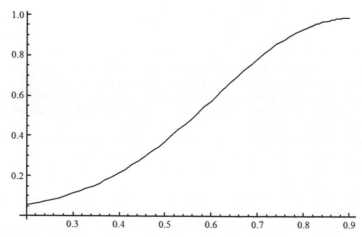

Figure 1.13 The fractional flow of water, f_w, versus saturation as determined from the above relative permeabilities.

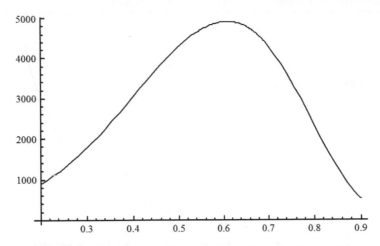

Figure 1.14   The position of each saturation, $z(Sw)$, as determined from the fractional flow.

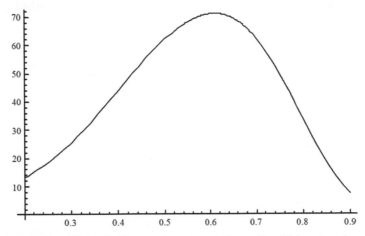

Figure 1.15   The velocity, $v(Sw)$, of each saturation contour is shown. Here higher saturation contours move faster than lower saturation contours a result that is physically impossible indicting that the equations are physically incorrect.

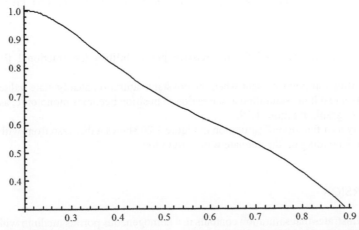

Figure 1.16   The relative permeability to oil, K_{ro}, versus saturation as calculated using Equations (1.44), (1.45) and (1.46) under dynamic conditions.

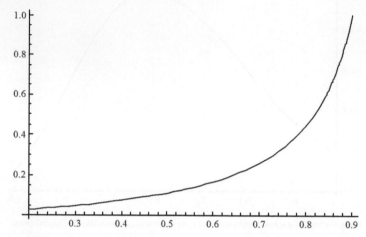

Figure 1.17   The relative permeability to water, K_{rw}, versus saturation as calculated using Equations (1.44), (1.45) and (1.46) under dynamic conditions.

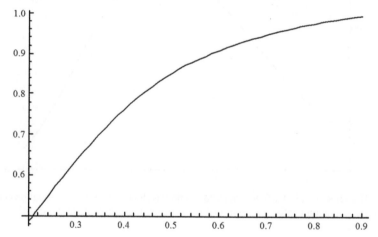

Figure 1.18   The fractional flow of water, f_w, versus saturation as determined from the above dynamic relative premeabilities.

When these values are used for the relative permeabilities the fractional flow becomes Figure 1.18.

Here the inflection point present when the constant saturation steady state values were used disappears. As a result the saturation as a function of position becomes monotonic, as can be seen from the $z(Sw)$ graph in Figure 1.19.

The velocity as a function of saturation in Figure 1.20 shows a disperse front with the amount of dispersion depending on the connate water saturation.

1.6   DISPERSION

Figure 1.20 illustrates that saturation contours in a homogeneous porous medium without viscous fingering (i.e. a one dimensional displacement) evolve to constant velocity waves. Furthermore this figure describes how these saturation contours are moving apart. This process of mechanical

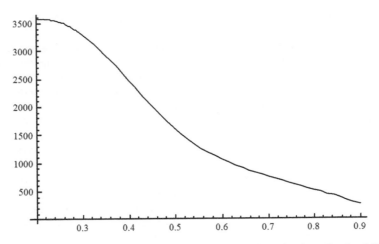

Figure 1.19   The position of each saturation, $z(Sw)$, as determined from the above fractional flow.

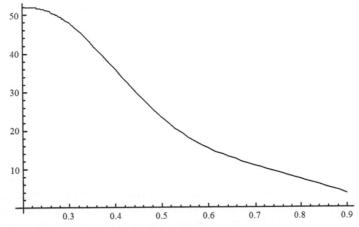

Figure 1.20   This graph illustrates that during frontal dispersion lower water saturation contours move faster than higher water saturation contours. This process is not allowed by Equations (1.47), (1.48) and the associated capillary pressure assumptions. Note that in many experiments the asymptote appearing at low water saturations can be much more pronounced than illustrated in the current graph and appear more like a water front.

mixing of the fluids at the pore scale is called dispersion. Equation (1.44) describes the change in saturation within a volume element in time due to a volume fraction one fluid flowing into the volume element and a volume fraction of the other fluid flowing out causing a change in saturation due to pore scale mixing of the fluids. The constraint of a static pressure difference at the megascale states the pore scale mixing is not occurring causing physical contradiction at a front and is the origin of the Buckely Leverett paradox. In real porous media two competing processes occur: dispersion and viscous fingering. Here viscous fingering is channeling of the displacing fluid bypassing the displaced fluid either due to heterogeneities in the medium or a less viscous fluid channeling through a more viscous fluid due to the associated mobility instability (c.f. Spanos 2002). Generally one of the two processes will dominate the other. Since dispersion enhances sweep efficiency and results in a greater amount of the porous medium being contacted by the injected fluid and a greater amount of the displaced being recovered it is generally the process one wishes to enhance.

1.7   FEW COMMENTS ABOUT THE ASSOCIATED THERMODYNAMICS

When two phase flow in porous media is considered saturation becomes a variable which is necessary to construct both a thermomechanical and thermodynamic description of flow at the megascale. As well when compressional motions are considered porosity also becomes variable which is required to construct both a self consistent thermomechanical and thermodynamic description of compressional motions at the megascale. Note in the case of compressible flow in a rigid porous medium one obtains two dynamic pressure equations and a dynamic saturation equation related to both the flow and compressional motions of the fluid phases (c.f. Spanos 2002 pg 133). This dynamic saturation equation is identical in form to the dynamic porosity equation which occurs for a porous matrix containing a single fluid phase. In the limit as compression goes to zero the two pressure equations an the saturation equation combine to form the dynamic pressure equation described in this chapter (c.f. Spanos 2002 pg 133 et seq.). Fluid flow requires that non equilibrium thermodynamics be used and that discussion is too complex for the current presentation, besides it is proprietary information currently used in Wavefront Technology Solutions reservoir simulator. As a demonstration the Onsager relations are constructed for the thermostatic compression of porous medium with a single fluid phase in the Appendix. There it is shown that one of the Onsager relations constrains the parameters in the porosity equation and thus for two fluid phases the parameters in the saturation equation and not the permeability terms. If these constraints are not satisfied then Newton's third law is violated and mass is not conserved.

1.8   CONCLUSIONS

Permeability and relative permeability are terms that are used rather loosely in descriptions of porous media and fluid flow in the earth. Permeability has a very specific physical meaning in megascopic laboratory experiments. This physical meaning is often lost when one goes to the field scale (gigascale) because at that scale Darcy's equation is no longer a physical equation. Relative permeability does not even have physical meaning as a parameter at the macroscale because: it does not account for the forces between the fluid phases at the pore scale which is required to describe fluid displacement processes; also an additional equation of motion is required to account for the dynamic role played by saturation during fluid displacements. As well it is shown that the constraints that Onsager's relations place on multiphase flow has been badly misinterpreted when they have been presented in the context of relative permeability. As a final comment, Onsager's relations in the context of non equilibrium thermodynamics has a profound influence in the description of multiphase fluid flow in the earth.

ACKNOWLEDGMENT

A number of helpful suggestions for modifying this chapter were made by Ramon Bentsen.

1.A   APPENDIX

Summary of the equations that describe thermostatic deformations:

1.A.1   *Solid properties*

Internal Energy

$$dU_s = \tau_{ik}^s du_{ik}'^s + T_s dS_s \tag{1.A1}$$

Solid stress

$$\tau_{ik}^s = -\phi_o K_s \alpha_s (T^s - T_o)\delta_{ik} + 2\mu_M \left(u_{ik}'^s - \frac{1}{3}\delta_{ik} u_{jj}'^s \right) + \phi_o K_s \delta_{ik} u_{jj}'^s \tag{1.A2}$$

where $\phi = 1 - \eta$ and the solid strain is given by

$$u_{ik}^{\prime s} = u_{ik}^s + \frac{1}{3}\frac{\phi - \phi_o}{\phi_o}\delta_{ik} \qquad (1.A3)$$

Solid internal energy

$$U_s(S, u_{ik}^{\prime s}) = \phi_o U_o^s(S_s) - \frac{T_o^s K_s \alpha_s}{c_v^s}(S_s - S_o)u_{jj}^{\prime s} + \mu_M \left(u_{ik}^{\prime s} - \frac{1}{3}\delta_{ik}u_{jj}^{\prime s}\right)^2 + \frac{1}{2}\phi_o K_{ad}^s u_{jj}^{\prime s\,2} \quad (1.A4)$$

In equilibrium thermodynamics entropy balance of the solid component must describe how the energy of the system is arranged at all scales. Thus

$$S_o^s(T^s) = -\phi_o \frac{dF_o^s(T^s)}{dT^s} \qquad (1.A5)$$

the entropy balance of the solid component is given by

$$S_s - S_o^s = \phi_o c_v^s \frac{T^s - T_o}{T_o} + \phi_o K_s \alpha_s u_{jj}^{\prime s} \qquad (1.A6)$$

and the heat capacity by

$$c_v^s = -T_o \frac{d^2 F_o^s(T_o)}{dT_o^2} \qquad (1.A7)$$

In the current formulation the Helmholtz free energy for the solid represents the total energy required to create the solid part of the system minus the heat obtained from the environment at temperature T^S. In deformations where granular motions are allowed, changes in porosity can also cause changes in entropy. For reversible motions of the matrix involving only elastic deformations of the matrix and thus only mechanical energy and heat the Helmholtz free energy for the solid may be written as

$$dF_s = \sigma_{ik}^s du_{ik}^{\prime s} + S_s dT^s \qquad (1.A8)$$

$$F_s(T^s, u_{ik}^{\prime s}) = \phi_o F_o^s(T^s) - \phi_o K_s \alpha_s(T^s - T_o)u_{jj}^{\prime s} + \mu_M \left(u_{ik}^{\prime s} - \frac{1}{3}\delta_{ik}u_{jj}^{\prime s}\right)^2 + \frac{1}{2}\phi_o K_s u_{jj}^{\prime s\,2} \quad (1.A9)$$

The enthalpy of the solid represents the solid internal energy of the system plus the work required to make room for the solid in an environment with constant pressure

$$dH_s = -u_{ik}^{\prime s} d\sigma_{ik}^s + T^s dS_s \qquad (1.A10)$$

$$H_s(S_s, u_{ik}^{\prime s}) = \phi_o U_o^s(S_s) - \left(\frac{T_o K_s \alpha_s}{c_v}\right)(S_s - S_o^s)u_{jj}^{\prime s} - \mu_M \left(u_{ik}^{\prime s} - \frac{1}{3}\delta_{ik}u_{jj}^{\prime s}\right)^2 + \frac{1}{2}\phi_o(K_{s(ad)} - K_s)u_{jj}^{\prime s\,2}$$
$$(1.A11)$$

The Gibbs Free Energy for the solid represents the solid internal energy of the system plus the work required to make room for it in an environment with constant pressure minus the heat obtained from an environment at temperature T^S

$$dG_s = -u_{ik}^{\prime s} d\sigma_{ik}^s + S_s dT^s \qquad (1.A12)$$

$$G_s(T^s, u_{ik}^{\prime s}) = \phi_o F_o^s(T^s) - \mu_M \left(u_{ik}^{\prime s} - \frac{1}{3}\delta_{ik}u_{jj}^{\prime s}\right)^2 - \frac{1}{2}\phi_o K_s u_{jj}^{\prime s\,2} \qquad (1.A13)$$

1.A.2 *Fluid Properties*

The equivalent potentials for the fluid are given by
  Internal Energy

$$dU_f = -\eta_o p_f du_{kk}^f + T^f dS_f \tag{1.A14}$$

$$U_f(S_f, u_{jj}'^f) = \phi_o U_o^f(S_f) - \frac{T_o \alpha_f}{c_p^f}(S_f - S_o)p_f + \frac{1}{2}\eta_o K_{f(ad)} u_{jj}'^{f2} \tag{1.A15}$$

Fluid compressive strain

$$u_{jj}'^f = \vec{\nabla} \cdot \vec{u}_f + \frac{1}{3}\frac{\eta - \eta_o}{\eta_o} \tag{1.A16}$$

Ambient Temperature

$$T_o = \eta_o \frac{dU_o^f(S_o)}{dS_o} \tag{1.A17}$$

Helmholtz Free Energy

$$dF_f = -\eta p_f du_{jj}^f - S_f dT^f \tag{1.A18}$$

$$F_f(T^f, u_{ik}'^f) = \phi_o F_o^f(T^f) + \eta_o K_f \alpha_f(T^f - T_o)u_{jj}'^f - \frac{1}{2}\eta_o K_f u_{jj}'^{f2} \tag{1.A19}$$

Thermostatic fluid entropy balance

$$S_f - S_o^f = \eta_o c_p^f \frac{T^f - T_o}{T_o} + \eta_o \alpha_f p_f \tag{1.A20}$$

$$S_o^f(T^f) = -\eta_o \frac{dF_o^f(T_o^f)}{dT_o^f} \tag{1.A21}$$

Enthalpy

$$dH_f = u_{jj}^f d(\eta p_f) - T^f dS_f \tag{1.A22}$$

$$H_f(S_f, u_{jj}'^f) = \eta_o U_o^f(S_f) - \left(\frac{T_o \alpha_f}{c_p^f}\right)(S_f - S_o)p_f + \frac{1}{2}\eta_o K_{f(ad)} u_{jj}'^{f2} + \eta_o p_f u_{jj}'^s \tag{1.A23}$$

Gibbs Free Energy

$$dG_f = u_{jj}'^f d(\eta p_f) - S_f dT^f \tag{1.A24}$$

$$G_f(T_f, \sigma_{ik}^f) = \eta_o F_o^f(T_f) - 2\eta_o \alpha_f(T_f - T_o)p_f - \eta_o p_f u_{jj}'^f \tag{1.A25}$$

1.A.3 *Reciprocity*

Reciprocity is the constraint imposed by the Onsager relations which require that the entropy production must be positive or in the present case of equilibrium thermo-statics that it must be zero.

The Onsager relations are constructed as follows: The empirical flux laws are expressed by the following relations.
  Temperature (heat)

$$\mathbf{J}_Q = -\kappa \nabla T \tag{1.A26}$$

Porosity variations (order) associated with quasi static deformations enter the equilibrium thermo-statics through the strains as described in Equations (1.A3) and (1.A16)

$$\mathbf{J}_\Pi = -D_s \nabla \eta \tag{1.A27}$$

Pressure (mass)

$$\mathbf{J}_M = -D_f \nabla p \tag{1.A28}$$

These quantities may also be expressed as

$$\mathbf{J}_Q = L_{QQ}\left(\frac{\nabla T}{T}\right) - L_{Q\Pi}\nabla\eta - L_{QM}\nabla p \tag{1.A29}$$

$$\mathbf{J}_\Pi = L_{\Pi Q}\left(\frac{\nabla T}{T}\right) - L_{\Pi\Pi}\nabla\eta - L_{\Pi M}\nabla p \tag{1.A30}$$

$$\mathbf{J}_M = L_{MQ}\left(\frac{\nabla T}{T}\right) - L_{M\Pi}\nabla\eta - L_{MM}\nabla p \tag{1.A31}$$

Now define

$$\boldsymbol{\Gamma}_Q = -\frac{\nabla T}{T} \tag{1.A32}$$

$$\boldsymbol{\Gamma}_\Pi = -\nabla\eta \tag{1.A33}$$

$$\boldsymbol{\Gamma}_M = -\nabla p \tag{1.A34}$$

and the coefficients in Eqs. (1.A29) to (1.A31) are given by

$$L_{QQ} = \frac{\partial\mathbf{J}_Q}{\partial\boldsymbol{\Gamma}_Q}, \quad L_{Q\Pi} = \frac{\partial\mathbf{J}_Q}{\partial\boldsymbol{\Gamma}_\Pi}, \quad L_{QM} = \frac{\partial\mathbf{J}_Q}{\partial\boldsymbol{\Gamma}_M}, \quad L_{Qq} = \frac{\partial\mathbf{J}_Q}{\partial\boldsymbol{\Gamma}_q} \tag{1.A35}$$

$$L_{\Pi Q} = \frac{\partial\mathbf{J}_\Pi}{\partial\boldsymbol{\Gamma}_Q}, \quad L_{\Pi\Pi} = \frac{\partial\mathbf{J}_\Pi}{\partial\boldsymbol{\Gamma}_\Pi}, \quad L_{\Pi M} = \frac{\partial\mathbf{J}_\Pi}{\partial\boldsymbol{\Gamma}_M}, \quad L_{\Pi q} = \frac{\partial\mathbf{J}_\Pi}{\partial\boldsymbol{\Gamma}_q} \tag{1.A36}$$

$$L_{MQ} = \frac{\partial\mathbf{J}_M}{\partial\boldsymbol{\Gamma}_Q}, \quad L_{M\Pi} = \frac{\partial\mathbf{J}_M}{\partial\boldsymbol{\Gamma}_\Pi}, \quad L_{MM} = \frac{\partial\mathbf{J}_M}{\partial\boldsymbol{\Gamma}_M}, \quad L_{Mq} = \frac{\partial\mathbf{J}_M}{\partial\boldsymbol{\Gamma}_q} \tag{1.A37}$$

$$L_{qQ} = \frac{\partial\mathbf{J}_q}{\partial\boldsymbol{\Gamma}_Q}, \quad L_{q\Pi} = \frac{\partial\mathbf{J}_q}{\partial\boldsymbol{\Gamma}_\Pi}, \quad L_{qM} = \frac{\partial\mathbf{J}_q}{\partial\boldsymbol{\Gamma}_M}, \quad L_{qq} = \frac{\partial\mathbf{J}_q}{\partial\boldsymbol{\Gamma}_q} \tag{1.A38}$$

According to Onsager's reciprocity relations

$$L_{M\Pi} = L_{\Pi M} \tag{1.A39}$$

thus

$$\frac{\partial\mathbf{J}_M}{\partial\boldsymbol{\Gamma}_\Pi} = \frac{\partial\mathbf{J}_\Pi}{\partial\boldsymbol{\Gamma}_M} \tag{1.A40}$$

which may be written as

$$\frac{\partial(-D_f\nabla p)}{\partial(-\nabla\eta)} = \frac{\partial(-D_s\nabla\eta)}{\partial(-\nabla p)} \tag{1.A41}$$

or

$$D_f\frac{\partial(\nabla p)}{\partial(\nabla\eta)} = D_s\frac{\partial(\nabla\eta)}{\partial(\nabla p)} \tag{1.A42}$$

where

$$D_f = \frac{v_{pf}^2}{2a_{pf}} \tag{1.A43}$$

$$D_s = \frac{v_{\eta M}^2}{2a_{\eta M}} \tag{1.A44}$$

$$v_{pf}^2 = \frac{K_f}{\rho_{pf}} \tag{1.A45}$$

$$v_{\eta M}^2 = \frac{K_\eta}{\rho_{\eta M}} \tag{1.A46}$$

$$a_{pf} = \frac{1}{2}\frac{\mu_f}{K}\frac{\eta_o(\alpha_2 - 1)}{\rho_{pf}} \tag{1.A47}$$

$$a_{\eta M} = \frac{\rho_{\eta s}}{\rho_{\eta M}}a_{\eta s} + \frac{\rho_{\eta f}}{\rho_{\eta M}}\frac{\alpha_2 K_M}{\alpha_1 K_f}a_{\eta f} \tag{1.A48}$$

$$K_\eta = \left(1 - \frac{\eta_o}{(1 - \eta_o)\alpha_1}\right)K_s + \frac{4}{3}\frac{\mu_M}{(1 - \eta_o)} \tag{1.A49}$$

$$\rho_{pf} = \rho_f - \frac{\rho_{12}}{\eta_o}(\alpha_2 - 1) \tag{1.A50}$$

$$\rho_{\eta M} = \rho_{\eta s} + \rho_{\eta f}\frac{\alpha_2 K_M}{\alpha_1 K_f} \tag{1.A51}$$

$$\rho_{\eta s} = \rho_s - \frac{\rho_{12}(\alpha_1 + 1)}{\alpha_1(1 - \eta_o)} \tag{1.A52}$$

$$\rho_{\eta f} = \rho_f - \rho_{12}(\alpha_1 + 1) \tag{1.A53}$$

$$K_M = K_s + \frac{4}{3}\frac{\mu_M}{(1 - \eta_o)} \tag{1.A54}$$

$$\alpha_1 = \frac{\eta_o - \delta_f}{\delta_s} \tag{1.A55}$$

$$\alpha_2 = \frac{\delta_f}{\delta_s} \tag{1.A56}$$

Under quasi static conditions the equations of motion yield

$$D_f\nabla^2 p = -CD_s\nabla^2\eta \tag{1.A57}$$

and

$$D_s\nabla^2\eta = -BD_f\nabla^2 p \tag{1.A58}$$

Upon substituting the relations

$$D_f\frac{\partial(\nabla p)}{\partial(\nabla\eta)} = -CD_s \tag{1.A59}$$

$$D_s\frac{\partial(\nabla\eta)}{\partial(\nabla p)} = -BD_f \tag{1.A60}$$

into Onsager's relations one obtains

$$CD_s = BD_f \tag{1.A61}$$

where

$$B = 2\frac{\alpha_2\rho_{pM}\eta_o a_{pM}}{\alpha_1\rho_{\eta M}a_{\eta M}K_f} \tag{1.A62}$$

and

$$C = \frac{\rho_{\eta f}K_f a_{\eta f}}{\eta_o\rho_{pf}a_{pf}} \tag{1.A63}$$

Therefore

$$\eta_o^2(1 - \alpha_2)\left(\alpha_2 - \frac{K_f}{(1 - \eta_o)K_M}\right) = \frac{1}{2}(\alpha_1 + 1)K_f^2\left(\alpha_1 - \frac{\eta_o K_s}{(1 - \eta_o)K_M}\right) \tag{1.A64}$$

This relation places constraints on the α's for equilibrium processes. Thus Onsagers relation places a constraint on α_2 to be in the range $(K_f/(1 - \eta_o)K_M) \leq \alpha_2 \leq 1$ which in turn constrains α_1 to essentially be a constant.

Similarly in non-equilibrium thermodynamics involving multiphase flow, Onsager's relations place new restrictions on the parameters in the equations for porosity and saturation in the context of a specific thermodynamic process.

REFERENCES

Anderson, T.B. & Jackson, R. (1967) Fluid mechanical description of fluidized beds. Equations of motion. *Industrial & Engineering Chemistry Fundamentals*, 6, 527–539.

Bear, J. (1972) *Dynamics of Fluids in Porous Media*. New York, Elsevier.

Bentsen, R.G. (1994) Effect of hydrodynamic forces on the pressure difference equation. *Transport in Porous Media*, 17, 121–135.

Buckley, S.E. & Leverett, M.C. (1942) Mechanics of fluid displacement in sand, *Trans. AIME*, 146, 107–116.

de la Cruz, V. & Spanos, T.J.T. (1983) Mobilization of Oil Ganglia, *AICHE Journal*, 29, 854–858.

Dullien, F.A.L. (1992) *Porous Media Fluid Transport and Pore Structure*. San Diego, Academic Press.

Hubbert, M.K. (1940) The Theory of Ground Water Motion. *Journal of Geology*, 48, 785–944.

Hubbert, M.K. (1956) Darcy law and the field equations of the flow of underground fluids. *Petroleum Transactions, AIME*, 207, 22–239.

Muscat, M. (1949) *Physical Principles of Oil Production*. New York, McGraw Hill.

Muscat, M. (1937) *The Flow of Homogeneous Fluids Through Porous Media*. New York, McGraw Hill.

Newman, S.P. (1977) Theoretical Derivation of Darcy's Law, *Acta Mechanica*, 25, 153–170.

Rose, W. (1949) Theoretical generalizations leading to the evaluation of relative permeability. *AIME*, 111–126.

Rose, W. (2000) Myths about later-day extensions of Darcy's Law. *Journal of Petroleum Science and Engineering*, 26, 187–198.

Scheidegger, A.E. (1974) *The Physics of Flow Through Porous Media*. Toronto, University of Toronto Press.

Slattery, J.C. (1967) Flow of viscoelastic fluids through porous media. *AICHE Journal*, 13, 1066–1071.

Slattery, J.C. (1970) Two phase flow through porous media. *AICHE Journal*, 16, 345–352.

Spanos, T.J.T. (2002) *The Thermophysics of Porous Media*. Boca Raton, Chapman and Hall/CRC.

Udey, N. & Spanos, T.J.T. (1993) The equations of miscible flow with negligible molecular diffusion. *Transport in Porous Media*, 10, 1–41.

Udey, N. (2009) Dispersion waves of two fluids in a porous medium. *Transport in Porous Media*, 79, 107–115.

Whittaker, S. (1967) Diffusion and dispersion in porous media. *AICHE Journal*, 13, 420–427.

Yuster, S.T. (1951) Theoretical consideration of multiphase flow in idealized capillary systems. *Proceedings of the Third World Petroleum Congress*, 2, 437–445.

CHAPTER 2

From upscaling techniques to hybrid models

I. Battiato & D.M. Tartakovsky

2.1 INTRODUCTION

Any mathematical model is an idealization of a real system at a specified scale. Assumptions and/or simplifications upon which such models are based enable their formulation, analytical and/or numerical treatment and, consequently, their use as predictive tools. The acceptance of a model derives from an optimal balance between simplicity and accuracy in capturing a system's behavior on the one hand and computational costs on the other. Different models might offer optimal performances, both in terms of fidelity and computation, in various regimes. A further complication in model selection arises when a scale at which predictions are sought is much larger than a scale at which governing equations and first principles are well defined. This situation is particularly common in analysis of flow and transport in porous media: typical scales of interest for predictions are often many orders of magnitude larger than a scale at which most biochemical processes take place. Such complex systems are of particular interest because of their ubiquitous nature: they characterize a variety of environments ranging from geologic formations to biological cells, and from oil reservoirs to nanotechnology products.

Flow and transport in porous media can be modeled at the pore- (microscopic) or Darcy- (macroscopic) scales. Equations that have a solid physical foundation and are based on the first principles (e.g., Stokes' equations for fluid flow and Fick's law of diffusion for solute transport) require the knowledge of pore geometry that is seldom available in real applications. While rapid advancements in computational power and imagine techniques bode well for the widespread use of pore-scale models at increasingly larger scales, computational domains that can be modeled with modern-day pore-scale simulations are still too small to be of any use for predictions at the field scale: the heterogeneity of most natural porous media (e.g., oil reservoirs and aquifers) and technology products (carbon nanotubes assemblies) and prohibitive computational costs render lattice-Boltzmann modeling (Leemput et al. 2007), smoothed particle hydrodynamics (Tartakovsky et al. 2007), molecular dynamics (Walther et al. 2004) and other pore-scale simulations impractical as a predictive tool at scales that are many orders of magnitude larger than the pore scale.

Macroscopic models (e.g., Darcy's law for fluid flow and an advection-dispersion equation for transport), which treat a porous medium as an "averaged" continuum, overcome these limitations by relying on phenomenological descriptions and a number of simplifications (e.g., spatial smoothness of pore-scale quantities, spatial periodicity of pore structures, and low degree of physical and chemical heterogeneity).

The ubiquitous presence of heterogeneities in natural systems might lead to a localized breakdown of such continuum models. Whenever a *localized* breakdown of continuum-scale models occurs, hybrid models must be used to attain an increased rigor and accuracy in predictions, while keeping computational costs in check. Hybrid simulations (Leemput et al. 2007, Tartakovsky et al. 2007) resolve a small reactive region with a pore-scale model that is coupled to its continuum counterpart in the rest of a computational domain.

In Section 2.2 we present a classification of the most common upscaling methods (Section 2.2.1), which allow one to derive macroscopic equations from their pore-scale counterparts. Sections 2.2.2 and 2.2.3 contain classical results of homogenization theory and their implications for applicability of macroscopic models. In Section 2.3 we identify the applicability range of macroscopic models for diffusive systems with nonlinear homogeneous reaction (Section 2.3.1) and advective-diffusive systems with nonlinear heterogeneous reactions (Section 2.3.4). In Section 2.4 we use volume averaging to construct the formalism of two types of hybrid algorithms, intrusive (Section 2.4.1) and non-intrusive (Section 2.4.5). Our major conclusions are presented in Section 2.5.

2.2 FROM FIRST PRINCIPLES TO EFFECTIVE EQUATIONS

2.2.1 *Classification of upscaling methods*

We consider a porous medium $\Omega = \Omega_s \cup \Omega_l$ consisting of a solid matrix Ω_s and a fluid-filled pore space Ω_l. A major goal of upscaling is to establish connections between pore- and continuum-scale descriptions of transport processes in Ω. Mathematical approaches to upscaling include the method of volume averaging (Whitaker 1999) and its modifications (Kechagia et al. 2002), generalizations of the method of moments (Shapiro & Brenner 1986, 1988, Shapiro et al. 1996), homogenization via multiple-scale expansions (Adler 1992), pore-network models (Acharya et al. 2005), and thermodynamically constrained averaging (Gray & Miller 2005).

Let u be a real-valued function on a pore-scale domain Ω_l that exhibits rapid spatial oscillations. It describes a certain physical quantity and satisfies a partial differential equation:

$$\mathscr{L}[u] = f \tag{2.1}$$

One can define the local average of u as:

$$\langle u \rangle(\mathbf{x}) = \frac{1}{|\mathscr{V}|} \int_{\mathscr{V}(\mathbf{x})} u(\mathbf{y}) \, d\mathbf{y} \tag{2.2}$$

In the method of volume averaging, the support volume $\mathscr{V}$ "is a small, but not too small, neighborhood of point $\mathbf{x}$ of the size of a representative elementary volume, REV (several hundred or thousand of pores)" (Hornung 1997, p. 1). The ambiguity in defining the size of an REV is typical. For example, in (de Marsily 1986, p. 15) "the size of the REV is defined by saying that it is

- sufficiently large to contain a great number of pores so as to allow us to define a mean global property, while ensuring that the effects of the fluctuations from one pore to another are negligible. One may take, for example, 1 cm^3 or 1 dm^3;
- sufficiently small so that the parameter variations from one domain to the next may be approximated by continuous functions, in order that we may use infinitesimal calculus."

A continuum-scale equation:

$$\overline{\mathscr{L}}[\langle c \rangle] = g, \tag{2.3}$$

is constructed by volumetric averaging of the original pore-scale equation (Eq. 2.1). The procedure is facilitated by the spatial averaging theorem, which enables one to exchange spatial integration and differentiation (Whitaker 1999):

$$\langle \nabla u \rangle = \nabla \langle u \rangle + \frac{1}{|\mathscr{V}|} \int_{A_{ls}} u \mathbf{n} \, dA, \tag{2.4}$$

where A_{ls} is the liquid-solid interface contained in $\mathscr{V}$ and $\mathbf{n}$ is the outward normal unit vector of A_{ls}.

Similar concepts are used in thermodynamically constrained averaging theory (Gray & Miller 2005), wherein thermodynamics is introduced into a constrained entropy inequality to guide the formation of closed macroscale models that retain consistency with microscale physics and thermodynamics.

In the homogenization theory by multiple-scale expansions (Hornung 1997), the volume $\mathcal{V}$ is the unit cell of a periodic porous medium Ω with period ϵ. A homogenized equation is obtained by determining the following limit:

$$\langle u \rangle = \langle \lim_{\epsilon \to 0} u_\epsilon \rangle, \tag{2.5}$$

where u_ϵ is the sequence (indexed by ϵ) of solutions of Eq. (2.1) with periodically oscillating coefficients. The limit is determined by utilizing a two-scale asymptotic expansion that "is an ansatz of the form:

$$u_\epsilon(\mathbf{x}) = u_0(\mathbf{x}, \mathbf{x}/\epsilon) + \epsilon u_1(\mathbf{x}, \mathbf{x}/\epsilon) + \epsilon^2 u_2(\mathbf{x}, \mathbf{x}/\epsilon) + \cdots \tag{2.6}$$

where each function $u_i(\mathbf{x}, \mathbf{y})$ in this series depends on two variables, $\mathbf{x}$ the macroscopic (or slow) variable and $\mathbf{y}$ the microscopic (or fast) variable, and is $\mathcal{V}$-periodic in $\mathbf{y}$ ($\mathcal{V}$ is the unit period). Inserting Eq. (2.6) into Eq. (2.1) satisfied by u_ϵ and identifying powers of ϵ leads to a cascade of equations for each term $u_i(\mathbf{x}, \mathbf{y})$. In general averaging with respect to $\mathbf{y}$ yields the homogenized equation for u_0. Another step is required to rigorously justify the homogenization result obtained heuristically with this two-scale asymptotic expansion" (Hornung 1997, p. 238).

Similar to the homogenization theory definition of average is that of the methods of moments, wherein the global ($\mathbf{x}$) and local ($\mathbf{y}$) variables "characterize the instantaneous position (configuration) of the Brownian particle in its phase space. Together the vectors ($\mathbf{x}, \mathbf{y}$) define a multidimensional phase space $\mathbf{x} \oplus \mathbf{y}$ within which convective and diffusive solute-particle transport processes occur. The domain of permissible values of $\mathbf{x}$ will always be unbounded; in contrast, the domain of permissible or accessible values of $\mathbf{y}$ will generally be bounded" (Brenner 1987, pp. 66–67), i.e., $\mathbf{y} \in \mathcal{V}$. In this case, a macroscopic transport equation is obtained for the probability density function of a Brownian particle (Brenner 1987, Equations 3.3–3.5):

$$\overline{P}(\mathbf{x}, t|\mathbf{y}') \stackrel{def}{=} \int_{\mathcal{V}} P(\mathbf{x}, \mathbf{y}, t|\mathbf{y}') \, d\mathbf{y} \tag{2.7}$$

where $P(\mathbf{x}, \mathbf{y}, t|\mathbf{y}') \equiv P(\mathbf{x} - \mathbf{x}', \mathbf{y}, t - t'|\mathbf{y}')$ with $\mathbf{x}' = \mathbf{0}$ and $t' = 0$ denotes the "conditional probability density that the Brownian particle is situated at position ($\mathbf{x}, \mathbf{y}$) at time t, given that it was initially introduced into the system at the position ($\mathbf{x}', \mathbf{y}'$) at some earlier time t' ($t > t'$)" (Brenner 1987, p. 68). "For sufficiently long times (i.e. 'long' relative to the time scale of evolution of the microscale transport process, but 'short' relative to the time scale of the macrotransport process) we expect that the particle(s) will loose memory of the initial position(s) $\mathbf{y}'$". Consequently, $\overline{P}(\mathbf{x}, t|\mathbf{y}') \approx \overline{P}(\mathbf{x}, t)$ and a fully macrotransport equation can be determined.

A number of other approaches to upscaling are reviewed by Brenner (1987). Even if based on different definitions of the averaging volume and on distinct mathematical tools, all upscaling methods require closure assumptions to decouple the average system behavior from the pore-scale information: the latter is exclusively incorporated into the upscaled equation through effective parameters that can be determined by laboratory experiments or numerical solution of a closure problem at the unit cell level.

2.2.2 *Flow: From Stokes to Darcy/Brinkman equations*

Single-phase flow of an incompressible Newtonian fluid in porous media in the pore-space Ω_l is described by the Stokes and continuity equations subject to the no-slip boundary condition on A_{ls}:

$$\mu \nabla^2 \mathbf{v} - \nabla p = 0, \quad \nabla \cdot \mathbf{v} = 0, \quad \mathbf{x} \in \Omega_l, \quad \mathbf{v} = \mathbf{0}, \quad \mathbf{x} \in A_{ls}, \tag{2.8}$$

where $\mathbf{v}(\mathbf{x})$ is the fluid velocity, p denotes the fluid dynamic pressure, and μ is the dynamic viscosity.

Upscaling of the Stokes equations (Eq. 2.8) at the pore-scale to the continuum scale has been the subject of numerous investigations, including those relying on multiple-scale expansions (Sanchez-Palencia & Zaoui 1989, Hornung 1997, Auriault & Adler 1995, Mikelić et al. 2006, Peter 2007, Marušić-Paloka & Piatnitski 2005, and references therein), volume averaging (Neuman 1977, and references therein), the method of moments, etc. These studies have demonstrated that Darcy's law, which was empirically established by Darcy (1856), and the continuity equation for $\langle \mathbf{v} \rangle$:

$$\langle \mathbf{v} \rangle = -\frac{\mathbf{K}}{\mu} \cdot \nabla \langle p \rangle, \qquad \nabla \cdot \langle \mathbf{v} \rangle = 0, \qquad \mathbf{x} \in \Omega, \tag{2.9}$$

provide an effective representation of the pore-scale Stokes flow (Hornung 1997, Eq. 4.7). Such upscaling procedures also enable one to formally define the permeability tensor $\mathbf{K}$ in Eq. (2.9) as the average of a "closure variable" $\mathbf{k}(\mathbf{y})$, i.e., $\mathbf{K} = \langle \mathbf{k}(\mathbf{y}) \rangle$. The latter is the unique solution of a local problem (e.g., Hornung 1997, pp. 46–47, Theorem 1.1 and Auriault & Adler 1995, Eq. 22) defined on a representative (unit) cell of the porous medium. "It is well admitted that the existence of continuum behaviors that are macroscopically equivalent to finely heterogeneous media needs a good separation of scales. If l and L are the characteristic lengths at the local and the macroscopic scale, respectively, their ratio should obey" (Auriault et al. 2005):

$$\epsilon = \frac{l}{L} \ll 1, \tag{2.10}$$

To describe flow through "hyperporous" media, Brinkman (1949) introduced a modification of Darcy's law:

$$\nabla \langle p \rangle = -\frac{\mu}{\mathbf{K}} \langle \mathbf{v} \rangle + \mu_e \nabla^2 \langle \mathbf{v} \rangle, \tag{2.11}$$

where μ_e is an effective viscosity "which may differ from μ" (Brinkman 1949). The *raison d'etre* for such a modification was the necessity of obtaining an equation that was valid in the high permeability limit ($|\mathbf{K}| \to \infty$) and that allowed for a direct coupling with the Stokes equations at interfaces separating Stokes flow (infinite permeability regions) and filtration flow (low permeability regions). In Brinkman's words, "this equation has the advantage of approximating Eq. (2.9) for low values of $\mathbf{K}$ and Eq. (2.8) for high values of $\mathbf{K}$".

After its introduction and its widespread use, an increasing research effort was devoted to the identification of domains of validity of both Darcy's & Brinkman's law (Lévy 1983, Auriault 2009, Durlofsky & Brady 2009, and references therein). Brinkman's intuition was mathematically proven later by Goyeau et al. (1997) and Auriault et al. (2005), who used respectively the method of volume averaging and multiple-scale expansions to demonstrate that Brinkman's equation represents a higher-order approximation of Darcy's law when the separation of scales is poor.

2.2.3 *Transport: From advection-diffusion to advection-dispersion equation*

Consider a fluid that contains a dissolved species $\mathcal{M}$, whose dimensional concentration $c(\mathbf{x}, t)$ [$mol L^{-3}$] at point $\mathbf{x} \in \Omega_l$ and time $t > 0$ changes due to advection, molecular diffusion, homogeneous reaction in the liquid phase and heterogeneous reaction at the solid-liquid interface A_{ls}. The first three phenomena are described by an advection-diffusion-reaction equation:

$$\frac{\partial c}{\partial t} + \mathbf{v} \cdot \nabla c = \nabla \cdot (\mathbf{D} \nabla c) + R(c), \qquad \mathbf{x} \in \Omega_l, \quad t > 0, \tag{2.12}$$

where the molecular diffusion coefficient $\mathbf{D}$ is, in general, a positive-definite second-rank tensor. If diffusion is isotropic, $\mathbf{D} = \mathscr{D} \mathbf{I}$ where $\mathscr{D}$ [$L^2 T^{-1}$] is the diffusion coefficient and $\mathbf{I}$ is the identity matrix. The source term $R(c)$ represents a generic homogeneous reaction. At the solid-liquid interface A_{ls} impermeable to flow, mass conservation requires that mass flux of the species $\mathcal{M}$ be balanced by net mass flux due to heterogeneous reaction, $Q(c)$:

$$-\mathbf{n} \cdot \mathbf{D} \nabla c = Q(c), \qquad \mathbf{x} \in A_{ls}, \tag{2.13}$$

In addition to Eq. (2.13), the flow and transport equations, Eq. (2.8) and Eq. (2.12), are supplemented with boundary conditions on the external boundary of the flow domain Ω. The upscaling of Eq. (2.12) and Eq. (2.13) leads to effective equations for the average concentration $\langle c \rangle$, generally written in the following form:

$$\frac{\partial \langle c \rangle}{\partial t} + \langle \mathbf{v} \rangle \cdot \nabla \langle c \rangle = \nabla \cdot (\mathbf{D}^* \nabla \langle c \rangle) + \overline{R}(\langle c \rangle) + \overline{Q}(\langle c \rangle), \qquad \mathbf{x} \in \Omega, \quad t > 0, \tag{2.14}$$

where $\mathbf{D}^*$ is a dispersion tensor and $\overline{R}(\langle c \rangle)$ and $\overline{Q}(\langle c \rangle)$ are effective reactive sources.

A significant research effort and ingenuity has been devoted to the upscaling of various functional forms of $R(c)$ and $Q(c)$ relevant to engineering, chemical, biochemical, hydrological, and other applications (Shapiro & Brenner 1986, 1988, Ochoa-Tapia et al. 1991, Shapiro et al. 1996, Wood & Ford 2007, Wood et al. 2007, van Noorden & Pop 2008, Hesse et al. 2009). Yet, very little, and only recent, attention has been paid to the identification of the applicability conditions of the upscaled models proposed by such a prolific research path.

While useful in a variety of applications, continuum models fail to capture experimentally observed transport features, including a difference between fractal dimensions of the diffusion and dispersion fronts (isoconcentration contours) (Maloy et al. 1998), long tails in breakthrough curves (Neuman & Tartakovsky 2009), and the onset of instability in variable density flows (Tartakovsky et al. 2008b). ADE-based models of transport of (bio-)chemically reactive solutes, which are the focus of our analysis, can significantly over-predict the extent of reactions in mixing-induced chemical transformations (Knutson et al. 2007, Wood et al. 2007, Tartakovsky et al. 2007, Li et al. 2006, Tartakovsky et al. 2008a, and references therein). These and other shortcomings stem from the inadequacy of either standard macroscopic models or their parametrizations or both. Upscaling from the pore-scale, on which governing equations are physically based and well defined, to the continuum scale, on which they are used for qualitative predictions, often enables one to establish the connection between the two modeling scales.

Upscaling approaches that rely on characteristic dimensionless numbers (e.g., the Damköhler and Péclet numbers) can provide quantitative measures for the validity of various upscaling approximations. Auriault and Adler (1995) used multiple scale expansions to establish the applicability range of advection-dispersion equation for a non-reactive solute in terms of Péclet number. Mikelić et al. (2006) provided a rigorous upscaled version of the Taylor dispersion problem with linear heterogeneous reaction. For flow between two parallel reacting plates they established the applicability range of the upscaled equation in terms of the Damköhler and Péclet numbers.

Nonlinearity of governing equations complicates the upscaling of most reactive transport phenomena. It requires a linearization and/or other approximations, whose accuracy and validity cannot be ascertained a priori. This is especially so for a large class of transport processes, such as mixing-induced precipitation, which exhibit highly localized reacting fronts and consequently defy macroscopic descriptions that are completely decoupled from their microscopic counterparts (Ochoa-Tapia et al. 1991, Auriault & Adler 1995, Kechagia et al. 2002).

In the following section, we present results from Battiato et al. (2009) and Battiato and Tartakovsky (2010) that generalize the analyses of Auriault and Adler (1995) and Mikelić et al. (2006) to nonlinear reactive processes. In Section 2.3.1 we consider a multicomponent system undergoing nonlinear homogeneous and linear heterogeneous reaction described by a system of coupled reaction-diffusion equations (RDEs); we specify key physical and (bio-)chemical assumptions that underpin this model and identify the Damköhler numbers for homogeneous and heterogeneous reactions as dimensionless parameters that control the phenomenon. We present the major results of volume averaging (Whitaker 1999) pore-scale equations to derive a system of upscaled RDEs that are commonly used to model mixing-induced precipitation on the continuum scale (e.g., Steefel et al. 2005, and the references therein). The goal here is to identify sufficient conditions for the macroscopic RDEs to be a valid descriptor of mixing-induced precipitation. To focus on the relative effects of nonlinear geochemical reactions and diffusion, we neglect advection.

In Section 2.3.4 we consider the advective-diffusive transport of a solute that undergoes a nonlinear heterogeneous reaction: after reaching a threshold concentration value, it precipitates on the solid matrix to form a crystalline solid. The relative importance of three key pore-scale transport mechanisms (advection, molecular diffusion, and reactions) is quantified by the Péclet (Pe) and Damköhler (Da) numbers. We use multiple-scale expansions to upscale a pore-scale advection-diffusion equation with reactions entering through a boundary condition on the fluid-solid interface, and to establish sufficient conditions under which macroscopic advection-dispersion-reaction equations (ADREs) provide an accurate description of the pore-scale processes. These conditions are summarized by a phase diagram in the (Pe, Da) space, parameterized with a scale-separation parameter that is defined as the ratio of characteristic lengths associated with the pore- and macro-scales.

2.3　APPLICABILITY RANGE OF MACROSCOPIC MODELS FOR REACTIVE SYSTEMS

2.3.1　*Diffusion-reaction equations: mixing-induced precipitation processes*

Consider a porous medium Ω that is fully saturated with an incompressible liquid at rest. The liquid, occupying the pore-space Ω_l, is a solution of two chemical (or biological) species M_1 and M_2 (with respective concentrations c_1 and c_2) that react to form an aqueous reaction product M_3. Whenever c_3, the concentration of M_3, exceeds a threshold value, M_3 undergoes a heterogeneous reaction and precipitates on the solid matrix, forming a precipitate $M_{4(s)}$. In general, this process of mixing-induced precipitation is fully reversible, $M_1 + M_2 \rightleftarrows M_3 \rightleftarrows M_{4(s)}$, and its speed is controlled by the reaction rates k_{12} $[L^3 mol^{-1} T^{-1}]$, k_p $[LT^{-1}]$, k_3 $[T^{-1}]$ and k_d $[mol T^{-1} L^{-2}]$ corresponding to the following reactions:

$$M_1 + M_2 \xrightarrow{k_{12}} M_3 \xrightarrow{k_p} M_{4(s)} \quad \text{and} \quad M_1 + M_2 \xleftarrow{k_3} M_3 \xleftarrow{k_d} M_{4(s)} \tag{2.15}$$

For bimolecular and unimolecular elementary reactions at constant temperature, the change in concentration is proportional to the product of the concentration of the reactants. Hence, the consumption and production rates, R_i^c with $i \in \{1, 2\}$ and R_3^p, of species M_i, $i \in \{1, 2\}$, and M_3, respectively, associated with the homogeneous reaction in Eq. (2.15) are typically concentration-driven and of the form $R_i^c = -R_3^p = -k_{12}\hat{c}_1\hat{c}_2 + k_3\hat{c}_3$. For the heterogeneous reaction, it is common to assume (Knabner et al. 1995, Duijn & Pop 2004, and references therein) that i) precipitation rate r_p is proportional to concentration $\hat{c}_3$, i.e., $r_p = k_p\hat{c}_3$; ii) dissolution rate r_d is constant, $r_d = k_d$; and iii) super-saturation index does not become large enough to support precipitation in the liquid phase, i.e., precipitation of M_3 occurs solely as an overgrowth on solid grains.

With these assumptions, the aqueous concentrations $\hat{c}_i(\hat{\mathbf{r}}, \hat{t})$ $[mol L^{-3}]$ at point $\hat{\mathbf{r}}$ and time $\hat{t}$ satisfy a system of reaction-diffusion equations (RDEs):

$$\frac{\partial \hat{c}_i}{\partial \hat{t}} = \mathscr{D}_i \hat{\nabla}^2 \hat{c}_i - k_{12}\hat{c}_1\hat{c}_2 + k_3\hat{c}_3 \quad \text{for } \hat{\mathbf{r}} \in \Omega_l, \hat{t} > 0 \quad i = 1, 2 \tag{2.16a}$$

$$\frac{\partial \hat{c}_3}{\partial \hat{t}} = \mathscr{D}_3 \hat{\nabla}^2 \hat{c}_3 + k_{12}\hat{c}_1\hat{c}_2 - k_3\hat{c}_3 \quad \text{for } \hat{\mathbf{r}} \in \Omega_l, \hat{t} > 0, \tag{2.16b}$$

subject to the boundary conditions on the (multi-connected) liquid-solid interface $\mathscr{A}_{ls}$:

$$\mathbf{n} \cdot \hat{\nabla}\hat{c}_i = 0, \quad i = 1, 2; \quad -\mathscr{D}_3\mathbf{n} \cdot \hat{\nabla}\hat{c}_3 = k_p(\hat{c}_3 - c_{eq}) \tag{2.17}$$

and the initial conditions:

$$\hat{c}_i(\mathbf{x}, 0) = c_{i0}(\mathbf{x}), \quad i = 1, 2, 3, \quad \Omega_l(0) = \Omega_{l0}, \tag{2.18}$$

when concentration of $M_{4(s)}$ is strictly positive. Here the hatted quantities have appropriate units (physical dimensions), $c_{eq} = k_d / k_p$ is the equilibrium concentration, $\mathscr{D}_i$ $[L^2 T^{-1}]$ $(i = 1, 2, 3)$ are the diffusion coefficients of the aqueous species M_1, M_2, and M_3, respectively. Due to precipitation and dissolution, the liquid-solid interface $\mathscr{A}_{ls}(\hat{t})$, with the outward normal unit vector $\mathbf{n}(\hat{t})$, evolves in time $\hat{t}$ with velocity $\mathbf{v}$ $[LT^{-1}]$, according to $\rho_c \mathbf{v} \cdot \mathbf{n} = k_p(\hat{c}_3 - c_{eq})$, where ρ_c $[mol_s L^{-3}]$ is the molar density of the precipitate. The dynamics of the interface $\mathscr{A}_{ls}(\hat{t})$, result from a modeling assumption about the dependence of $\mathbf{v}$ on precipitation/dissolution rates and mass conservation (van Noorden & Pop 2008).

To be specific, we consider a scenario in which two identical solvents (e.g., water), one containing M_1 with concentration $\hat{c}_{10}$ and the other containing M_2 with concentration $\hat{c}_{20}$, are brought in contact with each other at time $\hat{t} = 0$. Since reactants M_1 and M_2 are initially separated, no reactions took place and the initial concentration of reaction product M_3 is $\hat{c}_{30} = 0$. This is a typical situation, corresponding, for example, to injection of a solution of M_1 into a porous medium occupied by a solution of M_2 (Tartakovsky et al. 2008a).

The characteristic time scales associated with the chemical reactions Eq. (2.15) are $\tau_1 = \tau_2 = 1/k_{12}c_{10}$ for concentrations $\hat{c}_1$ and $\hat{c}_2$, and $\tau_3 = c_{eq}/k_{12}c_{10}^2$ for concentration $\hat{c}_3$. To simplify the presentation, we assume that the diffusion coefficients for reactants M_1 and M_2 and product $M_{3(l)}$ are the same, $\mathscr{D}_1 = \mathscr{D}_2 = \mathscr{D}_3 = \mathscr{D}$. Let us introduce dimensionless quantities:

$$t = \frac{\hat{t}}{\tau}, \quad q = \frac{c_{eq}}{c_{10}}, \quad c_i = \frac{\hat{c}_i}{c_{10}}, \quad c_3 = \frac{\hat{c}_3}{c_{eq}}, \quad K = \frac{k_3 c_{eq}}{k_{12}c_{10}^2}, \quad Da = \frac{l^2 k_{12}c_{10}}{\mathscr{D}}, \tag{2.19}$$

where $i = 1, 2$; l denotes a characteristic length scale associated with pore structure; and the Damköhler number Da is the ratio of diffusion and reaction time scales for species M_i ($i = 1, 2, 3$). RDEs Eq. (2.16) can now be written in a dimensionless form as:

$$\frac{\partial c_i}{\partial t} = \frac{l^2}{Da}\hat{\nabla}^2 c_i - c_1 c_2 + K c_3 \ (i = 1, 2), \qquad q\frac{\partial c_3}{\partial t} = \frac{ql^2}{Da}\hat{\nabla}^2 c_3 + c_1 c_2 - K c_3, \tag{2.20}$$

Following Tartakovsky et al. (2007), we define the Damköhler number for the precipitation/dissolution process as:

$$Da_{ls} = \frac{k_p l}{\mathscr{D}}, \tag{2.21}$$

This yields a dimensionless form of the boundary conditions on the liquid-solid interface $\mathscr{A}_{ls}$:

$$\mathbf{n} \cdot \hat{\nabla} c_i = 0 \ (i = 1, 2), \quad \mathbf{n} \cdot l\hat{\nabla} c_3 = Da_{ls}(1 - c_3). \tag{2.22}$$

We proceed by employing the local volume averaging (Whitaker 1999) to upscale the pore-scale equations, Eq. (2.20) and Eq. (2.22) to the macroscopic scale. Section 2.3.2 contains definitions of the averaging procedure. The results of the upscaling procedure are presented in form of Propositions in Section 2.3.3 (Battiato et al. 2009). The results are summarized in a phase diagram identifying sufficient conditions under which the upscaled (macroscopic) description is valid.

2.3.2 *Preliminaries*

Consider a portion of the porous medium $\mathscr{V} \in \Omega$ whose volume is $|\mathscr{V}|$ and characteristic radius $r_0 \gg l$, where l is the pore-geometry length scale. Let $B(\hat{\mathbf{x}}) \in \mathscr{V}$ denote the volume of the liquid phase contained in $\mathscr{V}$, which is centered at $\hat{\mathbf{x}} \in \Omega$. If a characteristic length-scale of the macroscopic domain Ω is L, then the size of the averaging volume $\mathscr{V}$ is selected to satisfy $l \ll r_0 \ll L$.

Following Whitaker (1999), we define superficial and intrinsic averages of a quantity $c(\hat{\mathbf{r}})$ with $\hat{\mathbf{r}} \in \Omega_l$ as:

$$\langle c \rangle(\hat{\mathbf{x}}) = \frac{1}{|\mathscr{V}|}\int_{B(\hat{\mathbf{x}})} c(\hat{\mathbf{r}})\, \mathrm{d}^3 r \quad \text{and} \quad \langle c \rangle_B(\hat{\mathbf{x}}) = \frac{1}{|B(\hat{\mathbf{x}})|}\int_{B(\hat{\mathbf{x}})} c(\hat{\mathbf{r}})\, \mathrm{d}^3 r, \tag{2.23}$$

respectively. The two averages are related through porosity $\phi \equiv |B|/|\mathscr{V}|$ by $\langle c \rangle = \phi\langle c \rangle_B$. The application of spatial averaging is facilitated by the spatial averaging theorem Eq. (2.4).

Let L_c, L_{c1} and L_ϕ denote characteristic length-scales associated with the macroscopic quantities $\langle c \rangle_B$, $\hat{\nabla}\langle c \rangle_B$ and ϕ, respectively. These scales are defined by (Whitaker 1999, p. 19):

$$\hat{\nabla} f_i(x) = \mathcal{O}\left(\frac{\Delta f_i}{L_i}\right), \qquad \Delta f_i(x) \equiv f_i\left(x + \frac{L_i}{2}\right) - f_i\left(x - \frac{L_i}{2}\right) \tag{2.24}$$

for $f_i = \{\langle c \rangle_B, \hat{\nabla}\langle c \rangle_B, \phi\}$ and $L_i = \{L_c, L_{c1}, L_\phi\}$, respectively. The notation $f = \mathcal{O}(g)$ denotes an order of magnitude estimate in the following sense (Kundu & Cohen 2008, p. 391):

$$\frac{|g|}{\sqrt{10}} \leq |f| \leq |g|\sqrt{10}. \tag{2.25}$$

2.3.3 *Upscaling via volume averaging*

In this section, we present results of Battiato et al. (2009) for the upscaling of the third equation in Eq. (2.20). The remaining two equations in Eq. (2.20) are upscaled in a similar fashion. Details of the derivation can be found in Battiato et al. (2009).

We assume that reactions in the fluid phase are much faster than precipitation on the solid phase, so that $\langle \partial c_3 / \partial t \rangle = \partial \langle c_3 \rangle / \partial t$. No assumptions are required for the upscaling of the linear term $\langle Kc_3 \rangle_B = K \langle c_3 \rangle_B$. The averaging procedure is presented below as a series of propositions. Their proofs are provided in Battiato et al. (2009).

Proposition 2.3.1. *Suppose that the following scale constraints hold:*

1) $l \ll r_0$,
2) $r_0^2 \ll \overline{L}^2$ *where* $\overline{L} = min\{L_{c1}, L_\phi\}$,
3) $\epsilon \ll 1$ *where* $\epsilon = l/L_c$,
4) $r_0 \ll L_c$,
5) $r_0^2 \ll L_c L_{c1}$.

Then the average of the Laplacian in Eq. (2.20) can be approximated by:

$$\langle \hat{\nabla}^2 c_3 \rangle = \phi \hat{\nabla}^2 \langle c_3 \rangle_B + \hat{\nabla}\phi \cdot \hat{\nabla}\langle c_3 \rangle_B + \frac{1}{|\mathcal{V}|}\hat{\nabla} \cdot \int_{A_{ls}} \tilde{c}_3 \mathbf{n}_{ls} \, dA - a_v \, Da_{ls} \frac{\langle c_3 \rangle_B - 1}{l} \tag{2.26}$$

where $a_v \equiv |\mathcal{A}_{ls}|/|\mathcal{V}|$ *and* $\tilde{c}_3$ *is such that* $c_3 = \langle c_3 \rangle_B + \tilde{c}_3$.

Proposition 2.3.2. *Suppose that the scale constraints 3)–5) of the Proposition 2.3.1 hold. Then the average of the reaction term in Eq. (2.20) can be approximated by:*

$$\langle c_1 c_2 \rangle = \phi \langle c_1 \rangle_B \langle c_2 \rangle_B. \tag{2.27}$$

Proposition 2.3.3. *Suppose that in addition to the constraints in Proposition 2.3.1 the following scale constraints hold:*

1) $a_v \approx l^{-1}$,
2) $t \gg Da$,
3) $l \ll L_\phi$.

Then, the concentration fluctuations $\tilde{c}_3$ satisfy a differential equation:

$$0 = \frac{ql^2}{Da}\hat{\nabla}^2\tilde{c}_3 + \frac{qa_v l}{\phi}\frac{Da_{ls}}{Da}(\langle c_3 \rangle_B - 1) + \tilde{c}_1 \langle c_2 \rangle_B + \tilde{c}_2 \langle c_1 \rangle_B + \tilde{c}_1\tilde{c}_2 - K\tilde{c}_3 \tag{2.28}$$

subject to the boundary conditions:

$$-\mathbf{n} \cdot \hat{\nabla}\tilde{c}_3 = \mathbf{n} \cdot \hat{\nabla}\langle c_3 \rangle_B + Da_{ls}\frac{\langle c_3 \rangle_B + \tilde{c}_3 - 1}{l}. \tag{2.29}$$

Boundary-value problems for fluctuations $\tilde{c}_1$ and $\tilde{c}_2$ are derived in a similar manner. Further progress requires an assumption of periodicity of the porous medium.

Proposition 2.3.4. *Suppose that in addition to the scale constraints imposed by Propositions 2.3.1 and 2.3.3 the porous medium is periodic with a unit cell characterized by* $\mathbf{n}(\hat{\mathbf{r}} + \hat{\mathbf{l}}_i) = \mathbf{n}(\hat{\mathbf{r}})$, *where* $\hat{\mathbf{l}}_i$ *with* $i = 1, 2, 3$ *represents the three lattice vectors describing a spatially periodic porous medium. Then concentration fluctuations are periodic,* $\tilde{c}(\hat{\mathbf{r}} + \hat{\mathbf{l}}_i) = \tilde{c}(\hat{\mathbf{r}})$, *and* $\langle c_3 \rangle_B$ *and* $\hat{\nabla} \langle c_3 \rangle_B$ *in Eqs. (2.28) and (2.29) are evaluated at the centroid.*

Proposition 2.3.5. *Suppose that in addition to the constraint imposed by Proposition 2.3.3 the following constraints hold:*

1) $Da_{ls} \ll \epsilon$.
2) $Da \ll 1$,

Then concentration fluctuations $\tilde{c}_3$ *can be represented in terms of the macroscopic variables as:*

$$\tilde{c}_3 = \hat{\mathbf{b}} \cdot \hat{\nabla} \langle c_3 \rangle_B + s \langle c_3 \rangle_B + \psi, \qquad (2.30)$$

where the closure variables $\hat{\mathbf{b}}$, s *and* ψ *are solutions of the boundary value problems (wherein* $j = 1, 2, 3$):

$$\hat{\nabla}^2 \hat{\mathbf{b}} - \frac{k_3}{\mathscr{D}} \hat{\mathbf{b}} = 0, \qquad -\mathbf{n} \cdot \hat{\nabla} \hat{\mathbf{b}} = \mathbf{n} \quad at\ A_{ls}, \qquad \hat{\mathbf{b}}(\hat{\mathbf{r}} + \hat{\mathbf{l}}_j) = \hat{\mathbf{b}}(\hat{\mathbf{r}}) \qquad (2.31)$$

$$\hat{\nabla}^2 s - \frac{k_3}{\mathscr{D}} s = -\frac{a_v Da_{ls}}{\phi l}, \qquad -\mathbf{n} \cdot \hat{\nabla} s = \frac{Da_{ls}}{l} \langle c_3 \rangle^l \quad at\ A_{ls}, \qquad s(\hat{\mathbf{r}} + \hat{\mathbf{l}}_j) = s(\hat{\mathbf{r}}); \qquad (2.32)$$

$$\hat{\nabla}^2 \psi - \frac{k_3}{\mathscr{D}} \psi = \frac{a_v Da_{ls}}{\phi l}, \qquad -\mathbf{n} \cdot \hat{\nabla} \psi = -\frac{Da_{ls}}{l} \quad at\ A_{ls}, \qquad \psi(\hat{\mathbf{r}} + \hat{\mathbf{l}}_j) = \psi(\hat{\mathbf{r}}). \qquad (2.33)$$

Combining the results from Propositions 2.3.1–2.3.5 with analogous results for $\langle c_1 \rangle_B$ and $\langle c_2 \rangle_B$, the volume averaging of Eq. (2.20) leads to a system of macroscopic equations:

$$\phi \frac{\partial \langle c_i \rangle_B}{\partial t} = \frac{\epsilon^2}{Da} \nabla \cdot (\phi \mathbf{D}_{\mathrm{eff}} \cdot \nabla \langle c_i \rangle_B) - \phi \langle c_1 \rangle_B \langle c_2 \rangle_B + \phi K \langle c_3 \rangle_B \qquad (i = 1, 2), \qquad (2.34)$$

$$\phi q \frac{\partial \langle c_3 \rangle_B}{\partial t} = \frac{q \epsilon^2}{Da} \nabla \cdot (\phi \mathbf{D}_{\mathrm{eff}} \cdot \nabla \langle c_3 \rangle_B) - q a_v l \frac{Da_{ls}}{Da} [\langle c_3 \rangle_B - 1] + \phi \langle c_1 \rangle_B \langle c_2 \rangle_B - \phi K \langle c_3 \rangle_B, \qquad (2.35)$$

where the effective diffusivity tensor $\mathbf{D}_{\mathrm{eff}}$ is defined as:

$$\mathbf{D}_{\mathrm{eff}} = \mathbf{I} + \frac{1}{|B|} \int_{A_{ls}} \mathbf{n}\hat{\mathbf{b}}\, dA. \qquad (2.36)$$

According to Proposition 2.3.5, a sufficient condition for the validity of the macroscopic description Eqs. (2.34)–(2.35), requires that $Da \ll 1$, which implies that on the pore scale the system is well-mixed with diffusion dominating reactions. Further insight is gained by relating different macroscopic diffusion and/or reaction regimes to the Damköhler number Da expressed in terms of the scale-separation parameter ϵ. (This is conceptually similar to the Auriault and Adler (1995) analysis of macroscopic dispersion equations, which identifies distinct transport regimes by expressing the Péclet number as powers of ϵ.) Interplay between the Damköhler number and ϵ determines whether macroscopic RDEs, Eqs. (2.34) and (2.35), are diffusion or reaction dominated. For $Da < \epsilon^2$, the macroscopic process is diffusion-driven and the nonlinear effects introduced by reactions are negligible. The two mechanisms are of the same order of magnitude in the region $\epsilon^2 < Da < \epsilon$, and reactions dominate diffusion if $\epsilon < Da < 1$. Mixing-induced precipitation, which is characterized by $Da \gg 1$, falls into the category of physical phenomena for which pore-scale or hybrid simulations are a priori necessary.

2.3.4   *Advection-diffusion-reaction equation*

Consider reactive transport in a porous medium whose characteristic length is L. Let us assume that the medium can be represented microscopically by a collection of spatially periodic "unit cells" with a characteristic length l, such that a scale parameter $\varepsilon \equiv l/L \ll 1$. Spatially periodic representations of micro-structures of porous media are routinely used to derive macroscopic properties and effective models of phenomena taking place in disordered media that lack such periodicity (Nitsche & Brenner 1989, Section 2). The unit cell $\hat{\mathcal{V}} = \hat{B} \cup \hat{G}$ consists of the pore space $\hat{B}$ and the impermeable solid matrix $\hat{G}$ that are separated by the smooth surface $\hat{A}_{ls}$. The pore spaces $\hat{B}$ of each cell $\hat{\mathcal{V}}$ form a multi-connected pore-space domain $\hat{B}^\varepsilon \subset \hat{\Omega}$ bounded by the smooth surface $\hat{A}_{ls}^\varepsilon$.

2.3.4.1   *Governing equations*

Single-phase flow of an incompressible fluid in the pore-space $\hat{B}^\varepsilon$ is described by the Stokes and continuity equations Eq. (2.8) subject to the no-slip boundary condition on $\hat{A}_{ls}^\varepsilon$. The fluid contains a dissolved species $\mathcal{M}$, whose molar concentration $\hat{c}_\varepsilon(\hat{\mathbf{x}}, \hat{t})$ $[mol L^{-3}]$ at point $\hat{\mathbf{x}} \in \hat{B}^\varepsilon$ and time $\hat{t} > 0$ changes due to advection, molecular diffusion, and a nonlinear heterogeneous reaction at the solid-liquid interface $\hat{A}_{ls}^\varepsilon$. The first two phenomena are described by the advection-diffusion equation Eq. (2.12).

Whenever the concentration $\hat{c}_\varepsilon$ exceeds a threshold value $\bar{c}$, a heterogeneous reaction $n\mathcal{M} \leftrightarrow \mathcal{N}_{(s)}$ occurs, in which n molecules of the solute $\mathcal{M}$ precipitate in the form of one molecule of a crystalline solid $\mathcal{N}_{(s)}$. At the solid-liquid interface $\hat{A}_{ls}^\varepsilon$ impermeable to flow, mass conservation requires that mass flux of the species $\mathcal{M}$ be balanced by the difference between the precipitation rate R_p and the dissolution rate R_d:

$$-\mathbf{n} \cdot \hat{\mathbf{D}} \hat{\nabla} \hat{c}_\varepsilon = R_p - R_d, \tag{2.37}$$

where $\mathbf{n}$ is the outward unit normal vector of $\hat{A}_{ls}^\varepsilon$. Following Knabner et al. (1995), we assume that $R_p = \hat{k}\hat{c}_\varepsilon^a$ and $R_d = \hat{k}\bar{c}^a$, where $\hat{k}$ $[L^{3a-2}T^{-1}mol^{1-a}]$ is the reaction rate constant, $a \in \mathbf{Z}^+$ is related to the order of reaction n (Morse & Arvidson 2002, Eq. 6), and the threshold concentration $\bar{c}$ represents the solubility product (Morse & Arvidson 2002). Mass conservation on the liquid-solid interface $\hat{A}_{ls}^\varepsilon$ yields a boundary condition (Morse & Arvidson 2002, Eq. 5):

$$-\mathbf{n} \cdot \hat{\mathbf{D}} \hat{\nabla} \hat{c}_\varepsilon = \hat{k}(\hat{c}_\varepsilon^a - \bar{c}^a), \qquad \hat{\mathbf{x}} \in \hat{A}_{ls}^\varepsilon, \qquad \hat{t} > 0. \tag{2.38}$$

In addition to Eq. (2.38), the flow and transport equations Eq. (2.8) and Eq. (2.12) are supplemented with proper boundary conditions on the external boundary of the flow domain $\hat{\Omega}$.

2.3.4.2   *Dimensionless formulation*

Let us introduce dimensionless quantities:

$$c_\varepsilon = \hat{c}_\varepsilon/\bar{c}, \quad \mathbf{x} = \hat{\mathbf{x}}/L, \quad \mathbf{v}_\varepsilon = \hat{\mathbf{v}}_\varepsilon/U, \quad \mathbf{D} = \hat{\mathbf{D}}/D, \quad p = \hat{p}l^2/\hat{\mu}UL, \tag{2.39}$$

where D and U are characteristic values of $\mathbf{D}$ and $\mathbf{v}_\varepsilon$, respectively. The scaling of pressure $\hat{p}$ ensures that the pressure gradient and the viscous term are of the same order of magnitude, as prescribed by Stokes equation (Auriault & Adler 1995, Eqs. 15 and 16). Furthermore, we define three time scales associated with diffusion $(\hat{t}_D)$, reactions $(\hat{t}_R)$ and advection $(\hat{t}_A)$ as:

$$\hat{t}_D = \frac{L^2}{D}, \qquad \hat{t}_R = \frac{L}{\hat{k}\bar{c}^{a-1}}, \qquad \hat{t}_A = \frac{L}{U}. \tag{2.40}$$

Ratios between these time scales define the dimensionless Damköhler $(Da = \hat{t}_D/\hat{t}_R)$ and Péclet $(Pe = \hat{t}_D/\hat{t}_A)$ numbers:

$$Da = \frac{L\hat{k}\bar{c}^{a-1}}{D} \quad \text{and} \quad Pe = \frac{UL}{D}. \tag{2.41}$$

Rewriting Eqs. (2.8), (2.12) and (2.38) in terms of the dimensionless quantities Eq. (2.39) and the dimensionless time $t = \hat{t}/\hat{t}_D$ yields a dimensionless form of the flow equations:

$$\varepsilon^2 \nabla^2 \mathbf{v}_\varepsilon - \nabla p = 0, \qquad \nabla \cdot \mathbf{v}_\varepsilon = 0, \qquad \mathbf{x} \in B^\varepsilon, \tag{2.42}$$

subject to:

$$\mathbf{v}_\varepsilon = \mathbf{0}, \qquad \mathbf{x} \in A_{ls}^\varepsilon \tag{2.43}$$

and a dimensionless form of the transport equation:

$$\frac{\partial c_\varepsilon}{\partial t} + \nabla \cdot (-\mathbf{D}\nabla c_\varepsilon + \mathrm{Pe}\mathbf{v}_\varepsilon c_\varepsilon) = 0, \qquad \mathbf{x} \in B^\varepsilon, \qquad t > 0, \tag{2.44}$$

subject to:

$$-\mathbf{n} \cdot \mathbf{D}\nabla c_\varepsilon = \mathrm{Da}(c_\varepsilon^a - 1), \qquad \mathbf{x} \in A_{ls}^\varepsilon, \qquad t > 0. \tag{2.45}$$

2.3.4.3   *Homogenization via multiple-scale expansions*

Homogenization aims to derive effective equations for average state variables that are representative of an averaging volume (e.g., Darcy scale). To this end, two types of local averages of a quantity $c(\mathbf{x})$ can be defined as in Eq. (2.23). We also define:

$$\langle \mathcal{A} \rangle_{ls} \equiv \frac{1}{|A_{ls}|} \int\limits_{A_{ls}(\mathbf{x})} \mathcal{A} \, d\mathbf{y}. \tag{2.46}$$

In the subsequent derivation of effective (continuum- or Darcy-scale) equations for average flow velocity $\langle \mathbf{v}(\mathbf{x}) \rangle$ and solute concentration $\langle c(\mathbf{x}, t) \rangle$, we employ the method of multiple-scale expansions (Auriault & Adler 1995, Hornung 1997).

The method of multiple-scale expansions introduces a fast space variable $\mathbf{y}$ and two time variables τ_r and τ_a:

$$\mathbf{y} = \frac{\mathbf{x}}{\varepsilon}, \qquad \tau_r = \mathrm{Da}\,t = \frac{\hat{t}}{\hat{t}_R}, \qquad \tau_a = \mathrm{Pe}\,t = \frac{\hat{t}}{\hat{t}_A}. \tag{2.47}$$

Furthermore, it represents the concentration $c_\varepsilon(\mathbf{x}, t)$ in Eq. (2.44) as $c_\varepsilon(\mathbf{x}, t) := c(\mathbf{x}, \mathbf{y}, t, \tau_r, \tau_a)$. The latter is expanded into an asymptotic series in powers of ε:

$$c(\mathbf{x}, \mathbf{y}, t, \tau_r, \tau_a) = \sum_{m=0}^{\infty} \varepsilon^m c_m(\mathbf{x}, \mathbf{y}, t, \tau_r, \tau_a), \tag{2.48}$$

wherein $c_m(\mathbf{x}, \mathbf{y}, t, \tau_r, \tau_a)$ $(m = 0, 1, \ldots)$ are $\mathcal{V}$-periodic in $\mathbf{y}$. Finally, we set:

$$Pe = \varepsilon^{-\alpha} \quad \text{and} \quad Da = \varepsilon^\beta, \tag{2.49}$$

with the exponents α and β determining the system behavior. For example, transport due to advection and dispersion at the pore scale is not homogenizable if $\alpha \geq 2$ (Auriault & Adler 1995, Section 3.5, Table 1).

Pore-scale reactive transport processes described by Eqs. (2.44)–(2.45) can be homogenized, i.e., approximated up to order ε^2 with an effective ADRE:

$$\phi \frac{\partial \langle c \rangle_B}{\partial t} = \nabla \cdot (\mathbf{D}^* \nabla \langle c \rangle_B - Pe \langle c \rangle_B \langle \mathbf{v} \rangle) - \varepsilon^{-1} \phi Da \mathcal{K}^* (\langle c \rangle_B^a - 1), \qquad \mathbf{x} \in \Omega, \tag{2.50}$$

provided the following conditions are met:

1) $\varepsilon \ll 1$,
2) $Pe < \varepsilon^{-2}$,
3) $Da/Pe < \varepsilon$,

4) $Da < 1$,

5) $\langle \chi \rangle_{ls} \approx \langle \chi \rangle_B$.

In Eq. (2.50), the dimensionless effective reaction rate constant $\mathcal{K}^*$ is determined by the pore geometry:

$$\mathcal{K}^* = \frac{|A_{ls}|}{|B|}, \qquad (2.51)$$

and the dispersion tensor $\mathbf{D}^*$ is given by:

$$\mathbf{D}^* = \langle \mathbf{D}(\mathbf{I} + \nabla_{\mathbf{y}}\chi) \rangle + \varepsilon\, Pe \langle \chi \mathbf{k} \rangle \nabla_{\mathbf{x}} p_0. \qquad (2.52)$$

The closure variable $\chi(\mathbf{y})$ has zero mean, $\langle \chi \rangle = \mathbf{0}$, and is defined as a solution of the local problem:

$$-\nabla_{\mathbf{y}} \cdot \mathbf{D}(\nabla_{\mathbf{y}}\chi + \mathbf{I}) + \varepsilon\, Pe\, \mathbf{v}_0 \nabla_{\mathbf{y}}\chi = \varepsilon\, Pe\, (\langle \mathbf{v}_0 \rangle_B - \mathbf{v}_0), \qquad \mathbf{y} \in B; \qquad (2.53a)$$

$$-\mathbf{n} \cdot \mathbf{D}(\nabla_{\mathbf{y}}\chi + \mathbf{I}) = 0, \qquad \mathbf{y} \in A_{ls}; \qquad (2.53b)$$

where $\mathbf{v}_0 = -\mathbf{k} \cdot \nabla_{\mathbf{x}} p_0$ and the pressure p_0 is a solution of:

$$\langle \mathbf{v} \rangle = -\mathbf{K} \cdot \nabla p_0, \qquad \nabla \cdot \langle \mathbf{v} \rangle = 0, \qquad \mathbf{x} \in \Omega. \qquad (2.54)$$

Details of the derivations can be found in Battiato and Tartakovsky (2010). Constraints 1)–4) ensure the separation of scales. While constraint 1) is almost always met in practical applications, the rest of them depend on the relative importance of advective, diffusive, and reactive mechanisms of transport. These conditions are summarized in the phase diagram in Fig. 2.1, where the line $\beta = 0$ refers to $Da = 1$ and the half-space $\beta > 0$ to $Da < 1$ because $\varepsilon < 1$; the line $\alpha = 2$ refers to $Pe = \varepsilon^{-2}$ and the half-space $\alpha < 2$ refers to $Pe < \varepsilon^{-2}$; the line $\alpha + \beta = 1$ refers to $Da/Pe = \varepsilon$; and the half-space underneath this line refers to $Da/Pe < \varepsilon$. Constraints 3) and 4) require that either diffusion or advection-diffusion dominate reactions at the pore scale. This allows one to decouple the pore- and continuum-scale descriptions. Constraint 5) is not required for scale separation, but facilitates the derivation of the effective parameters Eq. (2.51) and Eq. (2.52). This constraint allows one to interchange the surface and volume averages, $\langle c_1 \rangle_{ls} \approx \langle c_1 \rangle_B$, within errors on the order of ε^2.

The results above generalize the conclusions of the analysis of reactive-diffusive transport performed in Section 2.3.1, which relied on the method of volume averaging. While using different upscaling approaches, both analyses provide the same bound on the Damköhler number Da in the absence of advection. The effective reaction rate $\mathcal{K}^*$ for heterogeneous reactions Eq. (2.51) is likewise consistent with that obtained by Battiato et al. (2009). This suggests that the conditions for validity and breakdown of continuum models of reactive transport presented in the phase diagram in Fig. 2.1 are universal and independent of the upscaling method. Finally, these upscaling results justify the use of reaction terms similar to the one in Eq. (2.50) in continuum models of precipitation and dissolution processes in porous media (Lichtner & Tartakovsky 2003, Tartakovsky et al. 2009, Broyda et al. 2010).

2.4   HYBRID MODELS FOR TRANSPORT IN POROUS MEDIA

2.4.1   *Intrusive hybrid algorithm*

Having identified the criteria of the applicability of macroscopic equations, we present a hybrid algorithm that couples pore-scale simulations in a small domain Ω_p with continuum simulations elsewhere in the computational domain, Ω / Ω_p. The coupling is accomplished via an iterative procedure in a handshake region Ω_{pc}, where both pore-scale and continuum-scale descriptions are solved iteratively to ensure the continuity of state variables and their fluxes across the interface between Ω_p and the rest of the computational domain.

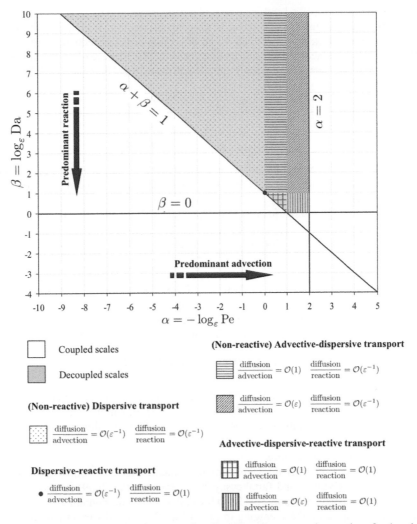

Figure 2.1   Phase diagram indicating the range of applicability of macroscopic equations for the advection-reaction-diffusion system Eqs. (2.44)–(2.45), in terms of *Pe* and *Da*. The grey region identifies the sufficient conditions under which the macroscopic equations hold. In the white region, macro- and micro-scale problems are coupled and have to be solved simultaneously. Also identified are different transport regimes depending on the order of magnitude of *Pe* and *Da*. Diffusion, advection, and reaction are of the same order of magnitude at the point $(\alpha, \beta) = (1, 0)$. After Battiato and Tartakovsky (2010).

This section contains a general formulation of flow and transport equations at the pore- and continuum-scales (Section 4.1.1), as well as an outline of the proposed hybrid algorithm (Section 2.4.1.2). Both the hybrid formulation and its numerical implementation are demonstrated in Section 2.4.2 by applying them to model Taylor dispersion in a planar fracture with chemically reactive walls. In Section 2.4.4, we use this well-studied problem to validate our hybrid algorithm via comparison with analytical solutions and two-dimensional pore-scale numerical simulations.

2.4.1.1   *Governing equations at the pore- and continuum-scale*

Consider reactive transport in a fully-saturated porous medium Ω^T. Within the pore space Ω_l contained in Ω^T, single-phase flow of an incompressible fluid is described by the Stokes and continuity equations Eq. (2.8), subject to the no-slip boundary condition on the solid-liquid interface. The flow is driven by boundary conditions imposed on $\partial\Omega^T$, the external boundary of Ω^T.

The fluid contains a dissolved species with molar concentration $c(\mathbf{x}, t)$ $[ML^{-3}]$ that undergoes advection, molecular diffusion and a linear heterogeneous reaction at the solid-liquid interface A_{sl}. The evolution of $c(\mathbf{x}, t)$ is described by an advection-diffusion equation Eq. (2.12) with $\mathbf{D} = \mathscr{D}\mathbf{I}$ and $R(c) = 0$, subject to the boundary condition Eq. (2.13) on the solid-fluid interface with $Q(c) = \mathscr{K}c$. Proper boundary conditions are applied on the external boundary $\partial\Omega^T$. This leads to:

$$\frac{\partial c}{\partial t} + \nabla \cdot (\mathbf{v}c) = \mathscr{D}\nabla^2 c, \tag{2.55a}$$

subject to:

$$-\mathbf{n} \cdot \mathscr{D}\nabla c = \mathscr{K}c. \tag{2.55b}$$

where $\mathscr{D}$ $[L^2T^{-1}]$ is the molecular diffusion coefficient, $\mathscr{K}$ $[LT^{-1}]$ is the reaction constant describing an interface reaction (e.g., linear microbial degradation), and $\mathbf{n}$ is the outward unit normal vector of A_{sl}.

Let $\overline{A}(\mathbf{x}, t)$ denote the volumetric average of a pore-scale quantity $A(\mathbf{x}, t)$ defined as:

$$\overline{A}(\mathbf{x}, t) \equiv \frac{1}{\phi|\mathscr{V}|} \int\limits_{\mathscr{V}(\mathbf{x})} A(\mathbf{y}, t)\,\mathrm{d}\mathbf{y}, \tag{2.56}$$

where ϕ is the porosity of a porous medium and the averaging volume $\mathscr{V}$ might or might not constitute a representative elementary volume (REV).

The upscaling of Eq. (2.55) by standard upscaling methodologies e.g., multiple-scale expansions or volumetric averaging (e.g., Battiato & Tartakovsky 2010, and references therein), leads to:

$$\phi\frac{\partial\overline{c}}{\partial t} + \phi\nabla \cdot (\mathbf{V}\overline{c}) = \nabla \cdot (\mathbf{D}^*\nabla\overline{c}) - K\overline{c}, \tag{2.57}$$

where $\mathbf{D}^*$ is the dispersion tensor, $\mathbf{V} = \phi\overline{\mathbf{v}}$ is Darcy's flux (Auriault & Adler 1995) and K is the effective reaction rate. As previously discussed, a number of simplifying approximations are required for Eq. (2.57) to be valid regardless of the choice of an upscaling technique.

2.4.1.2 General hybrid formulation

When Darcy's law is valid over the whole computational domain Ω^T but one or more of the sufficient conditions (Battiato and Tartakovsky 2010, Battiato et al. 2009) for the validity of the continuum-scale transport equation Eq. (2.57) break down in a sub-domain Ω_p of the computational domain Ω (Fig. 2.2), the averaging in Ω_p of Eq. (2.55) results in an integro-differential equation:

$$\phi\frac{\mathrm{d}\overline{c}}{\mathrm{d}t} + \overline{\nabla \cdot (\mathbf{v}c)} = \mathscr{D}\overline{\nabla^2 c}. \tag{2.58}$$

Here the averaging of Eq. (2.56) is defined over $\mathscr{V} \equiv \Omega_p(\mathbf{x}^\star)$ and $\mathbf{x}^\star$ is the centroid of Ω_p, i.e., the subdomain Ω_p shrinks to a point $\mathbf{x}^\star \in \Omega^T$.

Application of Gauss' theorem, boundary condition Eq. (2.55b) and the no-slip condition yield:

$$\phi\frac{\partial\overline{c}}{\partial t} = -\frac{1}{\phi|\Omega_p|}\int\limits_{A_{ll}^P} q_n\,\mathrm{d}s - \frac{1}{\phi|\Omega_p|}\int\limits_{A_{sl}^P} \mathscr{K}c\,\mathrm{d}s, \tag{2.59}$$

where $A_p = A_{ll}^P \cup A_{sl}^P$ is the bounding surface of Ω_p and consists of liquid-liquid (A_{ll}^P) and solid-liquid (A_{sl}^P) segments, $\mathrm{d}s$ is an infinitesimal element of A_p and $q_n = \mathbf{n} \cdot (\mathbf{v}c - \mathscr{D}\nabla c)$ is the flux through the liquid-liquid portion of the boundary, A_p. The right hand side of Eq. (2.59) depends on pore-scale quantities. It represents the fluxes exchanged at the boundary A_p between the pore- and continuum-scale descriptions. While multiscale approaches (Christie 1996, Efendief & Durlofsky 2003, Langlo & Espedal 1994, among others) aim to decouple the two descriptions by employing closure assumptions for the the unresolved flux, q_n, our goal is to preserve the nonlinearity of the

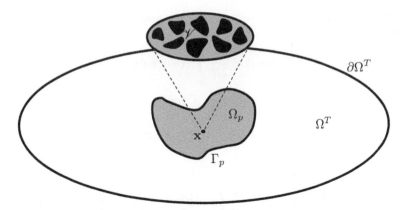

Figure 2.2   A schematic representation of the pore- and continuum-scale domains. After Battiato et al. (2010).

problem and to compute the unresolved normal flux q_n without any assumption on the microscale behavior.

This will allow us to bypass the assumptions needed for continuum-scale models. To this end, we obtain the pore-scale concentration $c(\mathbf{x}, t)$ in Eq. (2.59) by solving the transport problem Eq. (2.55) defined on Ω_p. The boundary condition Eq. (2.55b) is now defined on the union of all solid-liquid surfaces A_{sl} contained in Ω_p. On the fluid-fluid segments A_{ll}^p, mass conservation requires that $\mathbf{n} \cdot (\mathscr{D}\nabla c - \mathbf{v}c) = q_n$. The flux q_n, which represents a boundary condition for the pore-scale problem Eq. (2.55) and a source term for the continuum-scale equation Eq. (2.59), is unknown.

In summary, the hybrid pore-scale/continuum scale algorithm contains the three unknowns $(c, \overline{c}, q_n)$ that satisfy a system of coupled partial-differential equations (Battiato et al. 2010):

$$\phi\frac{\partial \overline{c}}{\partial t} + \phi\nabla \cdot (\mathbf{V}\overline{c}) = \nabla \cdot (\mathbf{D}\nabla\overline{c}) - K\overline{c}, \quad \mathbf{x} \in \Omega^T, \quad t > 0 \tag{2.60}$$

$$\phi\frac{\mathrm{d}\overline{c}}{\mathrm{d}t} = \frac{1}{\phi|\Omega_p|}\int\limits_{A_{ll}} q_n \, \mathrm{d}\mathbf{x} - \frac{1}{\phi|\Omega_p|}\int\limits_{A_{sl}} \mathscr{K}c \, \mathrm{d}\mathbf{x}, \quad \mathbf{x} = \mathbf{x}^\star, \quad t > 0 \tag{2.61}$$

$$\frac{\partial c}{\partial t} + \nabla \cdot (\mathbf{v}c) = \mathscr{D}\nabla^2 c, \qquad \mathbf{x} \in \Omega_p, \quad t > 0 \tag{2.62}$$

$$\mathbf{n} \cdot (\mathscr{D}\nabla c - \mathbf{v}c) = q_n, \qquad \mathbf{x} \in A_{ll}, \quad t > 0 \tag{2.63}$$

$$-\mathbf{n} \cdot \mathscr{D}\nabla c = \mathscr{K}c, \qquad \mathbf{x} \in A_{sl}, \quad t > 0, \tag{2.64}$$

supplemented by boundary conditions on the external domain $\partial\Omega^T$ and initial conditions.

2.4.2   *Taylor dispersion in a fracture with reactive walls*

Transport of a reactive solute by advection and diffusion in a fracture of width $2H$ that undergoes a first-order heterogeneous reaction at the walls of the channel is described by:

$$\frac{\partial c}{\partial t} + u(y)\frac{\partial c}{\partial x} - \mathscr{D}\left(\frac{\partial^2 c}{\partial x^2} + \frac{\partial^2 c}{\partial y^2}\right) = 0, \qquad (x, y) \in \Omega, \quad t > 0 \tag{2.65a}$$

$$-\mathscr{D}\frac{\partial c}{\partial y} = \mathscr{K}c, \qquad (x, y) \in A, \quad t > 0, \tag{2.65b}$$

where the flow domain $\Omega = \{(x, y) : x \in (0, \infty), |y| < H\}$ has the boundary $A = \{(x, y) : x \in (0, \infty), |y| = H\}$. A fully developed flow assumption in laminar regime yields to a Poiseuille's velocity profile for the pore-scale velocity $\mathbf{v} = (u, 0)^T$, $u(y) = u_m[1 - (y/H)^2]$, where u_m is the maximum velocity at the center of the fracture ($y = 0$). The average concentration $\bar{c}(x,t)$ in Eq. (2.56) is now defined as:

$$\bar{c}(x,t) \equiv \frac{1}{2H} \int_{-H}^{H} c(x,y,t)\,\mathrm{d}y. \tag{2.66}$$

It satisfies the Darcy-scale equation (Mikelić et al. 2006):

$$\frac{\partial \bar{c}}{\partial t} + U\frac{\partial \bar{c}}{\partial x} + K\bar{c} = D\frac{\partial^2 \bar{c}}{\partial x^2}, \qquad x \in (0, \infty), \quad t > 0, \tag{2.67a}$$

where

$$U = u_m \left(\frac{2}{3} + \frac{4\mathrm{Da}_y}{45}\right), \quad K = \frac{\mathscr{K}}{H}\left(1 - \frac{\mathrm{Da}_y}{3}\right), \quad D = \mathscr{D}\left(1 + \frac{8\mathrm{Pe}_y^2}{945}\right) \tag{2.67b}$$

and

$$\mathrm{Pe}_y = \frac{u_m H}{\mathscr{D}}, \qquad \mathrm{Da}_y = \frac{\mathscr{K}H}{\mathscr{D}}. \tag{2.67c}$$

The validity of Eq. (2.67) requires that L, a macroscopic characteristic length scale in the x direction, be much larger than H, i.e., $\epsilon = H/L \ll 1$; and places a number of constraints on the order of magnitude of Pe and Da (see the phase diagram in Battiato and Tartakovsky (2010)). Specifically, Eq. (2.67) fails for $\mathrm{Da}_y \geq 3$ as K changes sign for increasing positive values of $\mathscr{K}$ (i.e., increasing mass loss at the solid-liquid interface): this leads to the unphysical behavior of $K < 0$ (i.e., source) while mass is absorbed (degraded, etc) at the micro-scale (i.e., sink).

Whenever one or more of these constraints are violated in a small portion of the computational domain, Ω_p, Eq. (2.67) is no longer valid and its nonlocal counterpart Eq. (2.59) must be employed. For the problem under consideration, the latter takes the form:

$$\frac{\partial \bar{c}}{\partial t} = \mathscr{D}\frac{\partial^2 \bar{c}}{\partial x^2} - \frac{\mathscr{K}J_c}{2H} - \overline{u(y)\frac{\partial c}{\partial x}}, \qquad (x,y) \in \Omega, \tag{2.68}$$

where $\Omega_p = \{(x,y) : x \in (a,b), |y| < H\}$ corresponds to a single macroscale grid block, $J_c = c(x,H) + c(x,-H)$, and the pore-scale concentration $c(x,y,t)$ satisfies Eq. (2.65). Equation (2.68) is subject to boundary conditions at the internal boundary $A_p = \{(x,y) : x = a, b; y \in (-H, H)\}$:

$$\mathbf{n} \cdot (\mathbf{v}c - \mathscr{D}\nabla c) = q_n. \tag{2.69}$$

2.4.3 *Hybrid algorithm*

Solving the nonlinear coupled system Eqs. (2.65)–(2.69) reduces to finding zero q_n of an algebraic equation in the form $F(q_n) = 0$, where q_n is the unknown flux at the boundary of Ω_p. The hybrid pore-scale/continuum-scale algorithm can be formulated as follows.

- Initialization. At timestep T^N, c^N and $\bar{c}(t = T^N)$ are known.
- Guess for fluxes. Make a guess for q_n. This imposes the Robin conditions Eq. (2.69) on A_p.
- Pore-scale evolution. The pore-scale problem Eq. (2.65) is evolved from T^N to T^{N+1}. The source term.
- Evaluation of volume and boundary integrals. J_c and the volume integral in the right hand side of Eq. (2.68) are evaluated.
- Continuum-scale evolution. The continuum-scale concentration $\bar{c}$ is evolved from T^N to T^{N+1} by solving Eq. (2.68) in Ω_p and Eq. (2.67) in the remainder of the computational domain.

- Continuum-scale fluxes computation. Continuum-scale flux $\tilde{q}_n$ at the interface A_p is evaluated by differentiation of continuum-scale solution.
- Convergence check and iteration. Select an acceptable tolerance ϵ. If $|\tilde{q}_n - q_n| > \epsilon$, refine the guess of q_n and go to the second step. If $|\tilde{q}_n - q_n| \leq \epsilon$, then the convergence is reached. March forward in time ($N := N + 1$) and go to the first step.

2.4.4 Numerical results

A finite volume implementation of the problem of reactive transport through a fracture is discussed in Battiato et al. (2010). Spatial discretization of Eqs. (2.67) and (2.68) leads to an algebraic system of equations of the form $Ax = b$ with some of the coefficients $a_{i,j} = [A]_{i,j}$ and b_i dependent on pore-scale quantities. While such formulation naturally arises from upscaling techniques, its intrusive nature renders it less computationally appealing: at each macroscopic timestep, A must be evaluated and inverted.

Some of the results discussed in Battiato et al. (2010) are reported here for completeness. These include a hybrid validation for advective-diffusive transport in a fracture with uniform reaction rates. This setting admits an analytical solution and, hence, is used to analyze the accuracy of the hybrid algorithm relative to that of its continuum (upscaled) counterpart (Section 2.4.4.1). In Section 2.4.4.2, the reaction coefficient is taken to be highly heterogeneous. For this situation, a comparison of the hybrid solution with both a solution of the upscaled equation Eq. (2.67) and an averaged solution of the fully two-dimensional problem ("pore-scale simulations") is presented.

2.4.4.1 Hybrid validation
The macroscopic problem Eq. (2.67), subject to the initial and boundary conditions:

$$\bar{c}(x,0) = 1, \qquad \bar{c}(0,t) = 0, \qquad \frac{\partial \bar{c}}{\partial x}(\infty, t) = 0, \tag{2.70}$$

admits the unique solution

$$\bar{c}(x,t) = e^{-Kt} \left(1 - \frac{1}{\sqrt{\pi}} e^{Ux/D} \int_{\frac{x+Ut}{2\sqrt{Dt}}}^{+\infty} e^{-\eta^2} d\eta + \frac{1}{\sqrt{\pi}} \int_{\frac{x-Ut}{2\sqrt{Dt}}}^{+\infty} e^{-\eta^2} d\eta \right). \tag{2.71}$$

Battiato et al. (2010) use this exact solution to verify the accuracy of both the hybrid solution and the numerical solution of the continuum problem Eq. (2.67) for advective-diffusive transport (Fig. 2.3) and advective-diffusive-reactive transport with uniform reaction rates (Fig. 2.4). The agreement between analytical and hybrid solution is perfect, which is to be expected since all the necessary conditions for the validity of the macroscopic (averaged) transport equation Eq. (2.67) hold for these flow and transport regimes. The parameter values used in these and subsequent simulations are provided in Battiato et al. (2010).

2.4.4.2 Hybrid simulations for highly localized heterogeneous reaction
Effects heterogenous reaction coefficient are presented. We allow $\mathcal{K}$ to change by two orders of magnitude with a typical Damköhler number ranging from 0.03 to 2.8. Specifically $\mathcal{K} = \mathcal{K}_{in}$ for $x = 3.75$ and $\mathcal{K} = \mathcal{K}_{out}$ everywhere else, with $\mathcal{K}_{in} = 450$ and $\mathcal{K}_{out} = 5$. We show here that significant deviations from the "pore-scale" solution occur even for Da < 3. The results of our hybrid simulations are compared with that of the upscaled 1D equation and the average of the fully 2D solution. Figure 2.5 shows the continuum-scale concentration obtained from the upscaled 1D continuum-scale, hybrid simulations and the 2D pore-scale equations. At the location of high heterogeneity, the continuum-scale equation overestimates the concentration, with values that double the true concentration obtained from the pore-scale simulations. On the contrary, the hybrid simulation significantly improves the predictions.

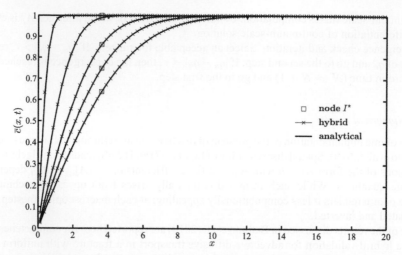

Figure 2.3   Advection-diffusion: Temporal snapshots of the average concentration $\bar{c}$ obtained analytically by Eq. (2.71) (solid line) and from hybrid simulation ($\times$) at times $t = 0.005$, $t = 0.05$, $t = 0.15$, $t = 0.25$, and $t = 0.395$ (from left to right). The box indicates the location where pore- and continuum-scales are coupled. After Battiato et al. (2010).

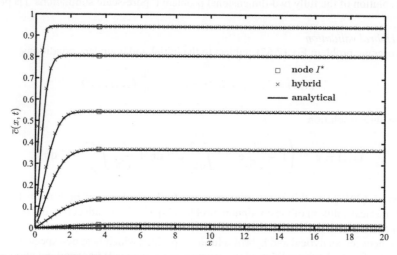

Figure 2.4   Advection-diffusion-reaction with homogeneous reaction rate: Temporal snapshots of the average concentration $\bar{c}$ obtained analytically by Eq. (2.71) (solid line) and from hybrid simulation ($\times$) at times $t = 0.001$, $t = 0.005$, $t = 0.015$, $t = 0.025$, $t = 0.05$, $t = 0.1$, and $t = 0.195$ (from top to bottom). The box indicates the location where pore- and continuum-scales are coupled. After Battiato et al. (2010).

2.4.5   *Non-intrusive hybrid algorithm*

Th hybrid algorithm developed in the previous section is intrusive in that it requires the modification of some of the coefficients of the system of discretized equations. Even though its formulation is quite general and can be applied to a variety of different numerical schemes, its implementation in legacy codes, in which discretized equations cannot be easily modified by the user, is challenging. Complex pore geometries introduce another complication.

Hence, a desirable feature of a hybrid algorithm is its portability and implementation in existing codes. This can be accomplished by eliminating the overlapping ("handshake") region and formulating appropriate conditions at the interfaces separating the two computational subdomains,

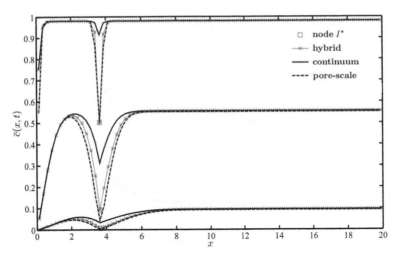

Figure 2.5 Advection-diffusion-reaction with localized reaction rate: Temporal snapshots of the average concentration $\bar{c}$ obtained from the 1D upscaled equation (solid line), from hybrid simulations ($- \times -$) and from the 2D pore-scale simulations (dashed line) at times $t = 0.0005$ (top), $t = 0.015$ (center) and $t = 0.06$ (bottom). The box indicates the location where pore- and continuum-scales are coupled. After Battiato et al. (2010).

while ensuring the continuity of state variables and fluxes. Within this framework, pore-scale simulations affect a continuum-scale solution through boundary conditions (and not as a modification of continuum-scale discretized equations): this will facilitate hybrid implementation for existing codes and/or software.

2.4.5.1 *Governing equations at the pore- and continuum-scale for advective-diffusive systems*

Consider advective-diffusive transport in a fully saturated porous medium Ω^T. Within the pore space $\Omega^T_{pore} \subset \Omega^T$, single-phase flow of an incompressible fluid is described by the Stokes and continuity equations Eq. (2.8). Flow equations are subject to the no-slip boundary condition on the solid-liquid interface $A^T_{s\ell}$, which is taken to be impermeable to flow. The flow is driven by boundary conditions imposed on $\partial\Omega^T$, the external boundary of Ω^T. The fluid contains a dissolved species with molar concentration $c(\mathbf{x}, t)$ that is advected and diffused in the system.

The evolution of the concentration $c(\mathbf{x}, t)$ of a tracer undergoing advection and diffusion is described by:

$$\frac{\partial c}{\partial t} + \nabla \cdot (\mathbf{v}c) = \mathscr{D}\nabla^2 c, \tag{2.72}$$

subject to a no-flux boundary condition on the solid-fluid interface $A^T_{s\ell}$

$$-\mathbf{n} \cdot \mathscr{D}\nabla c = 0, \tag{2.73}$$

and proper boundary conditions on $\partial\Omega^T$.

Let $\overline{A}(\mathbf{x}, t)$ denote the spatial average of a pore-scale quantity $A(\mathbf{x}, t)$ defined as in Eq. (2.56). Then, the spatial averaging of Eq. (2.72) leads to an upscaled equation:

$$\phi\frac{\partial\bar{c}}{\partial t} + \phi\nabla \cdot (\mathbf{V}\bar{c}) = \nabla \cdot (\mathbf{D}^*\nabla\bar{c}), \tag{2.74}$$

where $\mathbf{V}$ is the average macroscopic velocity and $\mathbf{D}^*$ is the dispersion coefficient.

2.4.5.2 *Derivation of coupling boundary conditions*

We are concerned with transport regimes in which the validity of the continuum-scale transport equation Eq. (2.74) breaks down in a subdomain $\Omega_p \subset \Omega^T_{pore}$ with boundary $\partial\Omega_p$ of the

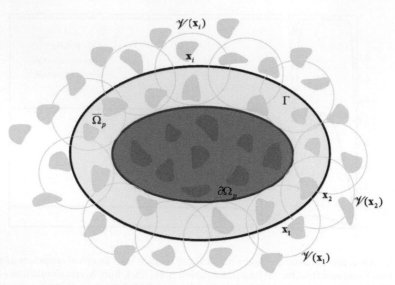

Figure 2.6   A schematic representation of the pore- and continuum-scale domains. The subdomain where continuum-scale representation breaks down is depicted in red. Its boundary is $\partial\Omega_p$. The boundary Γ is constructed as the locus of the centers of the family of averaging volumes $\mathscr{V}(\mathbf{x})$ whose envelope is $\partial\Omega_p$.

computational domain Ω. We define Γ to be the locus of the centers of the family of the averaging volumes $\mathscr{V}(\mathbf{x})$, whose envelope is $\partial\Omega_p$ as shown in Fig. 2.6. We denote $\overline{\Omega}_p$ the domain bounded by Γ. Let $A_{s\ell} = \overline{\Omega}_p \cap A_{s\ell}^T$.

Let $\overline{c}^{\leftarrow}$ denote the limiting value of $\overline{c}(\mathbf{x})$ as $\mathbf{x} \to \mathbf{x}^{\leftarrow} \in \Gamma$ from the *exterior* of $\overline{\Omega}_p$, and $\overline{c}^{\rightarrow} = \overline{c}(\mathbf{x}^{\rightarrow})$ as $\mathbf{x} \to \mathbf{x}^{\rightarrow} \in \Gamma$ from the *interior* of $\overline{\Omega}_p$. Since average concentration is a continuous function everywhere in Ω, it is continuous across Γ:

$$\overline{c}^{\leftarrow} = \overline{c}^{\rightarrow} \quad \text{for } |\mathbf{x}^{\rightarrow} - \mathbf{x}^{\leftarrow}| \to 0. \tag{2.75}$$

Let $\mathscr{V}^{in}(\mathbf{x}) := \mathscr{V}(\mathbf{x}) \cap \overline{\Omega}_p$ and $\mathscr{V}^{out}(\mathbf{x}) := \mathscr{V}(\mathbf{x}) \setminus \mathscr{V}^{in}(\mathbf{x})$ to form a partition of $\mathscr{V}$ where pore-scale is explicitly resolved and where only a continuum-scale representation exists, respectively (see Fig. 2.7). Then Eq. (2.75) can be written as:

$$\overline{c}^{\leftarrow} = \frac{1}{\phi|\mathscr{V}|} \int_{\mathscr{V}^{in}(\mathbf{x}^{\rightarrow})} c(\mathbf{y})\,\mathrm{d}\mathbf{y} + \frac{1}{\phi|\mathscr{V}|} \int_{\mathscr{V}^{out}(\mathbf{x}^{\rightarrow})} c(\mathbf{y})\,\mathrm{d}\mathbf{y}. \tag{2.76}$$

Expanding $c(\mathbf{y})$ into a Taylor series around the centroid $\mathbf{x}$, retaining the leading term, and substituting the result into Eq. (2.76) yields:

$$\overline{c}^{\leftarrow} = \frac{1}{\phi|\mathscr{V}|} \int_{\mathscr{V}^{in}(\mathbf{x}^{\rightarrow})} c(\mathbf{y})\,\mathrm{d}\mathbf{y} + \frac{|\mathscr{V}^{out}(\mathbf{x}^{\rightarrow})|}{\phi|\mathscr{V}|} \overline{c}^{\rightarrow} \tag{2.77}$$

In a similar manner, one can show that a flux continuity condition across Γ can be written as:

$$\mathbf{n} \cdot (\mathbf{D}^* \nabla \overline{c}^{\leftarrow} - \phi \mathbf{V} \overline{c}^{\leftarrow}) = \frac{1}{|\mathscr{V}|} \mathbf{n} \cdot \int_{\mathscr{V}^{in}(\mathbf{x}^{\rightarrow})} (-\mathscr{D}\nabla c + \mathbf{v}c)\,\mathrm{d}\mathbf{y} + q_n, \tag{2.78}$$

where

$$q_n(\mathbf{x}) := \frac{1}{|\mathscr{V}|} \mathbf{n} \cdot \int_{\mathscr{V}^{out}(\mathbf{x})} (-\mathscr{D}\nabla c + \mathbf{v}c)\,\mathrm{d}\mathbf{y} \tag{2.79}$$

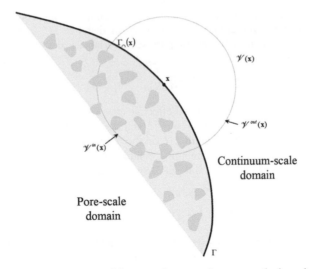

Figure 2.7 A schematic representation of the averaging procedure across the boundary separating pore- and continuum-scale representations. On the left of Γ pore-scale is fully resolved while on the right only a continuum-scale representation exists.

is an unknown flux through Γ. This flux serves as a coupling condition at the interface between pore- and continuum-scale subdomains.

The final form of the nonlinear coupled system of equations for the hybrid algorithm is:

$$\phi\frac{\partial\overline{c}}{\partial t} + \phi\nabla\cdot(\mathbf{V}\overline{c}) = \nabla\cdot(\mathbf{D}^*\nabla\overline{c}), \quad \mathbf{x}\in\Omega\setminus\overline{\Omega}_p, \tag{2.80}$$

$$\frac{\partial c}{\partial t} + \nabla\cdot(\mathbf{v}c) = \mathscr{D}\nabla^2 c, \quad \mathbf{x}\in\overline{\Omega}_p, \tag{2.81}$$

$$-\mathbf{n}\cdot\mathscr{D}\nabla c = 0, \quad \mathbf{x}\in A_{s\ell}, \tag{2.82}$$

$$\mathbf{n}\cdot(-\mathscr{D}\nabla c + \mathbf{v}c) = q_n, \quad \mathbf{x}\in\Gamma \tag{2.83}$$

$$\overline{c}^{\leftarrow} = \frac{1}{\phi|\mathscr{V}|}\int_{\mathscr{V}^{\mathrm{in}}(\mathbf{x})} c\,\mathrm{d}\mathbf{y} + \frac{|\mathscr{V}^{\mathrm{out}}|}{\phi|\mathscr{V}|}c^{\rightarrow}, \quad \mathbf{x}\in\Gamma, \tag{2.84}$$

$$\mathbf{n}\cdot(\mathbf{D}^*\nabla\overline{c}^{\leftarrow} - \phi\mathbf{V}\overline{c}^{\leftarrow}) = q_n + \frac{1}{|\mathscr{V}|}\mathbf{n}\cdot\int_{\mathscr{V}^{\mathrm{in}}(\mathbf{x}^{\rightarrow})}(-\mathscr{D}\nabla c + \mathbf{v}c)\,\mathrm{d}\mathbf{y}, \quad \mathbf{x}\in\Gamma. \tag{2.85}$$

The interfacial conditions Eqs. (2.84) and (2.85) are reminiscent of the macroscopic Dirichlet and Neumann boundary conditions derived by the method of volume averaging in Prat (1989). While similar in spirit, our conditions do not require a closure approximation, relying on pore-scale simulations instead.

Coupling conditions for the Taylor dispersion problem are derived as a special case of our more general formulation in the following section.

2.4.5.3 Taylor dispersion between parallel plates ($\mathscr{V}\cap\overline{\Omega}_p=\emptyset$)

Whenever the average of a pore-scale variable $\xi(\mathbf{x})$ with $\mathbf{x}=(x_1,x_2,x_3)\in\Omega_1\times\Omega_2\times\Omega_3$ is defined by integrating one of the independent variables x_i over Ω_i, $i=1,2,3$, the dimensionality of the correspondent continuum-scale equation is reduced. An example is the problem of Taylor dispersion between two infinite parallel plates separated by the distance $2H$. Now $\Omega_1 = (-\infty, +\infty)$, $\Omega_2 = (-H, H)$, and the average of a generic pore-scale variable is defined as:

$$\bar{\xi}(x) = \frac{1}{2H} \int_{-H}^{H} \xi(x, y) \, dy. \tag{2.86}$$

This allows one to derive a one-dimensional effective equation starting from a two-dimensional pore-scale problem. In such situation, $\mathcal{V} = \mathcal{V}^{\text{out}}$ and $\mathcal{V}^{\text{in}} = \emptyset$. The boundary Γ reduces to a point and to a vertical segment of length $2H$ on the continuum- and pore-scale subdomains, respectively. Since $\phi = 1$, the general coupling conditions for state variables and fluxes at the boundary Γ established by Eqs. (2.84) and (2.85) simplify to:

$$\bar{c}^{\leftarrow} = c^{\rightarrow}, \qquad\qquad \mathbf{x} \in \Gamma, \tag{2.87}$$

$$\mathbf{n} \cdot (\mathbf{D}^* \nabla \bar{c}^{\leftarrow} - \phi \mathbf{V} \bar{c}^{\leftarrow}) = q_n, \qquad \mathbf{x} \in \Gamma. \tag{2.88}$$

These conditions establish that pore-scale concentration and flux are constant along the boundary and equal to the continuum-scale value on the boundary exterior.

2.4.5.4 *Hybrid algorithm*

Solving the nonlinear coupled system Eqs. (2.80)–(2.85) reduces to finding zeros of a system of equations in the form:

$$F(q_n, c^{\rightarrow}) = 0, \quad G(q_n, c^{\rightarrow}) = 0, \tag{2.89}$$

where

$$F(q_n, c^{\rightarrow}) = \mathbf{n} \cdot (\mathbf{D}^* \nabla \bar{c}^{\leftarrow} - \phi \mathbf{V} \bar{c}^{\leftarrow}) - \frac{1}{|\mathcal{V}|} \int_{\mathcal{V}^{\text{in}}(\mathbf{x}^{\rightarrow})} \mathbf{n} \cdot (-\mathcal{D} \nabla c + \mathbf{v} c) \, dy - \frac{|\Gamma(\mathbf{x}^{\rightarrow})|}{|\mathcal{V}|} q_n, \tag{2.90}$$

$$G(q_n, c^{\rightarrow}) = \bar{c}^{\leftarrow} - \frac{1}{\phi |\mathcal{V}|} \int_{\mathcal{V}^{\text{in}}(\mathbf{x})} c \, dy - \frac{|\mathcal{V}^{\text{out}}|}{\phi |\mathcal{V}|} c^{\rightarrow}. \tag{2.91}$$

The hybrid pore-scale/continuum-scale algorithm can be formulated as follows

- Initialization. At timestep T^N, $c(t = T^N)$ and $\bar{c}(t = T^N)$ are known.
- Guess for flux. Make a guess for q_n. This imposes the Robin condition Eq. (2.83) at interface Γ.
- Pore-scale evolution. The pore-scale equation Eq. (2.81), supplemented with boundary condition Eq. (2.82), is evolved from T^N to T^{N+1}.
- Evaluation of boundary integrals. The right hand side of Eq. (2.91) is evaluated together with the integral term in Eq. (2.85). The latter imposes the Robin condition Eq. (2.85) at the interface Γ.
- Continuum-scale evolution. The continuum-scale concentration $\bar{c}$ is evolved from T^N to T^{N+1} by Eq. (2.80).
- Continuum-scale concentration evaluation. The function G is computed by means of Eq. (2.91).
- Convergence check and iteration. For a given tolerance ϵ, if $|G(q_n, c_\Gamma)| > \epsilon$, the Broyden method (or another root-finding algorithm) is used to refine the guess of q_n and go to step 2. If $|G(q_n, c_\Gamma)| \leq \epsilon$, then the convergence is reached. March forward in time ($N := N + 1$) and go to step 1.

2.5   CONCLUSIONS

Our study leads to the following major conclusions:

- While capable of describing many processes at a variety of different scales, macroscopic models might breakdown. We established conditions under which macroscopic reaction-diffusion equations (RDEs) provide an adequate averaged description of pore-scale processes. We showed that the range of applicability of macroscopic RDEs and various transport regimes can be described by a phase diagram in a space spanned by the dimensionless Damköhler number and a scale separation parameter. This was accomplished by upscaling a system of nonlinear diffusion-reaction equations at the pore-scale by means of volume averaging technique. For physical phenomena that do not satisfy such conditions, an upscaled (local) equation does not generally exist and integro-differential (non-local in space and time) alternatives or hybrid models must be used instead.

- The previous result was generalized by considering macroscopic advection-dispersion-reaction equations (ADREs). The method of multiple-scale expansion was used to upscale to the continuum (Darcy) scale a pore-scale advection-diffusion equation with nonlinear reactions entering through a boundary condition on the fluid-solid interfaces. The range of applicability of macroscopic ADREs can be described with a phase diagram in the (Da, Pe)-space (where Da and Pe are Damköhler and Péclet numbers, respectively). The latter is parametrized with a scale-separation parameter, defined as the ratio of characteristic lengths associated with the pore-and macro-scales. This phase diagram revealed that transport phenomena dominated at the pore-scale by reaction processes do not lend themselves to macroscopic descriptions and effective parameters do not generally exist. These results generalize our previous findings relative to RDEs and suggest that they are universal, i.e., independent of the choice of an upscaling technique.

- When the validity of continuum-scale models cannot be ascertained a priori in small portions of the computational domain, hybrid models that couple pore- and continuum-scale representations can be used. We developed a general intrusive hybrid algorithm to incorporate pore-scale effects into continuum models of reactive transport in fractured media. This formulation is based on overlapping the pore- and continuum-scale representations and therefore requires the modification of some coefficients in the discretized system of equations. We applied our algorithm to model Taylor dispersion in a planar fracture with chemically reactive walls. Existing analytical solutions served as validation. The hybrid model formulation reduces to a zero-finding algorithm for a vector function: this suggests its high applicability to a wide variety of problems and numerical schemes. The proposed method is capable of handling highly localized heterogeneities, which provides a considerable improvement in accuracy and enables one to properly capture the pore-scale physics.

- A desirable feature of a hybrid model is its ability to be easily incorporated into existing (legacy) codes/software. Even though not necessary to this purpose, a formalization that is not intrusive would render such a task much easier. Therefore we developed an alternate formalization for the hybridization that is nonintrusive and a priori does not require mesh refinement on the continuum-scale subdomain to match the mesh dimension on the pore-scale subdomain.

ACKNOWLEDGEMENT

This research was supported by the Office of Science of the U.S. Department of Energy (DOE) under the Scientific Discovery through Advanced Computing (SciDAC).

REFERENCES

Acharya, R.C., der Zee, S.E.A.T.M.V. & Leijnse, A. (2005) Transport modeling of nonlinearly adsorbing solutes in physically heterogeneous pore networks. *Water Resources Research*, 41, W02020.

Adler, P.M. (1992) *Porous Media: Geometry and Transports*. Butterworth-Heinemann, New York.

Auriault, J.L. (2009) On the domain of validity of Brinkman's equation. *Transport Porous Media*, 79, 215–223.

Auriault, J.L. & Adler, P.M. (1995) Taylor dispersion in porous media: Analysis by multiple scale expansions. *Advances in Water Resources*, 18 (4), 217–226.

Auriault, J.L., Geindreau, C. & Boutin, C. (2005) Filtration law in porous media with poor separation of scales. *Transport Porous Media*, 60, 89–108.

Battiato, I. & Tartakovsky, D.M. (2010) Applicability regimes for macroscopic models of reactive transport in porous media. *Journal of Contaminant Hydrology*, Available from: doi:10.1016/j.jconhyd.2010.05.005.

Battiato, I., Tartakovsky, D.M., Tartakovsky, A.M. & Scheibe, T.D. (2009) On breakdown of macroscopic models of mixing-controlled heterogeneous reactions in porous media. *Advances in Water Resources*, 32: 1664–1673.

Battiato, I., Tartakovsky, D.M., Tartakovsky, A.M. & Scheibe, T.D. (2011) Hybrid models of reactive transport in porous and fractured media. *Advances in Water Resources*, Available from: doi:10.1016/j.advwatres.2011.01.012.

Brenner, H. (1987) *Transport Processes in Porous Media*, New York, McGraw-Hill.

Brinkman, H.C. (1949) A calculation of the viscous force exerted by a flowing fluid on a dense swarm of particles. *Applied Science Research*, A1, 27–34.

Broyda, S., Dentz, M. & Tartakovsky, D.M. (2010) Probability density functions for advective-reactive transport in radial flow. *Stochastic Environmental Research and Risk Assessment*, Available from: doi:10.1007/s00477–010–0401–4.

Christie, M. (1996) Upscaling for reservoir simulation. *Journal of Petroleum Technology*, 48, 1004–1010.

Darcy, H. (1856) Les fontaines publiques de la Ville de Dijon, Paris, *Victor Darmon*.

de Marsily, G. (1986) *Quantitative Hydrogeology*. San Diego, California, Academic Press.

Duijn, C.J.V. & Pop, I.S. (2004) Crystal dissolution and precipitation in porous media: Pore-scale analysis. *Journal für die Reine und Angewandte Mathematik*, 577, 171–211.

Durlofsky, L.J. & Brady, J.F. (2009) Analysis of the Brinkman equation as a model for flow in porous media. *Physics of Fluids*, 30 (11), 3329–3341.

Efendief, Y. & Durlofsky, L.J. (2003) A generalized convection-diffusion model for subgrid transport in porous media. *Multiscale Modeling and Simulation*, 1 (3), 504–526.

Goyeau, B., Benihaddadene, T., Gobin, D., & Quintard, M. (1997) Averaged momentum equation for flow through a nonhomogeneous porous structure. *Transport Porous Media*, 28, 19–50.

Gray, W.G. & Miller, C.T. (2005) Thermodynamically constrained averaging theory approach for modeling flow and transport phenomena in porous medium systems: 1. Motivation and overview. *Advances in Water Resources*, 28 (2), 161–180.

Hesse, F., Radu, F.A., Thullner, M. & Attinger, S. (2009) Upscaling of the advection–diffusion–reaction equation with Monod reaction. *Advances in Water Resources*, 32, 1336–1351.

Hornung, U. (1997) *Homogenization and Porous Media*, New York, Springer.

Kechagia, P.E., Tsimpanogiannis, I.N., Yortsos, Y.C. & Lichtner, P.C. (2002) On the upscaling of reaction-transport processes in porous media with fast or finite kinetics. *Chemical Engineering Science*, 57 (13), 2565–2577.

Knabner, P., Duijn, C.J.V. & Hengst, S. (1995) An analysis of crystal dissolution fronts in flows through porous media. Part 1: Compatible boundary conditions. *Advances in Water Resources*, 18 (3), 171–185.

Knutson, C., Valocchi, A. & Werth, C. (2007) Comparison of continuum and pore-scale models of nutrient biodegradation under transverse mixing conditions. *Advances in Water Resources*, 30 (6–7), 1421–1431.

Kundu, P.K. & Cohen, I.M. (2008) *Fluid Mechanics*, 4th edition. San Diego, Elsevier.

Langlo, P. & Espedal, M.S. (1994) Macrodispersion for two-phase, immiscible flow in porous media. *Advances in Water Resources*, 17, 297–316.

Leemput, P., Vandekerckhove, C., Vanroose, W. & Roose, D. (2007) Accuracy of hybrid lattice Boltzmann/finite difference schemes for reaction diffusion systems. *Multiscale Modeling and Simulation* 6 (3), 838–857.

Lévy, T. (1983) Fluid flow through an array of fixed particles. *International Journal of Engineering Science*, 21, 11–23.

Li, L., Peters, C. & Celia, M. (2006) Upscaling geochemmical reaction rates sing pore-scale network modeling. *Advances in Water Resources*, 29, 1351–1370.

Lichtner, P.C. & Tartakovsky, D.M. (2003) Upscaled effective rate constant for heterogeneous reactions. *Stochastic Environmental Research and Risk Assessment*, 17 (6), 419–429.

Maloy, K.J., Feder, J., Boger, F. & Jossang, T. (1998) Fractal structure of hydrodynamic dispersion in porous media. *Physical Review Letter*, 61 (82), 2925.

Marušić-Paloka, E. & Piatnitski, A. (2005) Homogenization of a nonlinear convection-diffusion equation with rapidly oscillating coefficients and strong convection. *Journal of the London Mathematical Society*, 2 (72), 391–409.

Mikelić, A., Devigne, V. & Van Duijn, C.J. (2006) Rigorous upscaling of the reactive flow through a pore, under dominant Peclet and Damköhler numbers. *SIAM Journal on Mathematical Analysis*, 38 (4), 1262–1287.

Morse, J.W. & Arvidson, R.S. (2002) The dissolution kinetics of major sedimentary carbonate minerals. *Earth Science Reviews*, 58, 51–84.

Neuman, S.P. (1977) Theoretical derivation of Darcy's law. *Acta Mecanica*, 25, 153–170.

Neuman, S.P. & Tartakovsky, D.M. (2009) Perspective on theories of anomalous transport in heterogeneous media. *Advances in Water Resources*, 32 (5), 670–680.

Nitsche, L.C. & Brenner, H. (1989) Eulerian kinematics of flow through spatially periodic models of porous media. *Archive for Rational Mechanics and Analysis*, 107(3), 225–292.

Ochoa-Tapia, J.A., Stroeve, P. & Whitaker, S. (1991) Facilitated transport in porous media. *Chemical Engineering Science*, 46, 477–496.

Peter, M.A. (2007) Homogenization in domains with evolving microstructure. *Comptes Rendus Mécanique* 335: 357–362.

Prat, M. (1989) On the boundary conditions at the macroscopic level. *Transport Porous Media*, 4, 259–280.

Sanchez-Palencia, E. & Zaoui, A. (1989) Homogenization techniques for composite media. In: Sanchez-Palencia, E. & Zaoui, A. (eds.). *Lectures Delivered at the CISM International Center for Mechanical Sciences, July 1–5 1985, Udine, Italy*. Berlin, Heidelberg, Springer. *Lecture Notes in Physics*, 272, 1987.

Shapiro, M. & Brenner, H. (1986) Taylor dispersion of chemically reactive species: Irreversible first-order reactions in bulk and on boundaries. *Chemical Engineering Science*, 41 (6), 1417–1433.

Shapiro, M. & Brenner, M. (1988) Dispersion of a chemically reactive solute in a spatially periodic model of a porous medium. *Chemical Engineering Science*, 43 (3), 551–571.

Shapiro, M., Fedou, R., Thovert, J. & Adler, P.M. (1996) Coupled transport and dispersion of multicomponent reactive solutes in rectilinear flows. *Chemical Engineering Science*, 51 (22), 5017–5041.

Steefel, C.I., DePaolo, D.J. & Lichtner, P.C. (2005) Reactive transport modeling: An essential tool and a new research approach for the Earth sciences. *Earth Planetary Science Letters*, 240, 539–558.

Tartakovsky, A.M., Meakin, P., Scheibe, T.D. & West, R.M.E. (2007) Simulation of reactive transport and precipitation with smoothed particle hydrodynamics. *Journal of Computational Physics*, 222, 654–672.

Tartakovsky, A.M., Redden, G., Lichtner, P.C., Scheibe, T. D. & Meakin, P. (2008) Mixing-induced precipitation: Experimental study and multi-scale numerical analysis. *Water Resources Research*, 44, W06S04, doi:10.1029/2006WR005725.

Tartakovsky, A.M., Tartakovsky, D.M. & Meakin, P. (2008) Stochastic Langevin model for flow and transport in porous media. *Physical Review Letters*, 101(4), 044502. Available from: doi:10.1103/PhysRevLett.101.044502.

Tartakovsky, A.M., Tartakovsky, D.M., Scheibe, T.D. & Meakin, P. (2007) Hybrid simulations of reaction-diffusion systems in porous media. *SIAM Journal of Scientific Computing*, 30 (6), 2799–2816.

Tartakovsky, D.M., Dentz, M. & Lichtner, P.C. (2009) Probability density functions for advective-reactive transport in porous media with uncertain reaction rates. *Water Resources Research*, 45, W07414. Available from: doi:10.1029/2008WR007383.

van Noorden, T.L. & Pop, I.S. (2008) A Stefan problem modelling crystal dissolution and precipitation. *IMA Journal of Applied Mathematics* 73 (2), 393–411.

Walther, J.H., Werder, T., Jaffe, R.L. & Koumoutsakos, P. (2004) Hydrodynamic properties of carbon nanotubes. *Physical Review E*, 69, 062201.

Whitaker, S. (1999) *The Method of Volume Averaging*. Netherlands, Kluwer Academic Publishers.

Wood, B.D. & Ford, R.M. (2007) Biological processes in porous media: From the pore scale to the field. *Advances in Water Resources*, 30 (6–7), 1387–1391.

Wood, B.D., Radakovich, K. & Golfier, F. (2007) Effective reaction at a fluid-solid interface: Applications to biotransformation in porous media. *Advances in Water Resources*, 30 (6–7), 1630–1647.

CHAPTER 3

A tensorial formulation in four dimensions of thermoporoelastic phenomena

M.C. Suarez Arriaga

3.1 INTRODUCTION

Porous media are encountered in many man-made systems and in industrial processes such as in fuel cells, paper pulp drying, food production, diverse filtration methods, concrete, ceramics, moisture absorbents, textiles, paint drying, polymer composites, various wood applications, etc. There is also a growing interest in biological tissues, biomechanics of organic porous tissues and in engineering tissues. Porous media are obviously encountered in many natural systems involving multiphase flow, energy and solute transport in soils, aquifers, geothermal, oil and gas reservoirs. In this chapter attention is focused on the modeling of the thermo-poroelastic behavior of porous rocks.

Several factors affect the geomechanical behavior of porous crustal rocks containing fluids: porosity, pressure, and temperature, characteristics of the fluids, fissures, and faults. Rocks in underground systems (aquifers, geothermal and hydrocarbon reservoirs) are porous, compressible, and elastic. The presence of a moving fluid in the porous rock modifies its mechanical response. Its elasticity is evidenced by the compression that results from the decline of the fluid pressure, which can reduce the pore volume. This reduction of the pore volume can be the principal source of fluid released from storage. A rock mechanics model is a group of equations capable of predicting the porous medium deformation under different internal and external forces. In this paper, an original four-dimensional tensorial formulation of linear thermo-poroelasticity theory is presented. This formulation makes more comprehensible the linear Biot's theory, rendering the resulting equations more convenient to be solved using the Finite Element Method. To illustrate practical aspects of this model some classic applications are outlined and solved.

3.2 THEORETICAL AND EXPERIMENTAL BACKGROUND

In order to understand why we need four spatial dimensions to model poroelastic phenomena, we need first to understand, from a mathematical point of view, what a dimension is. In mathematical terminology two $\{e_1, e_2\}$ or three orthonormal vectors $\{e_1, e_2, e_3\}$ define a vectorial basis in the two- or in the three-dimensional space $\mathcal{R}^2$ or $\mathcal{R}^3$ respectively. For example, any vector $\mathbf{v}$ in $\mathcal{R}^3$ can be represented as a linear combination of the basic vectors: $\mathbf{v} = v_1 e_1 + v_2 e_2 + v_3 e_3$. We say that this basis generates the vectorial space $\mathcal{R}^3$. A basis of a higher dimensional vector space $\mathbf{V}$ is defined as a subset $\{e_1, e_2, \ldots, e_n\}$ of vectors in $\mathbf{V}$ that are linearly independent and generate $\mathbf{V}$. These vectors form a basis if and only if every $\mathbf{v} \in \mathbf{V}$ can be uniquely written as: $\mathbf{v} = v_1 e_1 + v_2 e_2 + \cdots + v_n e_n$; where the numbers $v_1, v_2, \ldots, v_n$ are the coordinates of vector $\mathbf{v}$. The number of basis vectors in $\mathbf{V}$ is called the dimension of $\mathbf{V}$. All these concepts are extended to vectorial spaces of real functions. An important property appears when the functions f in $\mathbf{V}$ are linear. Linearity means that $f(a\mathbf{u} + b\mathbf{v}) = af(\mathbf{u}) + bf(\mathbf{v})$, where a, b are real numbers and $\mathbf{u}, \mathbf{v}$ are arbitrary vectors. In this case it is possible to find an associated matrix $\mathbf{M}$ such that $f(\mathbf{v}) = \mathbf{M} \cdot \mathbf{v}$ (Lang 1969). This expression means that the value of f corresponding to any vector $\mathbf{v}$ is equal to the computation of the product **M** times **v**. In other words, the knowledge of **M** is enough to know all the values of the function f. And vice versa, for any given matrix **M**, it is possible to construct a function f that satisfies the previous relationship of linearity. The only restriction is that the determinant of **M** must be not equal to zero; this condition assures that the rows (or the columns) of **M** form a basis of the space **V**. The key concept is that the dimension of the space **V** is determined by the number of columns (or the number of rows) of this matrix **M**.

In classic elastic solids only the two Lame moduli, (λ, G) or Young's elastic coefficient and Poisson's ratio (E, ν), are sufficient to describe the relations between strains and stresses. In poroelasticity, we need five poroelastic moduli for the same relationships (Bundschuh & Suarez 2010), but only three of these parameters are independent. The Biot's field variables for an isotropic porous rock are the stress σ acting in the rock, the bulk volumetric strain ϵ_B, the pore pressure p_f and the variation of fluid mass content ζ. In a one-dimensional test (1D), the linear relationships among these variables are the experimental foundations of Biot's poroelastic theory (Biot & Willis 1957, Wang 2000):

$$\epsilon_B = \frac{\sigma}{K_B} + \frac{p_f}{H}, \quad \zeta = \frac{\sigma}{H} + \frac{p_f}{R} \Leftrightarrow \begin{pmatrix} \epsilon_B \\ \zeta \end{pmatrix} = \begin{pmatrix} K_B^{-1} & H^{-1} \\ H^{-1} & R^{-1} \end{pmatrix} \cdot \begin{pmatrix} \sigma \\ p_f \end{pmatrix} \tag{3.1}$$

where K_B, H, and R are poroelastic coefficients that are experimentally measured as follows (Wang 2000):

$$\epsilon_B = \frac{\Delta V_B}{V_B}, \quad C_B = \left(\frac{\Delta \epsilon_B}{\Delta \sigma} \right)_{p_f}, \quad C_B = \frac{1}{K_B}$$

$$\frac{1}{H} = \left(\frac{\Delta \epsilon_B}{\Delta p_f} \right)_\sigma = \left(\frac{\Delta \zeta}{\Delta \sigma} \right)_{p_f}, \quad \frac{1}{R} = \left(\frac{\Delta \zeta}{\Delta p_f} \right)_\sigma \tag{3.2}$$

Equation (3.1) contains an expression involving a (2×2) matrix. Therefore two is the spatial dimension of this one-dimensional experiment. Figure 3.1 illustrates all the parts forming a poroelastic medium. Here V_B is the bulk volume, consisting of the rock skeleton formed by the union of the volume of the pores V_Φ and the volume of the solid matrix V_S (Figure 3.1). The control volume is ΔV_B. The drained coefficients K_B and C_B are the bulk modulus and the bulk compressibility, respectively; $1/H$ is a poroelastic expansion coefficient, which describes how much ΔV_B changes when p_f changes while keeping the applied stress σ constant; $1/H$ also measures the changes of ζ when σ changes and p_f remains constant. Finally $1/R$ is an unconstrained specific storage coefficient, which represents the changes of ζ when p_f changes. Inverting the matrix Equation (3.1) and replacing the value of σ in ζ we obtain:

$$\sigma = K_B \epsilon_B - \frac{K_B}{H} p_f \Rightarrow \zeta = \frac{K_B}{H} \epsilon_B + \left(\frac{1}{R} - \frac{K_B}{H^2} \right) p_f$$

$$\Rightarrow \begin{pmatrix} \sigma \\ p_f \end{pmatrix} = \begin{pmatrix} \frac{C}{B} & -C \\ -C & M \end{pmatrix} \cdot \begin{pmatrix} \epsilon_B \\ \zeta \end{pmatrix} \tag{3.3}$$

Biot (1941) and Biot & Willis (1957) introduced three additional parameters, b, M and C, that are fundamental for the tensorial formulation herein presented (Bundschuh & Suarez 2010). $1/M$ is called the constrained specific storage, which is equal to the change of ζ when p_f changes measured at constant strain. The parameters M, C and B are expressed in terms of the three fundamental ones defined in Equation (3.2):

$$\frac{1}{M} = \left(\frac{\Delta \zeta}{\Delta p_f} \right)_{\epsilon_B} = \frac{1}{R} - \frac{K_B}{H^2} \Rightarrow M = \frac{RH^2}{H^2 - K_B R}; \quad C = \frac{K_B}{H} M; \quad B = \frac{R}{H} \tag{3.4}$$

B is called the Skempton coefficient, representing the change in p_f when σ changes for undrained conditions. Let $C_S = 1/K_S$ be the compressibility of the solid matrix. The Biot-Willis coefficient

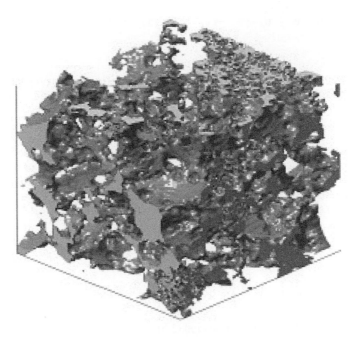

Figure 3.1   Skeleton of sandstone showing its pores and solid grains. Dimensions are ($3 \times 3 \times 3$ mm^3).

b is defined as the change of confining pressure p_k with respect to the fluid pressure change when the total volumetric strain remains constant:

$$b = \left(\frac{\partial p_k}{\partial p_f}\right)_{\epsilon_B} = 1 - \frac{K_B}{K_S} = \frac{C}{M} = \frac{K_B}{H} \tag{3.5}$$

The sign conventions are stress $\sigma > 0$ in tension and $\sigma < 0$ in compression; the volumetric strain $\epsilon_B > 0$ in expansion and $\epsilon_B < 0$ in contraction; the fluid content $\zeta > 0$ if fluid is added to the control volume ΔV_B and $\zeta < 0$ if fluid is extracted from ΔV_B; the pore pressure $p_f > 0$ if it is larger than the atmospheric pressure. The coefficient C represents the coupling of deformations between the solid grains and the fluid. The coefficient M is the inverse of the constrained specific storage, measured at constant strain (Wang 2000); this parameter characterizes the elastic properties of the fluid because it measures how the fluid pressure changes when ζ changes. These three parameters b, M and C are at the core of the poroelastic partial differential equations we introduce herein (Bundschuh & Suarez 2009).

3.3   MODEL OF ISOTHERMAL POROELASTICITY

Let $\mathbf{u}_s$ and $\mathbf{u}_f$ be the displacements of the solid and fluid particles; let vector $\mathbf{u} = \mathbf{u}_f - \mathbf{u}_s$ be the displacement of the fluid phase relative to the solid matrix respectively. Let ϵ_s, ϵ_f, φ_s, φ, V_s and V_f be the volumetric dilatations, porosities and volumes of each phase; ϵ_V is the volumetric deformation of the fluid phase relative to the solid phase. The mathematical expressions of these variables are:

$$\frac{\Delta V_s}{V_s} = \epsilon_s = \nabla \cdot \mathbf{u}_s; \quad \frac{\Delta V_f}{V_f} = \epsilon_f = \nabla \cdot \mathbf{u}_f$$

$$\epsilon_V = \epsilon_S - \epsilon_f; \quad \mathbf{u} = \mathbf{u}_f - \mathbf{u}_S \; \Rightarrow \; -\epsilon_V = -\nabla \cdot (\mathbf{u}_S - \mathbf{u}_f) = \nabla \cdot \mathbf{u} = \frac{\partial u_x}{\partial x} + \frac{\partial u_y}{\partial y} + \frac{\partial u_z}{\partial z}$$

(3.6)

Biot and Willis (1957) introduced the strain variable $\zeta(\mathbf{u}, t)$, defined in Equation (3.3), to describe the volumetric deformation of the fluid relative to the deformation of the solid with homogeneous porosity:

$$\zeta(\mathbf{u}, t) = \varphi \nabla \cdot (\mathbf{u}_S - \mathbf{u}_f) = \varphi \epsilon_S - \varphi \epsilon_f = \varphi \epsilon_V$$

(3.7)

The function ζ represents the variation of fluid content in the pore during a poroelastic deformation. The total applied stresses in the porous rock are similar to the equations of classic elasticity. However, we need to couple the effect of the fluid in the pores. The linear components of the global stresses in three dimensions, deduced experimentally by Biot (Biot 1941, Biot & Willis 1957, Wang 2000) are:

$$\sigma_{ij} = \lambda_U \epsilon_B \delta_{ij} + 2G \epsilon_{ij} - C \zeta \delta_{ij}$$

(3.8)

where:

$$\epsilon_{ij} = \frac{1}{2} \left(\frac{\partial u_i}{\partial x_j} + \frac{\partial u_j}{\partial x_i} \right), \quad \delta_{ij} = \begin{cases} 1, & \text{if } i = j \\ 0, & \text{if } i \neq j \end{cases}, \quad \lambda_U = \lambda + Cb; \quad \text{for } i, j = x, y, z$$

The fluid pressure is deduced from Equation (3.3):

$$p_f = \frac{K_B R H^2}{H^2 - K_B R} \left[\frac{\zeta}{K_B} - \frac{\epsilon_B}{H} \right]$$

(3.9)

We define a two-order tensor $\sigma_T = (\sigma_{ij})$ in four dimensions, which includes the bulk stress tensor σ_B acting in the porous rock and the fluid stress σ_F acting in the fluid inside the pores, positive in compression:

$$\sigma_T = \begin{pmatrix} \sigma_x & \sigma_{xy} & \sigma_{xz} & 0 \\ \sigma_{xy} & \sigma_y & \sigma_{yz} & 0 \\ \sigma_{xz} & \sigma_{yz} & \sigma_z & 0 \\ 0 & 0 & 0 & \sigma_f \end{pmatrix} = \begin{cases} \sigma_{ij} = (\lambda_U \epsilon_B - C \zeta) \delta_{ij} + 2G \epsilon_{ij} & i, j = x, y, z \\ \sigma_f = p_f = M \zeta - C \epsilon_B; \end{cases}$$

(3.10)

This tensorial equation becomes identical to the Hookean solids equation, when the rock has zero porosity and $b = 0$. From Equations (3.8), (3.9) and (3.10), we deduce that:

$$\sigma_T = \sigma_B + \sigma_F = (\sigma_{ij})$$

$$= \epsilon_B \begin{pmatrix} \lambda_U & 0 & 0 & 0 \\ 0 & \lambda_U & 0 & 0 \\ 0 & 0 & \lambda_U & 0 \\ 0 & 0 & 0 & -C \end{pmatrix} + 2G \begin{pmatrix} \epsilon_x & \epsilon_{xy} & \epsilon_{xz} & 0 \\ \epsilon_{xy} & \epsilon_y & \epsilon_{yz} & 0 \\ \epsilon_{xz} & \epsilon_{yz} & \epsilon_z & 0 \\ 0 & 0 & 0 & 0 \end{pmatrix} - \zeta \begin{pmatrix} C & 0 & 0 & 0 \\ 0 & C & 0 & 0 \\ 0 & 0 & C & 0 \\ 0 & 0 & 0 & -M \end{pmatrix}$$

(3.11)

$$\tau_{ij} = \lambda \epsilon_B \delta_{ij} + 2G \epsilon_{ij} \; \Rightarrow \; \sigma_{ij} = \tau_{ij} - b p_f \delta_{ij}$$

(3.12)

Tensor τ_{ij} is called the Terzaghi (1943) effective stress that acts only in the solid matrix; $b p_f$ is the pore-fluid pressure. Since there are no shear tensions in the fluid, the pore fluid pressure affects only the normal tensions $\sigma_i (i = x, y, z)$. The functions σ_{ij} are the applied stresses acting in the porous rock saturated with fluid. The solid matrix (τ_{ij}) supports one portion of the total applied tensions in the rock and the fluid in the pores ($b p_f$) supports the other part. This is a maximum for soils, when $b \approx 1$ and is minimum for rocks with very low porosity where $b \approx 0$.

For this reason, b is called the effective stress coefficient. Inverting the matrices of Equations (3.8) and (3.9), we arrive to the following tensorial form of the poroelastic strains:

$$\epsilon_{ii} = \frac{\sigma_{ii}}{2G} - \frac{3\nu}{E}\sigma_M + \frac{p_f}{3H}; \quad \epsilon_{ij} = \frac{\sigma_{ij}}{2G}; \quad \zeta = \frac{\sigma_M}{H} + \frac{p_f}{R} = \frac{C\sigma_M + K_U p_f}{MK_U - C^2} \tag{3.13}$$

$$\sigma_M = \frac{\sigma_{xx} + \sigma_{yy} + \sigma_{zz}}{3} = K_B \epsilon_B - bM\zeta; \quad K_B = \lambda + \frac{2}{3}G; \quad K_U = K_B + b^2 M \tag{3.14}$$

The coefficient K_U is the undrained bulk modulus, which is related to the previous defined coefficients. Note that both tensorial Equations (3.10) and (3.14) only need four basic poroelastic constants. The presence of fluid in the pores adds an extra tension due to the hydrostatic pressure, which is identified with the pore pressure, because it is supposed that all the pores are interconnected. This linear theory is appropriate for isothermal, homogeneous, and isotropic porous rocks.

3.4 THERMOPOROELASTICITY MODEL

The equations of non-isothermal poroelastic processes are deduced using the Gibbs thermoporoelastic potential or available enthalpy per unit volume and the energy dissipation function of the skeleton (Coussy 1991). Analytic expressions are constructed in terms of the stresses, the porosity, the pore pressure, and the density of entropy per unit volume of porous rock. As we did for the isothermal poroelasticity, we can write in a single four-dimensional tensor the thermoporoelastic equations relating stresses and strains (Bundschuh & Suarez 2010). We have for the pore pressure:

$$p - p_0 = M(\zeta - \zeta_0) - C\epsilon_B - M\varphi(\gamma_\varphi - \gamma_f)(T - T_0) \tag{3.15}$$

The volumetric thermal dilatation coefficient γ_B $[1/K]$ measures the dilatation of the skeleton and γ_φ $[1/K]$ measures the dilatation of the pores:

$$\gamma_B = \frac{1}{V_B}\left(\frac{\partial V_B}{\partial T}\right)_{p_k}, \quad \gamma_\varphi = \frac{1}{V_\varphi}\left(\frac{\partial V_\varphi}{\partial T}\right)_{p_f} = \frac{1}{\varphi}\left(\frac{\partial \varphi}{\partial T}\right)_{p_f} \left[\frac{1}{K}\right] \tag{3.16}$$

The fluid bulk modulus K_f and the thermal expansivity of the fluid γ_f $[1/K]$ are defined as follows:

$$\frac{1}{K_f} = C_f = \frac{1}{\rho_f}\left(\frac{\partial \rho_f}{\partial p}\right)_T; \tag{3.17}$$

$$\gamma_f = \frac{1}{V_f}\left(\frac{\partial V_f}{\partial T}\right)_{p_f} = -\frac{1}{\rho_f}\left(\frac{\partial \rho_f}{\partial T}\right)_{p_f} \tag{3.18}$$

The term p_k is the confining pressure. Expanding the corresponding functions of the Gibbs potential and equating to zero the energy dissipation we obtain the 4D thermoporoelastic equations, which include the thermal tensions in the total stress tensor (Bundschuh & Suarez 2010):

$$\sigma_{ij} - \sigma_{ij}^0 = (\lambda \epsilon_B - b(p - p_0))\delta_{ij} + 2G\epsilon_{ij} - K_B \gamma_B (T - T_0) \tag{3.19}$$

In this case, an initial reference temperature T_0 and an initial pore pressure p_0 are necessary because both thermodynamic variables T and p are going to change in non-isothermal processes occurring in porous rocks. The fluid stress is deduced in a similar way:

$$\sigma_f = p_f = M(\zeta - \zeta_0) - C\epsilon_B - M\varphi(\gamma_\varphi - \gamma_f)(T - T_0) \tag{3.20}$$

3.5   DYNAMIC POROELASTIC EQUATIONS

The formulation we introduced herein is very convenient to be solved using the Finite Element Method. The fundamental poroelastic differential equation is the tensorial form of Newton's second law in continuum porous rock dynamics:

$$\nabla \cdot \sigma_T + \mathbf{F} = \rho \frac{\partial^2 \mathbf{u}}{\partial t^2}; \quad \nabla \cdot \sigma_T = \mathbf{L}^T \cdot \sigma_T; \quad \sigma_T = \mathbf{C}_B \epsilon_T; \quad \epsilon_T = \mathbf{L} \cdot \mathbf{u} \qquad (3.21)$$

The terms σ_T and ϵ_T are the equivalent vectorial form of tensorial Equation (3.19) and $\mathbf{C}_B$ is the matrix of poroelastic constants. While $\mathbf{F}$ is the body force acting on the rock and the tensor differential operator $\mathbf{L}$ is given by:

$$\mathbf{L}^T = \begin{pmatrix} \partial_x & 0 & 0 & \partial_y & \partial_z & 0 & \partial_x \\ 0 & \partial_y & 0 & \partial_x & 0 & \partial_z & \partial_y \\ 0 & 0 & \partial_z & 0 & \partial_x & \partial_y & \partial_z \end{pmatrix} \Rightarrow \mathbf{L} \cdot \begin{pmatrix} u_x \\ u_y \\ u_z \end{pmatrix} = \epsilon_T = \begin{pmatrix} \epsilon_x & \epsilon_y & \epsilon_z & \epsilon_{xy} & \epsilon_{xz} & \epsilon_{yz} & \zeta \end{pmatrix} \qquad (3.22)$$

where $\mathbf{u} = (u_x, u_y, u_z)$ is the displacement vector of Equation (3.6). Using the operator $\mathbf{L}$ in Equation (3.22), the dynamic poroelastic equation becomes:

$$(\mathbf{L}^T \cdot \mathbf{C}_B \cdot \mathbf{L}) \cdot \mathbf{u} + \mathbf{F} = \rho \frac{\partial^2 \mathbf{u}}{\partial t^2} \qquad (3.23)$$

3.6   THE FINITE ELEMENT METHOD IN THE SOLUTION OF THE THERMOPOROELASTIC EQUATIONS

Equation (3.23) includes Biot's poroelastic theory. It can be formulated and numerically solved using the Finite Element Method (FEM). Let Ω be the bulk volume of the porous rock, and let $\partial\Omega$ be its boundary, $\mathbf{u}$ is the set of admissible displacements in Eq. (3.22); f_b is the volumetric force and f_s is the force acting on the surface $\partial\Omega$. After doing some algebra we arrive to a FEM fundamental equation for every element V^e in the discretization:

$$\mathbf{K}^e \cdot \mathbf{d}^e + \mathbf{M}^e \cdot \frac{\partial^2 \mathbf{d}^e}{\partial t^2} = \mathbf{F}^e; \quad e = 1, M \qquad (3.24)$$

$\mathbf{d}^e$ is a vector containing the displacements of the nodes in each V^e. Equation (3.24) approximates the displacement $\mathbf{u}$ of the poroelastic rock. $\mathbf{F}^e$ is the vector of total nodal forces. $\mathbf{K}^e$ and $\mathbf{M}^e$ are the stiffness and equivalent mass matrices for the finite element V^e. The mathematical definitions of both matrices are (Liu & Quek 2003):

$$\mathbf{K}^e = \int_{V^e} \mathbf{B}^T \cdot \mathbf{C}_B \cdot \mathbf{B} \, dV; \quad \mathbf{B} = \mathbf{L} \cdot \mathbf{N}; \quad \mathbf{M}^e = \int_{V^e} \rho \mathbf{N}^T \cdot \mathbf{N} \, dV; \quad e = 1, M \qquad (3.25)$$

where $\mathbf{N}$ is the matrix of shape functions that interpolate the displacements (Liu & Quek 2003). Matrix $\mathbf{B}$ is called the strain poroelastic matrix.

3.7   SOLUTION OF THE MODEL FOR PARTICULAR CASES

This section contains two brief illustrations of the deformation of an aquifer (Leake & Hsieh 1997) and the form that a temperature change can affect its poroelastic deformation. In the first example, we assume cold water at 20°C (1000 kg/m³). After, we consider a higher temperature of 250°C

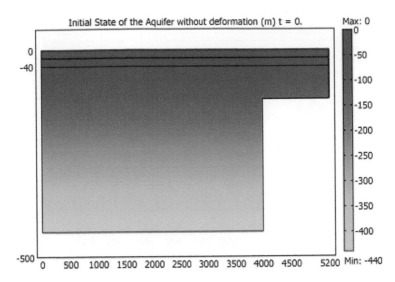

Figure 3.2   Simplified geometry of the aquifer and the impermeable bedrock in the basin. Initial state.

(50 bar). The model was programmed using COMSOL-Multiphysics (2009). Results are shown in Figures 3.4 to 3.7. Three sedimentary layers overlay impermeable bedrock in a basin where faulting creates a bedrock step (BS) near the mountain front (Fig. 3.2). The sediment stack totals 420 m at the deepest point of the basin ($x = 0$ m) but thins to 120 m above the step ($x > 4000$ m). The top two layers of the sequence are each 20 m thick. The first and third layers are aquifers; the middle layer is relatively impermeable to flow. Water obeys Darcy's law for head $h(K_X, K_Y)$ are the hydraulic conductivities and S_S is the specific storage:

$$\frac{\partial}{\partial x}\left(K_X\frac{\partial h}{\partial x}\right) + \frac{\partial}{\partial y}\left(K_Y\frac{\partial h}{\partial y}\right) + q_V = S_S\frac{\partial h}{\partial t} \tag{3.26}$$

3.8   DISCUSSION OF RESULTS

As given by the problem statement, the materials here are homogeneous and isotropic within a layer. The flow field is initially at steady state, but pumping from the lower aquifer reduces hydraulic head by 6 m per year at the basin center (under isothermal conditions). The head drop moves fluid away from the step. The fluid supply in the upper reservoir is limitless. The period of interest is 10 years. The corresponding FE mesh has 2967 elements excluding the bedrock step (Fig. 3.3). The rock is Hookean, poroelastic and homogeneous. For the computations, data of Table 3.1 were used. In the first example for the Biot-Willis coefficient we assume that $b = 0.3$; in the second example $b = 1.0$.

The two examples presented herein were solved using the Finite Element Method for a well-known problem of linked fluid flow and solid deformation near a bedrock step in a sedimentary basin described in a previous publication (Leake & Hsieh 1997). The problem concerns the impact of pumping for a basin filled with sediments draping an impervious fault block. In the first example, we considered the water in the aquifer to be cold, at 20°C. In the second example, the water is geothermal fluid, at 250°C. The basin is composed of three layers having a total depth of 500 m and is 5000 m long in both cases. The Darcy's law (Eq. 3.26) for water is coupled to the rock deformation via Equations (3.11) and (3.15) through the porosity φ, which is implicit in the storage coefficient S_S:

$$S_S = \rho_f g\left(C_B + \varphi C_f\right) \tag{3.27}$$

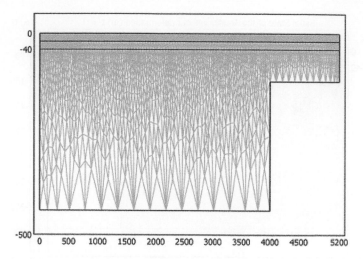

Figure 3.3   The mesh of the basin with 2967 elements.

Table 3.1   Numerical values of the parameters used in the simulations.

Hydraulic conductivity upper and lower aquifers	$K_X = 25$ m/day	Poroelastic storage coefficient, upper aquifer	$S_S = 1.0 \times 10^{-6}$
Hydraulic conductivity confining layer	$K_Y = 0.01$ m/day	Poroelastic storage coefficient, lower aquifer	$S_S = 1.0 \times 10^{-5}$
Biot-Willis coefficient (cold water at 20°C)	$b = 0.3$	Biot-Willis coefficient (hot water at 250°C)	$b = 1.0$
Young's modulus	$E = 8.0 \times 10^8$ Pa	Poisson's ratio	$\nu = 0.25$

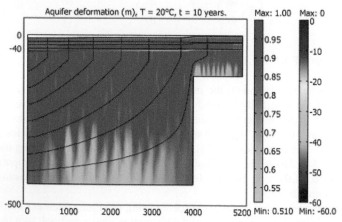

Figure 3.4   Poroelastic deformation of the basin for the BS problem with cold water (20°C). Streamlines represent the fluid to porous rock coupling.

where g (9.81 m/s^2) is gravity acceleration, ρ_f (1000 kg/m^3) is the water density, C_B (0.22 × 10^{-9} Pa^{-1}) is the bulk rock compressibility and C_f (0.4 × 10^{-9} Pa^{-1}) is the compressibility of water. All units are in the SI. Figures 3.4 and 3.5 show simulation results of the basin for years 1, 2, 5, and 10, respectively. The second simulation (Fig. 3.5) corresponds to a coupled thermoporoelastic deformation when the water in the aquifer is under geothermal conditions (fluid density

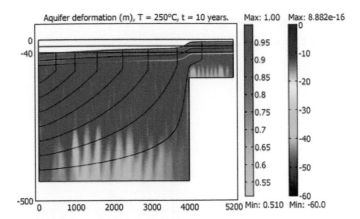

Figure 3.5   Poroelastic deformation of the basin for the BS problem with hot water (250°C). Streamlines represent the fluid to porous rock coupling.

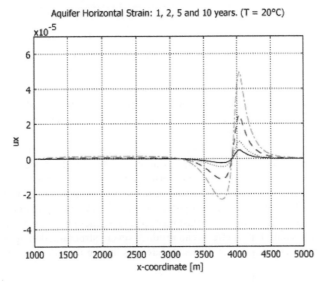

Figure 3.6   Horizontal strain at the basin with a BS for cold water (20°C).

of 800.4 kg/m³, temperature of 250°C, and pressure of 50 bar). Figures 3.6 and 3.7 compare the horizontal strains in both cases respectively, illustrating the evolution of lateral deformations that compensate for the changing surface elevation above the bedrock step. Note that vertical scales are different in both examples for clarity, except in Figures 3.4 and 3.5.

3.9 CONCLUSIONS

- All crustal rocks forming geothermal reservoirs are poroelastic and the fluid presence inside the pores affects their geomechanical properties. The elasticity of aquifers and geothermal reservoirs is evidenced by the compression resulting from the decline of the fluid pressure, which can shorten the pore volume. This reduction of the pore volume can be the principal source of fluid released from storage.
- The immediate physical experience shows that the supply or extraction of heat produces deformations in the rocks. Any variation of temperature induces a thermo-poroelastic behavior that influences the elastic response of porous rocks.

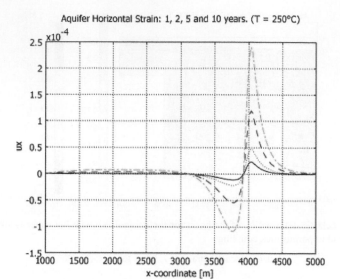

Figure 3.7   Horizontal strain at the basin with geothermal water (250°C).

- We introduced herein a general tensorial thermoporoelastic model that takes into account both the fluid and the temperature effects in linear porous rock deformations, and presenting two practical examples solved with finite elements.
- The second example illustrates the influence of temperature changes on the poroelastic strains. For cold water, the estimated value of ϵ_z is about -1.5×10^{-4}, while for hot water ϵ_z is -7.5×10^{-4}. Therefore, the poroelastic deformations are much higher in geothermal reservoirs than in isothermal aquifers. In the first case the bulk modulus of water $K_w = 0.45$ GPa, corresponding to $T = 250°$C. For cold aquifers $K_w = 2.5$ GPa approximately.
- Water bulk modulus affect other poroelastic coefficients, including the expansivity of rocks, which is relatively small, but its effects can produce severe structural damages in porous rocks subjected to strong temperature gradients, as happens during the injection of cold fluids.

REFERENCES

Biot, M.A. (1941) *General theory of three-dimensional consolidation. Journal of Applied Physics*, 12, 155–164.

Biot, M.A. & Willis, D.G. (1957) *The elastic coefficients of the theory of consolidation. Journal of Applied Mechanics*, 24, 594–601.

Bundschuh, J. & Suarez, M.C. (2010) *Introduction to the Numerical Modeling of Groundwater and Geothermal Systems: Fundamentals of Mass, Energy and Solute Transport in Poroelastic Rocks*. Leiden, Taylor & Francis CRC Press.

COMSOL Multiphysics software, (2009) Version 3.5a, Earth Science Module. Stockholm, Sweden, COMSOL AB.

Coussy, O. (1991) *Mecanique des Milieux Poreux*. Paris, Ed. Technip.

Leake, S.A. & Hsieh, P.A. (1997) *Simulation of deformation of sediments from decline of ground-water levels in an aquifer underlain by a bedrock step*. US Geological Survey Open File Report 97-47.

Liu, G.R. & Quek, S. (2003) *The Finite Element Method – A Practical course*. Bristol, Butterworth – Heinemann.

Terzaghi, K. (1943) *Theoretical Soil Mechanics*. New York, John Wiley.

Wang, H.F. (2000) *Theory of Linear Poroelasticity – With Applications to Geomechanics and Hydrogeology*. Princeton University Press, 287 pp.

Section 2:
Flow and transport

CHAPTER 4

New method for estimation of physical parameters in oil reservoirs by using tracer test flow models in Laplace space

J. Ramírez-Sabag, O.C. Valdiviezo-Mijangos & M. Coronado

4.1 INTRODUCTION

An important and useful tracer test application is on the estimation of reservoir properties such as porosity, dispersivity, fracture width, block size, rock adsorption, layer thickness or oil saturation. In order to obtain estimations of these parameters, mathematical tracer transport models are fitted to field tracer breakthrough data. Data fitting requires the solution of the so called inverse problem. The inverse problem consists in employing regression methods to determinate the free model parameters that provide the best field data fitting. When solving the inverse problem some difficulties associated to non-linearity might appear, thus finding a solution to the inverse problem can frequently result arduous and cumbersome. In this paper, an alternative approach to solve the inverse problem is analysed.

In the last decades many models and methodologies have been developed to describe tracer flow in porous media and to interpret tracer test results (Dai and Samper 2004, Ramírez-Sabag et al. 2005, Coronado et al. 2007). It should be noticed that many of the analytical solutions for tracer transport that are available obtained make use of the Laplace transform technique. The models include a large variety of circumstances such as continuous or instantaneous tracer injections in homogeneous and fractured reservoirs, with uniform linear or radial flow patterns. Analytical solutions are frequently given only in Laplace domain, because the inverse Laplace transform can not be analytically evaluated. In this case, the solution in real time is obtained numerically by standard inverse Laplace algorithms, such as that by de Hoog (1982). Thus, determining reservoir properties from tracer field data traditionally requires the employment of a non-linear optimization method coupled with numerical inverse Laplace transform techniques. This procedure becomes computer intensive and frequently faces numerical instability problems. In this work we introduce a new optimization procedure to fit analytical models in Laplace domain. This scheme requires the transformation of the original tracer breakthrough data into equivalent Laplace space data.

4.2 NUMERICAL LAPLACE TRANSFORMATION OF SAMPLE DATA

A primary problem consists in the translation of the original tracer concentration data, shown in Table 4.1, into the Laplace domain. This has been done previously by Roumboutsos and Stewart (1988) in regards to well test data. They first joined data pairs by straight lines and then analytically translated the line equation into the Laplace space. Wilkinson (1992) proposed a method to estimate the pressure transients as a parameter estimation in Laplace domain. In a subsequent paper, Bourgeois and Horne (1993) presented a mathematical background that justifies the use of Laplace space in well-test analysis. In this case, the whole time domain was considered as a single unit. A different numerical scheme was later introduced by Onur and Reynolds (1998) which consists in dividing the original data set into three sections that display different behaviour. These techniques have been applied to the specific curve shapes found in well tests. However, a direct application of this algorithm fails when applied to tracer test data. Therefore, following this idea, a new approach should be developed using information of the behaviour of the tracer test functions. The algorithm we propose for tracer data divides the data set into two parts. A first part uses the Roumboutsos and Stewart segmented straight lines approach and a second one which employs a decreasing exponential function. The way to aboard the problem in the interpretation of tracer test is completely new.

In order to introduce the new approach, it should be observed that tracer tests yield a collection of sampling times and their corresponding tracer concentration values as it is shown in Table 4.1. Typical tracer breakthrough curves exhibit an asymmetric bell-like shape as that illustrated in Figure 4.1. As mentioned before, the data set is divided into two regions as displayed in this figure. The first part comprehends the initial times and the peak, and the second contains the long tail. The cut-off time that divide the regions is denoted by t_p. This value should be chosen externally according to the specific data structure.

The Laplace transform of tracer concentration $C(t)$ is defined as

$$\mathcal{L}[C(t)] = \tilde{C}(s) = \int_0^\infty e^{-st} C(t)\, dt \tag{4.1}$$

Table 4.1　Tracer breakthrough in real space.

Time	Concentration
t_0	C_0
t_1	C_1
.	.
.	.
.	.
t_N	C_N

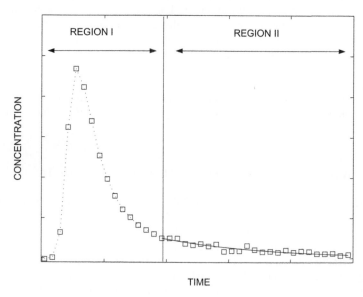

Figure 4.1　Tracer breakthrough data and its division into two regions.

As pointed out previously, this integral is divided into two sections. In the first part the aforementioned Roumboutsos algorithm is used, while in the second part a decreasing exponential is to be set. By this means Eq. (4.1) is written as

$$\tilde{C}(s) = \int_0^{t_p} e^{-st} C(t) \, dt + \int_{t_p}^{\infty} e^{-st} C(t) \, dt \tag{4.2}$$

The function $\tilde{C}(s)$ is written as

$$\tilde{C}(s) = \tilde{C}_I(s) + \tilde{C}_{II}(s), \tag{4.3}$$

where

$$\tilde{C}_I(s) = \int_0^{t_p} e^{-st} C(t) \, dt$$
$$\tilde{C}_{II}(s) = \int_{t_p}^{\infty} e^{-st} C(t) \, dt \tag{4.4}$$

By assuming a linear interpolation between each two consecutive points, i.e. a straight line in each segment (as illustrated in Region I of Figure 4.1), and after applying the Roumboutsos procedure to $\tilde{C}_I(s)$ it follows that

$$\tilde{C}_I(s) = \frac{C_0 - C_1 e^{-st}}{s} + \frac{d_0(1 - e^{-st_1})}{s^2}$$
$$+ \sum_{i=1}^{p} \left[\frac{C_i e^{-st_i}}{s} + \frac{d_i(e^{-st_i} - e^{-st_{i+1}})}{s^2} + \frac{C_{i+1} e^{-st_{i+1}}}{s} \right] \tag{4.5}$$

where

$$d_i = \frac{C_{i+1} - C_i}{t_{i+1} - t_i}$$

and p is the index of the time point associated to t_p. On the other hand, by setting

$$C(t) = a \exp[-\lambda(t - t_r)]$$

for $t \geq t_p$ it yields

$$\tilde{C}_{II}(s) = \frac{a e^{-\lambda t_r} e^{-t_p(\lambda+s)}}{\lambda + s} \tag{4.6}$$

The exponential function contains three parameters a, λ and t_r. The parameter a is introduced to satisfy concentration continuity at $t = t_p$, this means $a = C(t_p)/\exp[-\lambda(t_p - t_r)]$. The other two parameters λ and t_r are introduced to fit the tail data properly. They are determined by minimizing the following auxiliary objective function

$$OF_{aux}(\lambda, t_r) = \sum_{j=p}^{N} [C_j - a e^{-\lambda(t_j - t_r)}]^2 \tag{4.7}$$

After the application of this expression, a curve as the one that is displayed in Region II of Figure 4.1 follows. The tracer concentration function $\tilde{C}(s)$ in Laplace space given in Eq. (4.3) is a smooth function of s, and it is valid for the whole domain, $0 \leq s \leq \infty$. The procedure developed to obtain this expression will be called Modified Roumboutsos algorithm in this paper.

It should be pointed out that the main advantage in performing data fitting in Laplace space is that analytical tracer model solutions in real space are not required, thus the numerical inversions from Laplace to real domain are therefore unnecessary.

It should be mentioned that a different procedure was also tested. It was based on fitting the whole data set by a polynomial, and then translate analytically the polynomial into the Laplace domain. However, this alternative did not work properly, since too large data dispersion in Laplace domain appears, as reported in Ramírez-Sabag and Morales-Matamoros (2004).

4.3   THE LAPLACE DOMAIN OPTIMIZATION PROCEDURE

The general procedure we follow has been partially presented in the works by Wilkinson (1992) and by Onur and Reynolds (1998). The original tracer breakthrough data are incorporated in the Laplace domain using the function $\tilde{C}(s)$ from Eq. (4.3) after substitution of Eq. (4.5) and Eq. (4.6). This function should be fitted by a tracer transport model defined in Laplace space, $\tilde{M}(s; \bar{\alpha})$. This model is a function of s and the parameter set $\bar{\alpha} = \{\alpha_1, \alpha_2, \ldots\}$. In order to apply an optimization method we select a discrete collection of points $s_0, s_1, s_2, \ldots, s_M$ and search for a minimum value of the whole difference set $\{\tilde{C}(s_i) - \tilde{M}(s_i; \bar{\alpha})\}$ by adjusting the parameter values α. In this work an objective function made of a sum of squared weighted differences is employed, it is

$$OF(\bar{\alpha}) = \sum_{i=1}^{N} [\omega_i \tilde{C}(s_i) - \omega_i \tilde{M}(s_i; \bar{\alpha})]^2 \tag{4.8}$$

where ω_i is the weighting factor, which according to Onur and Reynolds (1998) is chosen as $\omega_i = s$. Here, $s_i \tilde{C}(s_i)$, is the *observed* concentration and $s_i \tilde{M}(s_i; \bar{\alpha})$ the corresponding model concentration.

As we can see, the objective function OF does not involve any numerical inverse Laplace transform, making this approach very attractive, since it avoids the numerical dispersion that regularly appear while numerically evaluating Laplace transforms. The relevance of this feature will be analysed by comparing the new procedure against the classical procedure that performs the parameter evaluation in real space. The traditional procedure is described in the next section.

4.4   THE REAL DOMAIN OPTIMIZATION PROCEDURE

The traditional method consists in minimizing the squared differences of the model and experimental data as it was done for example by Dai and Samper (2004) or Ramírez-Sabag et al. (2005). The corresponding objective function is

$$REOF(\bar{\alpha}) = \sum_{i=1}^{N} [C_i - M(t_i; \bar{\alpha})]^2 \tag{4.9}$$

where $REOF$ is the real space objective function, N is the number of data and $M(t; \bar{\alpha})$ is the tracer breakthrough model in real space. In this case, the model $M(t; \bar{\alpha})$ is not known in real space, but only in the Laplace space, then a numerical Laplace inversion should be performed in order to optimize $REOF$ the results of Eq. (4.9).

4.5   THE OPTIMIZATION METHOD

The next step in parameter estimation consists in minimizing the objective function in Eq. (4.8). The optimization method we will use is that by Nelder and Mead (1965), since this method has been proved to yield excellent results in tracer test problems (Ramírez-Sabag et al. 2005). The Nelder-Mead method (1965) is an algorithm developed to minimize nonlinear functions by a direct search, that is say, by evaluating the objective function in diverse points. Calculation of the derivative is not required here. The evaluation points are the vertexes of a *simplex*. A simplex is a polytope of $n + 1$ vertices in an n-dimensional space, for example, a line segment in a line, a triangle in a plane, a tetrahedron in a three-dimensional space and so on. The simplex is step-wise modified while searching for a minimum in the objective function. At each step the objective function is calculated at all vertices. A new simplex is generated by eliminating the vertex with the largest function value, and by applying one of the following operations: reflection, contraction, expansion or reduction. The procedure is repeated and a sequence of simplexes with always lower objective function values is obtained. The process continues until the diameter of the simplex becomes smaller than the specified tolerance.

4.6 THE VALIDATION PROCEDURE

We will apply the new method in two standard well known pulse tracer transport models, one for the so called homogeneous reservoirs and the other for fractured reservoirs. In the homogeneous case (non-fractured) both analytical expressions, for the tracer breakthrough concentration in real as well as in the Laplace space are available. To validate our method we will use synthetic data generated by those models plus adding a Gaussian random noise with zero mean and small standard deviation.

In the following subsection we describe briefly the models to be used in the pulse injection cases treated here, however it should be pointed out that our methodology can be also applied to other one tracer transport models.

4.6.1 *Employed mathematical models*

The models we will use to validate our procedure are regularly applied in the design and interpretation of the tracer test. The first one corresponds to homogeneous reservoirs and the other to fractured reservoirs.

4.6.1.1 *The tracer transport model for homogeneous formations*

The tracer advective-dispersive transport model for homogeneous reservoirs to be used was developed by Kreft and Zuber (1978). A pulse is injected in $x = 0$ at initial time and the tracer breakthrough concentration at point x and time t is given by

$$C_h(x_D, t_D) = \frac{C_0 x_D}{\sqrt{4\pi t_D^3 / Pe}} \exp\left[-\frac{Pe(x_D - t_D)}{4t_D}\right] \tag{4.10}$$

where $x_D = x/L$ and $t_D = tu/L$ are dimensionless distance and time respectively. Here u is a constant speed and L is a characteristic length, which can be set equal to the distance between injection and production wells. Pe is the Peclet number defined as $Pe = uL/D$, where D is a constant dispersion coefficient. The factor C_0 takes into account the amount of injected tracer. This model in the Laplace space is written as

$$\tilde{C}_h(x_D, s) = C_0 \exp\left[\frac{x_D Pe}{2}(1 - \sqrt{1 + 4s/Pe})\right]. \tag{4.11}$$

4.6.1.2 *The tracer transport model for fractured formations*

A standard model to describe tracer transport in a porous formation with fractures, Malosewski and Zuber (1985) and Ramírez-Sabag (1988), is given in Laplace space by

$$\tilde{C}_f(x_D, s) = C_0 \exp\left[\frac{x_D Pe}{2}(1 - \sqrt{1 + 4\sigma(s)/Pe})\right] \tag{4.12}$$

where

$$\sigma(s) = s + \beta\sqrt{(s)}, \qquad \beta = \frac{\phi_2 L}{w}\sqrt{\frac{R_a D_2}{Pe D_1}} \tag{4.13}$$

C_0 is a factor related to the total tracer injected mass. The factor β describes the matrix fracture coupling, and involves diverse parameters: $2w$ is the effective fracture width, R_a is the delay factor due to the adsorption on and desorption in the rock, D_1 and D_2 are the dispersion coefficient of fracture and matrix, respectively, and ϕ_2 is the porosity of the matrix. When the coupling parameter is not present, $\beta = 0$, the homogeneous model in Eq. (4.11) is recovered. In the next section, we describe how synthetic data are generated and we compare the results obtained by our method with those obtained by the traditional method.

4.6.2 *Generation of synthetic data*

We use synthetic data in order to analyse the capacity of the new procedure to evaluate model parameters in Laplace domain. In this way, we have a controlled environment in both physical parameters and the data. We can therefore straightforwardly compare the procedure outcome with the original parameter values. In the case of homogeneous reservoirs, synthetic data were generated by using the model of Eq. (4.10) together with Gaussian random noise having a zero mean, $\mu = 0$, and standard deviation $\sigma = 0.001$. The physical parameters values used are: $x_D = 1$, $Pe = 10$ and $C_0 = 1$. In the case of fractured reservoirs we use the model in Eq. (4.12), adding a random noise with zero mean and a standard deviation $\sigma = 0.005$. Here, the chosen physical parameters are: x_D, $Pe = 28$, $\beta = 1$ and $C_0 = 1$. The data in real space for fractured reservoir are obtained by numerical inversion of Eq. (4.12). The amount of synthetic points obtained in each case was 40; this is a representative situation of a typical field tracer test. In space domain, the s values were chosen from 1 to 10 with a step of 0.1. In the forthcoming part of the work, we apply the new methodology to the synthetic data previously generated.

4.6.3 *Result with synthetic data*

In the homogeneous case we generate synthetic tracer breakthrough data employing Eq. (4.10) and Gaussian noise. After exploring the curve data, depicted as square points in Figure 4.2, we choose the tail section as those points with greater times than $t_D = 1.75$, which corresponds to the 30th time point. By using Eq. (4.7) the optimization procedure yields the fitting parameter values displayed in the second column of Table 4.2.

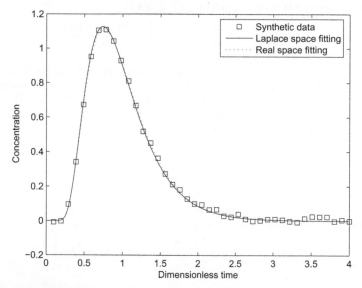

Figure 4.2   Tracer breakthrough curve using the parameter values obtained by data fitting in Laplace and real domains for the homogeneous case with synthetic data.

Table 4.2   Curve tail adjusting parameters.

	Homogeneous Model	Fractured Model
a	1.03×10^{-2}	5.54×10^{-2}
λ	6.56×10^{-2}	2.09×10^{-2}
t_p	1.76	1.76

The curve fitting in Laplace domain according to the new procedure presented in Section 4.3 is shown in Figure 4.3. The initial parameter values $x_D = 0.7$ and $Pe = 7$ are used. Data are represented as circles, and the Laplace space model, Eq. (4.11), as a solid line. A good fitting is achieved. The corresponding model parameters are displayed in Table 4.3. The second column contains the real space data fitting using Nelder-Mead optimization algorithm. The third column presents the parameters value obtained by Laplace domain fitting. Parameter results are very similar, as it can also be noticed in Figure 4.2, where real space concentration in terms of real time is displayed. The solid line and the broken line represent the fitting in Laplace and in real space respectively. A remarkable similarity can be appreciated. The difference between the original parameter values with those obtained by the optimization in Laplace space is 2% for x_D, 3% for Pe and less than 1% for C_0. The corresponding differences according to the optimization performed in real space is less than 1%. As shown in Table 4.3, the objective function takes a small value in both cases, indicating that a good fit has been achieved. The fact that the objective function value in Laplace domain is lower in size does not imply that the match is better, since the objective functions are different and they are defined in a distinct space.

The parameter values estimated using the model for fractured reservoirs previously described are shown in Table 4.4. The starting parameters values employed were: $x_D = 0.8$, $Pe = 20$ and $\beta = 2$. Data fitting in Laplace space is shown in Figure 4.4. A good fit is observed in Figure 4.4.

It can be seen the data fitting in real space in Figure 4.5. As in the previous case a good fit is achieved.

The difference between the original and the obtained parameters using the fitting in Laplace domain is 9% for x_D, 17% for Pe, 27% for β and 7% for C_0. The difference between the original

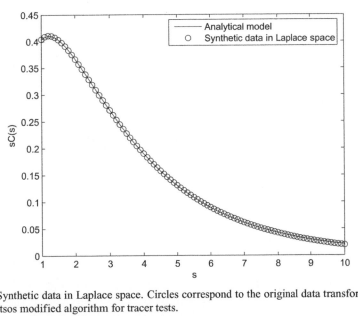

Figure 4.3   Synthetic data in Laplace space. Circles correspond to the original data transformed by using the Roumboutsos modified algorithm for tracer tests.

Table 4.3   Parameter values for the homogeneous case.

Physical parameter	Real domain	Laplace domain
x_D	1.0028	0.9896
Pe	10.0311	10.3453
C_0	0.993	1.0018
Objective Function Value	7.19×10^{-2}	5.00×10^{-3}

Table 4.4 Parameter values for the fractured case using synthetic data.

Physical parameter	Real domain	Laplace domain
x_D	0.9996	0.9036
Pe	27.4842	32.8673
β	1.005	1.2702
C_0	1.0024	1.0651
Objective Function Value	2.3338×10^{-2}	3.1958×10^{-3}

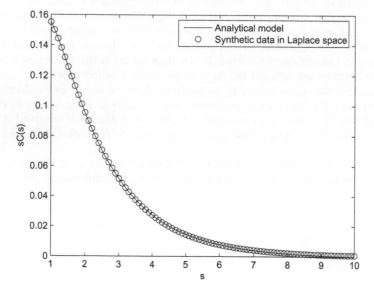

Figure 4.4 The same as Figure 4.3 for the fractured case.

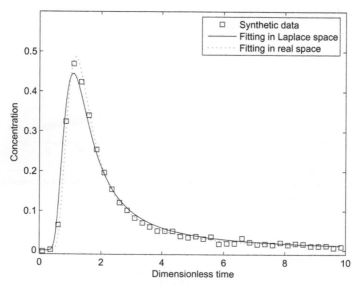

Figure 4.5 Tracer breakthrough curve obtained by Laplace space fitting (solid line) and real space fitting using numerical Laplace inversion (broken line) and data (squares) for synthetic data using the fractured reservoir model.

parameters values used to generate the data in real space and those obtained by fitting them in real space is 1%, 2%, 1% and 1% respectively. The parameters obtained after making optimization in real space are closer to the actual values than those obtained by optimization in Laplace space. However, performing optimization in real space without having a model in real space requires a great number of numerical Laplace inversions, which can give place to long computer times and numerical dispersion.

4.7   RESERVOIR DATA CASES

In this part of the work we will apply the new methodology to reservoir tracer tests. Here, some field data published in the literature for both homogeneous and fractured reservoirs are employed. We will compare the results obtained by the new method against those obtained by the traditional procedure in real space.

4.7.1   *A homogeneous reservoir (Loma Alta Sur)*

To test our method, we use data obtained in a tracer test carried out in the Loma Alta Sur, Argentina (Badessich et al. 2005). Tracer breakthrough data are displayed as squares in Figure 4.6. The tail section is chosen for larger times than $t_D = 1.5$, which corresponds to the 14th data point.

The distance between the injector and production well is 142 m. The average transit time is calculated using the first moment of the tracer arrival curve, which is 19.75 days. This is the time used to calculate the tracer velocity and the normalized time t_p. The curve tail is matched to the exponential function, and the parameters obtained are shown in Table 4.5.

Applying the whole procedure described before and using $x_D = 0.8$ and $Pe = 4$ as optimization starting parameters, the model parameters displayed in Table 4.6 follow. The new variable E is a measure of the effective tracer fraction that arrives at the production well. It is just a scale factor to fit the curve high. Figure 4.7 presents the fitting in the Laplace domain. In Figure 4.6, tracer data are displayed as squares; the curves using the parameters obtained from the Laplace domain fitting (solid line) and the real domain fitting (broken line) are plotted as well.

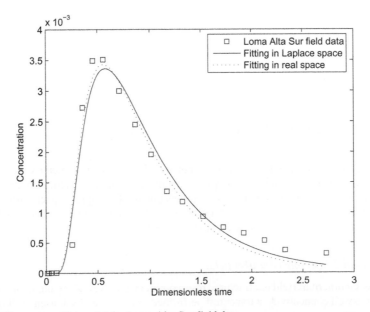

Figure 4.6   The same as Figure 4.5 for Loma Alta Sur field data.

Table 4.5   Curve tail adjusting parameters.

Adjusting parameter	Homogeneous model Loma Alta Sur
a	1.21×10^3
λ	9.66×10^1
t_p	1.25
Initial Point	14

Table 4.6   Model parameters for Loma Alta Sur Field test.

Model parameter	Real domain	Laplace domain
x_D	0.9501	1.0092
Pe	5.6307	5.05
C_0/E	3.4667×10^{-3}	3.4639×10^{-3}
Objective Function Value	3.7618×10^{-4}	3.6537×10^{-4}

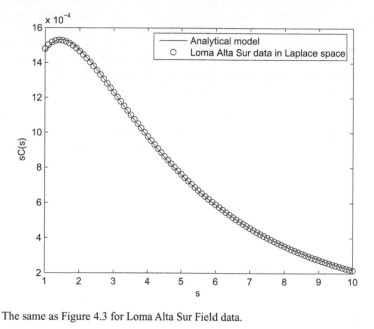

Figure 4.7   The same as Figure 4.3 for Loma Alta Sur Field data.

As we are dealing with field data, there are not actual reference parameter values to compare the fitting results with, we therefore simply confront the parameter values obtained using the Laplace domain fitting against the parameter values obtained using real space fitting previously published (Ramírez-Sabag et al. 2005). A difference is found which is less than 6% for x_D, less than 12% for the Peclet number and less than 0.1% for C_0/E.

4.7.2   A fractured reservoir (Wairakei field)

The Wairakei Geothermal field is a fractured reservoir located in New Zealand. In order to analyse reservoir inter-well connectivity, a tracer test performed as described by Jensen (1983). We select data from the tracer response at a specific production well. The distance between the production well and the injector is 210 m. The selected characteristic time is 0.214 days, which correspond to the first arrival time.

When applying the new procedure, we first evaluate the parameters for the exponential function at the tail region. With these values, we have all the elements required by the algorithm to transform the real space data to the Laplace domain. By taking the region cut-off time at $t_p = 15$ the cut-off point result the 87th data point. The obtained parameters are presented in Table 4.7. The model parameters values determined by means of our methodology are presented in the third column of Table 4.8. The starting optimization iteration values were $x_D = 1$, $Pe = 80$ and $\beta = 2$.

By comparing the fitting in Laplace versus in real domain, it is found a difference of less than 2% for x_D, less than 33% for Pe, for β, less than 7% for C_0/E. The curve adjustments are shown in Figures 4.8 and 4.9, for the Laplace and real domain respectively. It can be observed that data

Table 4.7 Adjusted parameters for the tail region of the Wairakei data.

Adjusting parameter	Fractured model Wairakei
a	7.83×10^3
λ	3.03×10^{-2}
t_p	1.85
Initial Point	87

Table 4.8 Model parameter values obtained for the Wairakei field.

Model parameter	Real domain	Laplace domain
x_D	1.3251	1.3006
Pe	88.6029	59.4767
β	1.804	1.9566
C_0/E	7.3575×10^4	7.8116×10^4
Objective Function Value	9.4983×10^2	3.4983×10^0

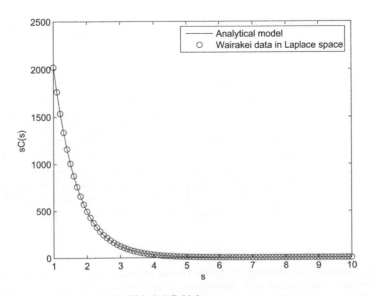

Figure 4.8   The same as Figure 4.5 for Wairakei field data.

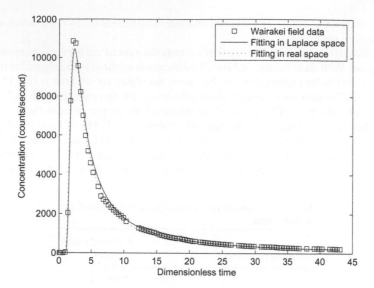

Figure 4.9   The same as Figure 4.3 for Wairakei field data.

fitting in Figure 4.9 in Laplace space is remarkably good and practically equivalent to real space fitting.

4.8   SUMMARY AND CONCLUDING REMARKS

A new methodology to solve the inverse problem involved in fitting pulse tracer breakthrough data to mathematical models for tracer transport has been introduced. The procedure consists in determining the model parameters by data fitting in the Laplace domain instead of the real domain. The method is particularly valuable when the available analytical models provide mathematical expression for the tracer concentration only in the Laplace domain, not in the time domain. In this case large computing operations associated to numerical Laplace inversion, as well as numerical dispersion can be avoided.

The procedure requires a translation of the tracer breakthrough data to Laplace space by what we called Modified Roumboutsos algorithm that divides the time domain into two regions. The first region comprehends the tracer concentrations peak and the second one the tracer curve tail. In this method an equivalent collection of data points in Laplace space is generated and matched to analytical tracer transport models. The procedure has been tested in synthetic data obtained for a classical tracer transport model for homogeneous (non-fractured) and for fractured porous media. A similar value for the model parameters obtained by both procedures (Laplace domain fitting and real space fitting) is gotten. The global parameter differences are smaller than 3% in the homogeneous case and less than 27% in the fractured case scratch it out. The procedure was also applied in two field tracer tests, one performed in a oil reservoir and the other performed in a geothermal reservoir. Both cases yield good results; this means less than 12% in the first case and less than 33% in the second case. This error is relatively small in field tracer tests where usually large uncertainties are present. Finally, it should be mentioned that special attention must be paid in cases when tracer breakthrough data show large dispersion around the average curve.

REFERENCES

Badessich M.F., Swinnen I. & Somaruga C. (2005) The use of tracer to characterize highly heterogeneous fluvial reservoirs: Field results. *SPE International Symposium on Oilfield Chemistry, 2-4 February, Houston, TX, Society of Petroleum Engineering library record 92030*. Available from: http://www.spe.org.

Bourgeois M. & Horne R. (1993) Well-test-model recognition with Laplace space. *SPE 1993 Formation Evaluation, March 1993, 22682-PA*. Available from: http://www.spe.org.

Coronado M., Ramírez-Sabag J. & Valdiviezo-Mijangos O. (2007) On the boundary condition in tracer transport models for fractured porous underground formations. *Revista Mexicana de Física*, 53 (4), 260–269.

Dai Z. & Samper J. (2004) Inverse problem of multicomponent reactive chemical transport in porous media: Formulation and applications. *Water Resources Research*, 40, W07407. Available from: doi: 10.1029/2004WR003248.

de Hoog F.R., Knigth J. & Stokes A. (1982) And improved method for numerical inversion of Laplace Transforms. *SIAM Journal of Scientific and Statistical Computing*, 3: 357–366.

Jensen C.L. (1983) *Matrix diffusion and its effect on the modeling of tracer returns from the fractured geothermal reservoir at Wairakei, New Zeland, SGP-TR-71. Report of the Stanford Geothermal Program.* Stanford CA, Stanford University.

Kreft A. & Zuber A. (1978) On the physical meaning of the dispersion equation and its solution for different initial and boundary condition. *Chemical Engineering Science*, 33, 1471–1480.

Maloszweski P. & Zuber A. (1985) On the theory of tracer experiments in fissured rock with porous matrix. *Journal of Hydrology*. 79, 333–358.

Nelder J.A. & Mead R. (1965) A simplex method for simplex minimization. *The Computer Journal*, 7, 308–313.

Onur M. & Reynolds A.C. (1998) Numerical Laplace transformation of sampled data for well-test analysis. *SPE, Annual Technical Conference and Exhibition, Denver, 6–9 October, 1996 No. 36554*. Available from: http://www.spe.org.

Ramírez-Sabag J. (1988) A model to predict the tracer flow in naturally fractured geothermal reservoirs. M. Eng. Thesis (in Spanish), México, School of Engineering, National University of Mexico.

Ramírez-Sabag and Morales-Matamoros D. (2004), Manual para Resolver el Problema Inverso en Pruebas de Trazadores en el Dominio de Laplace. México, *Instituto Nacional del Derecho de Autor* 03-2004-022611541100-01.

Ramírez-Sabag J., Valdiviezo-Mijangos O. & Coronado M. (2005) Inter-Well tracer tests in oil reservoirs using different optimization methods: A field case. *Geofsica Internacional*, 44 (1), 113–120.

Roumboutsos A. & Stewart G. (1988) A direct deconvolution or convolution algorithm for well test analysis. *SPE Annual Technical Conference and Exhibition, 2–5 October, Houston. Society of Petroleum Engineering library record 18157*. Available from: http://www.spe.org.

Wilkinson D.J. (1992) Pressure transient parameter estimation in the Laplace domain. *SPE The 67th Annual Technical Conference and Exhibition, 4–5 October, Washington. Society of Petroleum Engineering library record 24729*. Available from: http://www.spe.org.

CHAPTER 5

Dynamic porosity and permeability modification due to microbial growth using a coupled flow and transport model in porous media

M.A. Díaz-Viera & A. Moctezuma-Berthier

5.1 INTRODUCTION

Oil fields in its initial operation stage are produce by using its natural energy, which is known as primary recovery. As the reservoir loses energy it requires the injection of gas or water in order to restore or maintain the reservoir pressure, this stage is called secondary recovery. When the secondary recovery methods become ineffective it is necessary to apply other more sophisticated methods such as steam injection, chemicals, microorganisms, etc. These methods are known as tertiary or enhanced oil recovery (EOR). Some important oil fields in Mexico are entering the third stage.

For an optimal enhanced oil recovery method design a variety of laboratory tests under controlled conditions are required. They will contribute to understand the fundamental recovery mechanisms for a given EOR method in a specific reservoir. The laboratory tests commonly have a number of drawbacks, which include among others, that they are very sophisticated, expensive and largely unrepresentative of the whole range of phenomena involved. A proper modeling of the laboratory tests would be decisive in the interpretation and understanding of recovery mechanisms and in obtaining the relevant parameters for the subsequent implementation of enhanced recovery processes at the well and the reservoir scale.

In this paper we present a flow and transport model which was implemented using the finite element method within the environment of the COMSOL Multiphysics to numerically simulate, analyze and interpret biphasic oil-water displacement and multicomponent transport processes in porous media at core scale.

From a methodological point of view the four model development stages (conceptual, mathematical, numerical and computational) will be shown. The procedure is illustrated through a model example developed to simulate a microbial enhanced oil recovery (MEOR) method.

In a first stage, a multiphase flow (Diaz-Viera et al. 2008) and a multicomponent transport (Lopez-Falcon et al. 2008) models were simultaneously but separately developed. In particular, the flow model, which accounts for a two-phase water-oil flow, took into consideration the relative permeability and capillary pressure curves, the effects of gravity and the dynamic modification of porosity and permeability due to the phenomena of adsorption/desorption (clogging/declogging) of microorganisms, by using the Kozeny-Carman equation for the porosity-permeability relationship (Carman 1956).

Similarly, the transport model was accomplished for three phases (rock-water-biofilm) and three components (planktonic organisms, sessile microorganisms and nutrients). This model include physical-chemical-biological phenomena such as advection, diffusion, dispersion, adsorption, desorption, growth and decay of microorganisms. Adsorption of nutrients is implemented through a linear adsorption isotherm.

Finally, the previously developed models were coupled yielding a qualitatively new model. The performance of the new flow and transport model was numerically evaluated and validated in a study case. This case describes the oil displacement by the injection of water, followed by the injection of water with microorganisms and nutrients.

Here, this model will be exploded as a research tool to investigate the impact on the flow behavior of the porosity-permeability dynamic variation due to the clogging/declogging phenomena that occur during microbial enhanced oil recovery processes.

5.2 THE FLOW AND TRANSPORT MODEL

5.2.1 *Conceptual model*

The following hypothesis, postulations and conditions are being considered to be satisfied by the flow and transport model.

1. There are four phases: water ($\alpha = W$, $\mathbf{v}^W \geq 0$), oil ($\alpha = O$), as liquid hydrocarbon ($\mathbf{v}^O \geq 0$ and $\mathbf{v}^O \neq \mathbf{v}^W$), biofilm ($\alpha = B$) and rock ($\alpha = R$), both static ($\mathbf{v}^B = \mathbf{v}^R = 0$). Here, α denotes phases, and $\mathbf{v}$ is the velocity vector.
2. There are five components: water ($\gamma = w$), only in the water phase, oil ($\gamma = o$), only in the oil phase, rock ($\gamma = r$), only in the rock phase, microorganisms ($\gamma = m$) in the water phase (planktonic) and in the biofilm phase (sessile), and nutrients ($\gamma = n$) in the water phase (fluents) and in the rock phase (adsorbed). Here, γ denotes components.
3. The rock (porous matrix) and the fluids are incompressible.
4. The porous medium is fully saturated, i.e., the porous volume is completely filled.
5. The fluid phases stay separated inside the pores.
6. All phases are in thermodynamical equilibrium.
7. The porous medium is considered homogeneous and isotropic, which means that its properties are invariant with location and size, and direction independent, respectively. In particular, at the initial state the porosity is constant ($\phi = const$) and its permeability is the same in all directions ($\mathbf{k} = k\mathbf{I}$), but the dynamical porosity and permeability variation due to clogging/declogging (adsorption/desorption) processes is allowed.
8. The microorganisms and nutrients dispersive flux follows the Fick's law: $\boldsymbol{\tau}_\gamma^W(\mathbf{x}, t) = \phi S^W \mathbf{D}_\gamma^W \cdot \nabla c_\gamma^W$; $\gamma = m, n$, where S^W is the water saturation (volume fraction of water in the porous space), c_γ^W is the concentration of the component γ in water (mass of the component γ per volume of water), and

$$(D_\gamma^W)_{ij} = (\alpha_T)_\gamma^W |\mathbf{v}^W| \delta_{ij} + ((\alpha_L)_\gamma^W - (\alpha_T)_\gamma^W) \frac{v_i^W v_j^W}{|\mathbf{v}^W|} + \tau (D_d)_\gamma^W \delta_{ij}; \qquad (5.1)$$

are the hydrodynamic dispersion tensor components ($i, j = 1, 2, 3$) (Bear 1972), where (α_L) and (α_T) are the transversal and longitudinal dispersivity of component γ in water, respectively. $|\mathbf{v}^W|$ is the Euclidean norm of the water velocity vector, δ_{ij} is the Kronecker delta function, τ is the tortuosity and $(D_d)_\gamma^W$ is the molecular diffusion coefficient of the component γ in water.
9. Microorganisms and nutrients have biological interaction. The Monod growth rate function is used (Gaudy 1980): $\mu = \mu_{\max}(c_n^W / (K_{m/n} + c_n^W))$, where $\mu_{\max}$ is the maximum specific growth rate, $K_{m/n}$ is the Monod constant for nutrients, c_n^W is the nutrients concentration in water. Linear decay models for planktonic and sessile microorganisms are used, the processes are given by $\kappa_d \phi S^W c_m^W$ and $\kappa_d \rho_m^B \sigma$, respectively. Here κ_d is the specific decay rate of cells and we assume that it is the same for both, ρ_m^B is the density of microorganisms, σ is the volume fraction of sessile microorganisms, and c_m^W is the concentration of planktonic microorganisms.
10. Microorganisms form a biofilm phase. A quasilinear adsorption (Corapcioglu & Haridas 1984) and an irreversible limited desorption processes (Lappon & Fogler 1996) are assumed. The adsorption and desorption terms employed are $\kappa_a^m \phi c_m^W$ and $\kappa_r^m \rho_m^B (\sigma - \sigma_{irr})$ for $\sigma \geq \sigma_{irr}$ and 0 for $\sigma < \sigma_{irr}$, respectively. Here κ_a^m and κ_r^m are the corresponding adsorption and desorption

rate coefficients, and σ_{irr} is the minimum sessile cell concentration, which accounts for cells that are irreversibly adsorbed within the biofilm phase.

11. Nutrients and rock have physico-chemical interaction as a linear adsorption isotherm $\hat{c}_n^R = \kappa_a^n c_n^W$, where $\hat{c}_n^R$ is the adimensional nutrient concentration adsorbed within the porous medium.

In Table 5.1 a summary of the intensive properties associated with the mass of each components by phases is given.

Table 5.1   Intensive properties associated with the mass of the components by phases.

Phase (α)	Component (γ)	Intensive Properties (ψ_γ^α)
Water (W)	Water (w)	$\phi S^W \rho_w^W$
	Microorganisms (m)	$\phi S^W c_m^W$
	Nutrients (n)	$\phi S^W c_n^W$
Oil (O)	Oil (o)	$\phi S^O \rho_o^O$
Biofilm (B)	Microorganisms (m)	$c_m^B \equiv \rho_m^B \sigma$
Rock (R)	Rock (r)	$\rho_{r_b}^R \equiv (1-\phi)\rho_{r_p}^R$
	Nutrients (n)	$c_n^R \equiv \rho_{r_b}^R \hat{c}_n^R$

5.2.2   Mathematical model

From the assumptions given in the conceptual model we can derive the corresponding mathematical model. Here we will apply the axiomatic formulation for continuum systems (Allen et al. 1988) for deriving the equations of the mathematical model.

The axiomatic formulation adopts a macroscopic approach, which considers that the material systems are fully occupied by particles. A continuum system is made of a particle set known as material body, see Figure 5.1. The continuum system approach works with the volume average of the body properties and consequently there is a volume called representative elementary volume (REV) over which the average of the properties is valid.

The continuum approach formulation basically consists in establishing a one to one correspondence between the extensive properties $E(t)$, which are the volume integrals of the intensive properties, for example: the mass, and the intensive properties $\psi(\mathbf{x}, t)$, which are the physical properties by unit of volume, for example: the mass density. Both properties are related by the following expression:

$$E(t) \equiv \int_{B(t)} \psi(\mathbf{x}, t)\, d\mathbf{x} \tag{5.2}$$

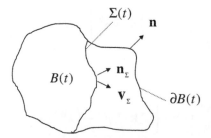

Figure 5.1   General scheme of a material body: $B(t)$ is the body with boundary $\partial B(t)$, where $\mathbf{n}$ is the outer normal vector, $\Sigma(t)$ is the discontinuity surface with a normal vector $\mathbf{n}_\Sigma$ and velocity $\mathbf{v}_\Sigma$.

The previous relationship allows us to express the global balance equation of an extensive property $E(t)$, Eq. (5.3), as the local balance equations of an intensive property $\psi(\mathbf{x}, t)$, Eqs. (5.4) and (5.5).

- Global Balance Equation

$$\frac{dE(t)}{dt} = \int_{B(t)} g(\mathbf{x}, t)\, d\mathbf{x} + \int_{\Sigma(t)} g_\Sigma(\mathbf{x}, t)\, d\mathbf{x} + \int_{\partial B(t)} \boldsymbol{\tau}(\mathbf{x}, t) \cdot \mathbf{n}\, d\mathbf{x} \tag{5.3}$$

where $g(\mathbf{x}, t)$ is the source term in the body $B(t)$; $g_\Sigma(\mathbf{x}, t)$ is the source term at the discontinuity surface $\Sigma(t)$; and $\boldsymbol{\tau}(\mathbf{x}, t)$ is the vector flux through the boundary $\partial B(t)$.

- Local Balance Equations

$$\frac{\partial \psi}{\partial t} + \nabla \cdot (\psi \mathbf{v}) = g + \nabla \cdot \boldsymbol{\tau}; \qquad \forall \mathbf{x} \in B(t) \tag{5.4}$$

$$[\![\psi(\mathbf{v} - \mathbf{v}_\Sigma) - \boldsymbol{\tau}]\!] \cdot \mathbf{n}_\Sigma = g_\Sigma; \qquad \forall \mathbf{x} \in \Sigma(t), \tag{5.5}$$

where $\mathbf{v}$ is the particle velocity and $[\![f]\!] \equiv f_+ - f_-$ is the jump of the function f across the surface $\Sigma(t)$. Here, f_+ and f_- are uniquely defined once the normal vector $\mathbf{n}_\Sigma$ to the surface $\Sigma(t)$ is chosen.

In the case when we have several phases and multiple components it is necessary to apply the previous procedure to each component in each phase, obtaining as many equations as components by phases we have. For the resulting system of equations it is required to specify certain constitutive laws which are related with the nature of the problem to be modeled, for example, the Darcy law in the case of a model of flow through a porous media. These constitutive relationships allow to link the intensive properties among them and define the flux and source terms. Even more, sometimes after this operation complementary relationships are necessary to obtain a closed equation system. Finally, the model is completed when sufficient initial and boundary conditions are specified. In this manner we get a well posed problem, which means that there exists one and only one solution for the problem.

5.2.2.1 *The governing equations of the flow and transport model*
Applying the previously described axiomatic formulation for continuum systems and considering the assumptions established in the conceptual model, we can obtain the governing equations system for the flow and transport model:

- The flow equations
  The flow model is based on the oil phase pressure and total velocity formulation given by Chen (Chen et al. 2006, Chen 2007) in which the capillary pressure, i.e., pressure difference between phases, and relative permeability curves are taken in account. The equations of the biphasic flow model are the following:
  Pressure equation (p^O):

$$-\nabla \cdot \left\{ \lambda \mathbf{k} \cdot \nabla p^O - \left(\lambda^W \frac{dp_c^{OW}}{dS^W} \right) \mathbf{k} \cdot \nabla S^W \right\} - \nabla \cdot \{(\lambda^O \rho^O + \lambda^W \rho^W)|\mathbf{g}|\mathbf{k} \cdot \nabla z\}$$
$$= q^O + q^W - \frac{\partial \phi}{\partial t} \tag{5.6}$$

Saturation equation (S^W):

$$\phi \frac{\partial S^W}{\partial t} - \nabla \cdot \left\{ \lambda^W \mathbf{k} \cdot \nabla p^O - \left(\lambda^W \frac{dp_c^{OW}}{dS^W} \right) \mathbf{k} \cdot \nabla S^W \right\}$$
$$- \nabla \cdot \{(\lambda^W \rho^W |\mathbf{g}|)\mathbf{k} \cdot \nabla z\} + \left(\frac{\partial \phi}{\partial t} \right) S^W = q^W \tag{5.7}$$

Phase velocities ($\mathbf{u}^\alpha$):

$$\mathbf{u}^W = -\lambda f^W \mathbf{k} \cdot \nabla p^O + \lambda f^W \left(\frac{dp_c^{OW}}{dS^W}\right) \mathbf{k} \cdot \nabla S^W - \lambda f^W \rho^W |\mathbf{g}| \mathbf{k} \cdot \nabla z;$$

$$\mathbf{u}^O = -\lambda f^O \mathbf{k} \cdot \nabla p^O - \lambda f^O \rho^O |\mathbf{g}| \mathbf{k} \cdot \nabla z \tag{5.8}$$

Here, $\mathbf{u}^\alpha$ is the Darcy velocity

$$\mathbf{u}^\alpha = -\frac{k_r^\alpha}{\nu^\alpha} \mathbf{k} \cdot (\nabla p^\alpha + \rho^\alpha |\mathbf{g}| \nabla z); \quad \alpha = O, W \tag{5.9}$$

where ϕ is the porosity, $\mathbf{k}$ is the absolute permeability tensor, S^α is the saturation, ν^α is the viscosity, ρ^α is the density, p^α is the pressure, k_r^α is the relative permeability, and q^α is the source term, for each phase $\alpha = O, W$, $|\mathbf{g}|$ is the absolute value of the gravity acceleration, p_c^{OW} is the capillary oil-water pressure and z is the vertical coordinate.

For convenience, the following additional notation was introduced: $\lambda^\alpha = k_r^\alpha / \nu^\alpha$ are the phase mobility functions, $\lambda = \sum \lambda^\alpha$ is the total mobility, $f^\alpha = \lambda^\alpha / \lambda$ are the fractional flow functions, so that $\sum f^\alpha = 1$ and the total velocity is defined as $\mathbf{u} = \sum \mathbf{u}^\alpha$, where $\alpha = O, W$.

In the present model, the constitutive relationships for the relative permeability functions are represented by the Brooks-Corey equation (Brooks & Corey 1964):

$$k_r^W = S_e^{\frac{2+3\theta}{\theta}} \quad k_r^O = (1 - S_e)^2 \left(1 - S_e^{\frac{2+\theta}{\theta}}\right); \tag{5.10}$$

where S_e is the effective or normalized saturation, which is defined as:

$$S_e = \frac{S^W - S_r^W}{1 - S_r^W - S_r^O}$$

Here, S_r^W and S_r^O are the residual saturations for water and oil, respectively. As long as θ characterizes the pore size distribution.

As the constitutive relationship for the oil-water capillary pressure was used the Brooks-Corey function (Brooks & Corey 1964):

$$p_c^{OW}(S^W) = p_t \left(\frac{S^W - S_r^W}{1 - S_r^W - S_r^O}\right)^{(-1/\theta)} \tag{5.11}$$

where p_t is the entry or left threshold pressure assumed to be proportional to $(\phi/k)^{1/2}$. Consequently, we can express the oil-water capillary pressure derivative as follows:

$$\frac{dp_{cow}}{dS_w}(S_w) = -\frac{p_t}{\theta(1 - S_{rw} - S_{ro})} \left(\frac{S_w - S_{rw}}{1 - S_{rw} - S_{ro}}\right)^{-(1+\theta)/\theta} \tag{5.12}$$

- The transport equations
The multicomponent transport model was derived based on the revision of the models introduced in (Corapcioglu & Haridas 1984) and in (Chang et al. 1992), respectively. The transport model, that is presented here, is constituted by a system of three fully coupled nonlinear partial differential equations.
Planktonic microorganisms equation (c_m^W):

$$\frac{\partial(\phi S^W c_m^W)}{\partial t} + \nabla \cdot (c_m^W \mathbf{u}^W - \phi S^W \mathbf{D}_m^W \cdot \nabla c_m^W) = (\mu - \kappa_d - \kappa_a^m)\phi S^W c_m^W + \kappa_r^m \rho_m^B(\sigma - \sigma_{irr}) \tag{5.13}$$

Sessile microorganisms equation (σ):

$$\frac{\partial(\rho_m^B \sigma)}{\partial t} = (\mu - \kappa_d)\rho_m^B \sigma + \kappa_a^m \phi S^W c_m^W - \kappa_r^m \rho_m^B(\sigma - \sigma_{irr}) \tag{5.14}$$

Nutrients equation (c_n^W):

$$\frac{\partial\{(\phi S^W + \rho_{r_b}^R \kappa_a^n)c_n^W\}}{\partial t} + \nabla \cdot (c_n^W \mathbf{u}^W - \phi S^W \mathbf{D}_n^W \cdot \nabla c_n^W) = -\frac{\mu}{Y_{m/n}}(\phi S^W c_m^W + \rho_m^B \sigma) \quad (5.15)$$

- Porosity and permeability relationships
  The porosity modification due to the clogging/declogging processes is taking in account by the following expression given in (Chang et al. 1992):

$$\phi(\mathbf{x}, t) = \phi_0 - \sigma(\mathbf{x}, t) \quad (5.16)$$

  where ϕ is the actual porosity, ϕ_0 is the initial porosity and σ is the volume fraction occupied by sessile microorganisms.

  Whereas, the permeability modification is expressed as a function of the porosity by the Kozeny-Carman equation (Carman 1956):

$$k(\mathbf{x}, t) = k_0 \frac{(1 - \phi_0)^2}{\phi_0^3} \frac{\phi^3}{(1 - \phi)^2} \quad (5.17)$$

  where k and k_0 are the actual and initial permeability, respectively.

- Initial and boundary conditions
  Initial conditions:

$$p^O(t_0) = p_0^O, \ S^W(t_0) = S_0^W; \quad c_m^W(t_0) = c_{m0}^W, \ \sigma(t_0) = \sigma_0, \ c_n^W(t_0) = c_{n0}^W \quad (5.18)$$

  Boundary conditions:
  The boundary conditions specified below are considering a particular domain, which is a cylindrical porous column (also known in the jargon of oil recovery "a core") where fluid is injected at the bottom and produced at the top:
  1. Inlet conditions (constant rate) at the bottom of the core

$$\mathbf{u}^O \cdot \mathbf{n} = \mathbf{u}^W \cdot \mathbf{n} = \mathbf{u}_{in}^W \cdot \mathbf{n}; \quad -[c_\gamma^W \mathbf{u}_{in}^W - \phi S^W \mathbf{D}_\gamma^W \cdot \nabla c_\gamma^W] \cdot \mathbf{n} = c_{\gamma_{in}}^W \mathbf{u}_{in}^W \cdot \mathbf{n}, \ \gamma = m, n$$

$$(5.19)$$

  where $\mathbf{u}_{in}^W$ is the water injection velocity, and $c_{\gamma_{in}}$ are injected concentrations for microorganisms ($\gamma = m$) and nutrients ($\gamma = n$), respectively.
  2. Outlet conditions (constant pressure, p_{out}^O) at the top of the core

$$p^O = p_{out}^O, \ \frac{\partial S^W}{\partial \mathbf{n}} = 0; \quad \frac{\partial c_\gamma^W}{\partial \mathbf{n}} = 0, \ \gamma = m, n \quad (5.20)$$

  where p_{out}^O is oil production pressure.
  3. No-flow conditions for other boundaries.

5.2.3   *Numerical model*

The numerical model consists in making the appropriate choice of the numerical methods in terms of accuracy and efficiency for the solution of the mathematical model.

In this case the resulting problem is a non linear system of partial differential equations with initial and boundary conditions. For the numerical solution we apply the following methods:

- For the temporal derivatives it was used a backward finite difference discretization of second order resulting a full implicit scheme in time.

- For the rest of differential operators in space derivatives a Galerkin standard finite element discretization was applied, where the quadratic Lagrange polynomials were employed for base and weighting functions.
- An unstructured mesh with tetrahedral elements in 3D was used.
- For the linearization of the non linear equation system the iterative Newton-Raphson method was applied.
- For the solution of the resulting algebraic system of equations a variant of the LU direct method for non symmetric and sparse matrices was utilized.
- The general procedure for coupling both the flow and transport models is sequential and is executed iteratively in the following manner:
  1. The flow model is solved and it is obtained: saturations S^α, pressures p^α and phase velocities $\mathbf{u}^\alpha$ for $\alpha = O, W$,
  2. The transport model is solved and it is obtained: components concentrations c_m^W, c_n^W and σ,
  3. The porosity and permeability are modified according with the Equations (5.16) and (5.17).
  4. When a given accuracy is achieved the procedure stops, but in the opposite case the iterative procedure should continue by going to Step 1.

5.2.4 *Computational model*

Once the numerical procedure for the solution of the mathematical model have been established its computational implementation is required. The realization of the flow model was performed using the COMSOL Multiphysics software by the PDE mode in general coefficient form for the time-dependent analysis (COMSOL Multiphysics 2007a). The implementation of the transport model was also carried out in the COMSOL Multiphysics but using the Earth Science Module (COMSOL Multiphysics 2007b).

5.3 NUMERICAL SIMULATIONS

In this section, the dynamic modification of the porosity and permeability, and its influence on the flow behavior due to the clogging/declogging processes that occur during microbial enhanced oil recovery are studied. To this purpose, a set of numerical experiments are carried out. Basically, these numerical simulations consist in solving the flow and transport model established in Section 5.2.2.1 using data of the reference test given in the Table 5.2 modifying one or more parameter. For each case, it can be observed the spatial and temporal variation in the porosity and permeability and hence assess its potential impact on the flow behavior.

5.3.1 *Reference study case description: a waterflooding test in a core*

The reference study case is a waterflooding test in a core that can be conventionally divided in two experiment sections that are sequentially performed. In the first section the oil is displaced by water. While, in the second one water is injected with microorganisms and nutrients. The data used here to simulate both experiment sections are taken from the published literature (Hoteit & Firoozabadi 2008, Chang et al. 1991). The reference study case can be described as follows:

- First experiment section: secondary recovery by water injection
  This test step is a typical one for a secondary recovery process. Initially, a Berea sandstone core (a cylindrical porous medium fragment) of 0.25 m of length and 0.04 m of diameter with homogeneous porosity ($\phi_0 = 0.2295$) and isotropic permeability ($k_0 = 326$ mD) set in vertical position, is fully saturated with water. To establish the appropriate initial conditions for a secondary recovery, the water is displaced by oil injection until the residual saturation of water is achieved, which means that the injection is stopped when it is impossible to recover

Table 5.2   Data for the flow and transport model taken from (Hoteit & Firoozabadi 2008, Chang et al. 1991).

Parameter	Notation	Value	Unit
Core length	L	0.25	m
Core diameter	d	0.04	m
Initial Porosity	ϕ_0	0.2295	
Initial Permeability	k_0	326	mD
Water viscosity	ν^W	1.0×10^{-3}	Pa·s
Oil viscosity	ν^O	7.5×10^{-3}	Pa·s
Water density	ρ^W	1.0	g/cm^3
Oil density	ρ^O	0.872	g/cm^3
Residual water saturation	S_r^W	0.2	
Residual oil saturation	S_r^O	0.15	
Oil production pressure	p_{out}^O	10	kPa
Brooks-Corey parameter	θ	2	
Entry pressure	p_t	10	kPa
Water injection velocity	u_{in}^W	1	ft/day
Nutrients dispersion	D_n^W	0.0083	ft^2/day
Microorganisms dispersion	D_m^W	0.0055	ft^2/day
Injected nutrients concentration	$c_{n_{in}}^W$	2.5	lb/ft^3
Injected microorganisms concentration	$c_{m_{in}}^W$	1.875	lb/ft^3
Microorganisms density	ρ_m^B	1000	kg/m^3
Irreducible biomass fraction	σ_{irr}	0.003	
Maximum specific growth	μ_{max}	8.4	day^{-1}
Affinity coefficient	$K_{m/n}$	0.5	lb/ft^3
Production coefficient	$Y_{m/n}$	0.5	
Decay coefficient	κ_d	0.22	day^{-1}
Adsorption coefficient	κ_a^m	25	day^{-1}
Desorption coefficient	κ_r^m	37	day^{-1}

additional water. Finally, a secondary recovery process is accomplished by the displacement of oil with the water flooding. The water is injected from the lower side of the core with a constant velocity of one foot per day (3.53×10^{-6} m/s), while the oil and water are produced at a constant pressure (10 kPa) in the opposite side of the core during 200 hours until a steady state is obtained. A summary of the data is given in Table 5.2.

- Second experiment section: enhanced oil recovery by water injection with microorganisms and nutrients

In the second step of the test, microorganisms and nutrients are injected simultaneous and continuously to the Berea sandstone core through the water phase during 24 hours until a steady state is obtained. The intention of this test consists in evaluating the additional oil production that can be recovered by mechanical effects because of the microbial activity (MEOR). As in the first step of the test data are those from Table 5.2.

5.3.2   *Modeling of secondary recovery by water injection*

For the numerical simulation of this secondary recovery test it is enough to apply the biphasic flow model described in Section 5.2.2.1 by the Equations (5.6) and (5.7). The finite element discretization mesh for the numerical solution of this initial and boundary valued problem is shown in the Figure 5.2, which is made of 1,702 tetrahedral elements and consequently, the problem has a total of 6,232 unknowns.

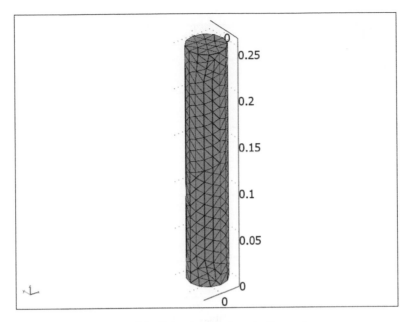

Figure 5.2   A 3D perspective view of the finite element discretization mesh.

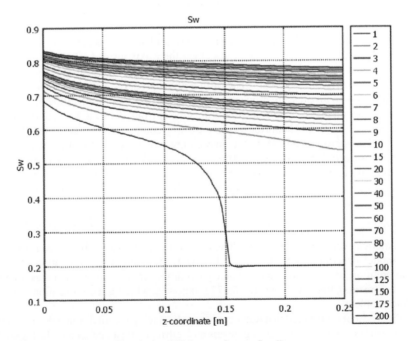

Figure 5.3   Water saturation profiles during 200 hours of water flooding.

Figure 5.3 displays water saturation profiles for various times from 1 to 200 hours of water flooding along the z-coordinate axe. It can be observed that a water front through the porous medium is formed. The oil displaced by the water front is recovered at the upper end of the core. The head of water breaks through the top of the core before the two hours.

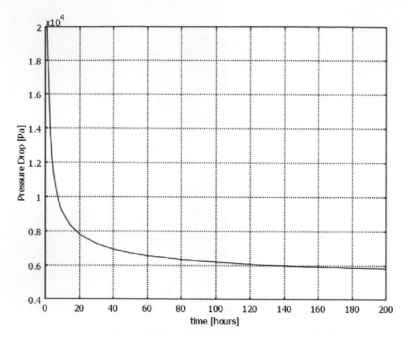

Figure 5.4   Oil pressure drop (in Pascals) as function of time (in hours) during 200 hours of water flooding.

It is pertinent to note here that although the model is implemented in 3D, in this work all figures which involve profiles are presented as values along the z axis instead of presenting the average value of the cross section since due to the cylindrical symmetry of the model this representation is considered ilustrative of the global behavior of a given property along the z coordinate.

In the Figure 5.4 an oil pressure drop curve for a water flooding during 200 hours is shown. It can be seen that the oil pressure drop downfalls drastically from 20 kPa to 8 kPa during the first 20 hours, whereas in the next 180 hours the oil pressure drop continues downfalling but slowly up to a value of 5.77 kPa, approximately. Here. the oil pressure drop is defined as the difference between the inlet oil pressure and the outlet oil pressure, i.e., $\delta p \equiv p_{in}^{O} - p_{out}^{O}$. It is worth to note that the oil recovery curve of Figure 5.5 represents a typical behavior of a recovery process where the porous medium is strongly oil wet. The total recovery of oil is approximately 74%, see Figure 5.5.

5.3.3   *Modeling of enhanced recovery by water injection with microorganisms and nutrients*

For the numerical simulation of this EOR test it is necessary to apply the coupled flow and transport model described in Section 5.2.2.1. The finite element discretization mesh for the numerical solution of this initial and boundary condition problem is the same as above (1702 tetrahedral elements), but now there are 15,577 unknowns, which is a computational much more demanding problem.

In the Figure 5.6 it is observed that water continues displacing the oil through the core during the 24 hours of water flooding with microorganisms and nutrients. It can be seen in the Figure 5.7 that there is a slightly increase in the oil pressure drop from 5680 to 6185 Pa. This repressurization phenomenon is associated with the modification of porosity and, consequently, with the modification of the permeability due to the biomass growth. This proves the principle that the biomass could be used to redirect the flow and to increase the oil sweep efficiency. A marginal additional oil recovery of about 0.2% is obtained, see Figure 5.8. Stationary state is established around the 6 hours, where nutrients are almost completely consumed, while at the 24 hours approximately an asymptotic value in the concentration of microorganisms is reached (Fig. 5.9). The maximum

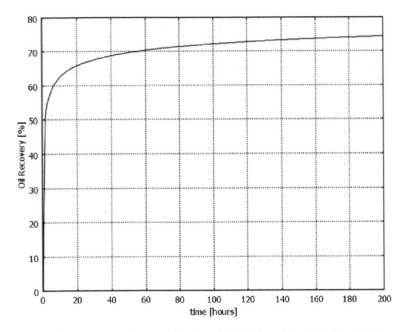

Figure 5.5   Oil recovery curve (in percent) as function of time (in hours) during 200 hours of water flooding.

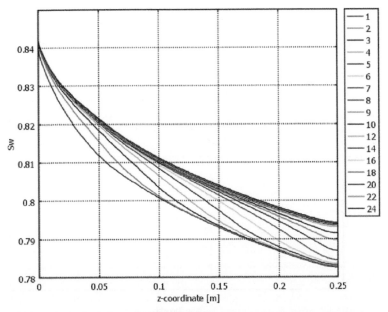

Figure 5.6   Water saturation distribution during 24 hours of water flooding with microorganisms and nutrients.

planktonic and sessile microorganisms concentration values are achieved in $c_m^W = 48.85$ kg/m^3 and $\sigma = 1.1\%$ at 0.074 m and 0.041 m, respectively, see Figures 5.9 and 5.10. The variations in σ are reflected directly in porosity and permeability changes, where the minimum values are achieved in a zone around 0.041 m from the core base. A more deep research of this fact will be presented in the following subsection.

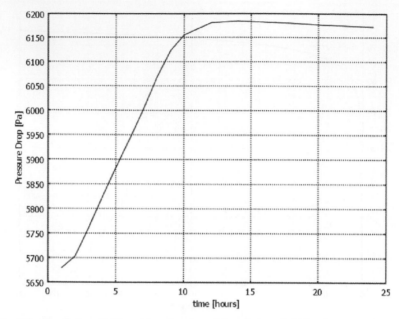

Figure 5.7   Oil pressure drop (in Pascals) as function of time (in hours) during 24 hours of water flooding with microorganisms and nutrients.

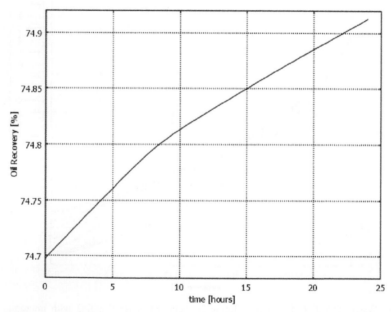

Figure 5.8   Oil recovery curve (in percent) as function of time (in hours) during 24 hours of water flooding with microorganisms and nutrients.

5.3.4   *Porosity and permeability modification due to microbial activity*

Here, it is investigated how the porosity and permeability modification due to the variation in the spatial and temporal biomass distribution along the core affects the flow behavior. To this purpose, seven numerical cases are simulated which represent variants of the test described in

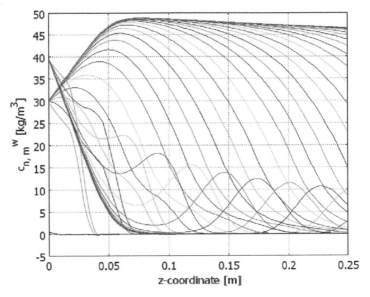

Figure 5.9   Nutrients and planktonic microorganisms profiles along the vertical axis at multiple sequential one-hour times (starting at $40 \, kg/m^3$ and $30 \, kg/m^3$, respectively in $z = 0$).

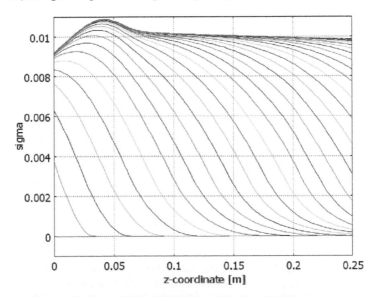

Figure 5.10   Sessile microorganisms profiles along the vertical axis at multiple sequential one-hour times.

Section 5.3.1. The data of each case are given in the Table 5.2, with the exception of the parameter values specified in Table 5.3. The Case A listed in Table 5.3 is exactly the same problem discussed in the previous subsection and it will be used as the reference problem.

In the Table 5.4 the results of the numerical simulations are summarized in terms of the maximum sessile microorganisms concentration value (σ_{max}), its position along the vertical axis (z_{max}), the minimum oil pressure drop (δp_{min}), the maximum oil pressure drop (δp_{max}) and their difference ($\delta p_{max} - \delta p_{min}$), for the cases given in Table 5.3 are displayed.

In the Figure 5.11 it is evident that the porosity (ϕ) and permeability (k) spatio-temporal distributions resemble the form of the sessile microorganisms concentration (σ). This is because

Table 5.3   Parameter values for each case.

Cases	κ_a^m [day^{-1}]	κ_r^m [day^{-1}]
A	25	37
B	8	37
C	25	10
D	8	10
E	100	37
F	25	80
G	100	80

Table 5.4   Summary of the numerical simulations for each case.

Cases	σ_{max}	z_{max} [m]	δp_{min} [Pa]	δp_{max} [Pa]	$\delta p_{max} - \delta p_{min}$ [Pa]
A	0.011	0.041	5680	6185	505
B	0.0058	0.045	5712	5975	263
C	0.05	0.005	5650	6900	1250
D	0.024	0.01	5700	6200	500
E	0.031	0.025	5600	7150	1550
F	0.0065	0.06	5700	6020	320
G	0.016	0.04	5650	6460	810

there is a linear relationship between porosity and σ, and since the Kozeny-Carman porosity-permeability dependence relationship is almost linear. In all cases, the region of the core that is more affected by the clogging, i.e., with the highest values of σ, is exactly the same region with the lowest values of porosity and permeability, and vice versa.

From a preliminary examination of the Table 5.4 it can be noticed that there is a direct relation between σ_{max} values and pressure drop increments. This fact can be explained as follows: a preferential reduction in the porosity and permeability in the bottom side of the core linked with the clogging process yields an increase of the inlet oil pressure, since the porous space for the fluids displacement is reduced. A more deep analysis of the results in the Table 5.4 allows to clearly identify two groups of cases with consistent behavior, one group formed by the Cases C and E in which $\kappa_a^m \gg \kappa_r^m$, and other one including Cases B and F in which $\kappa_a^m \ll \kappa_r^m$. In the first group the most intensive clogging and re-pressuring processes with the less oil recovery occur, whereas in the second one the opposite behavior is presented. For the Cases A, D and G whose κ_a^m and κ_r^m values are of the same order, their behavior is not completely clear and it is not possible to distinguish which is the predominant process.

It is easy to see in the Figure 5.12 that the water velocity presents certain fluctuations in the region close to the inlet side of the core which are mainly due to numerical instabilities, without any physical ground. However the velocity field is well defined in the rest of the core, except in the vicinity of the outlet side, where a small disturbances are emerged. In despite of the mentioned perturbations the water velocity remains in average around a value 3.53×10^{-6} m/s (one foot per day) which is the inlet velocity.

5.4   FINAL REMARKS

In this work a quite general flow and transport model in porous media was implemented using the standard formulation of the finite element method to simulate laboratory tests at core scale and under controlled conditions. Firstly, this model was successfully applied to a reference study

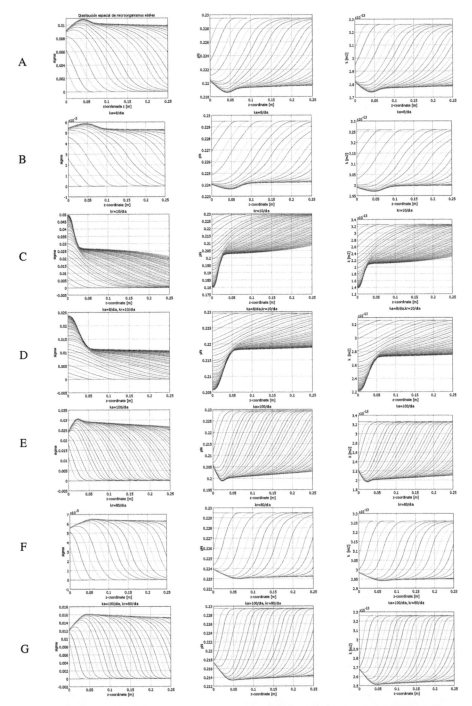

Figure 5.11   Profile modifications for the seven cases analyzed. The first column displays the sessile microorganism concentration (σ) along the path length, the second and the third columns show the corresponding porosity (ϕ) and permeability (k) profiles, respectively. Profiles are displayed with an hour difference during a 24 hours period of water flooding with microorganisms and nutrients for each case of Table 5.3.

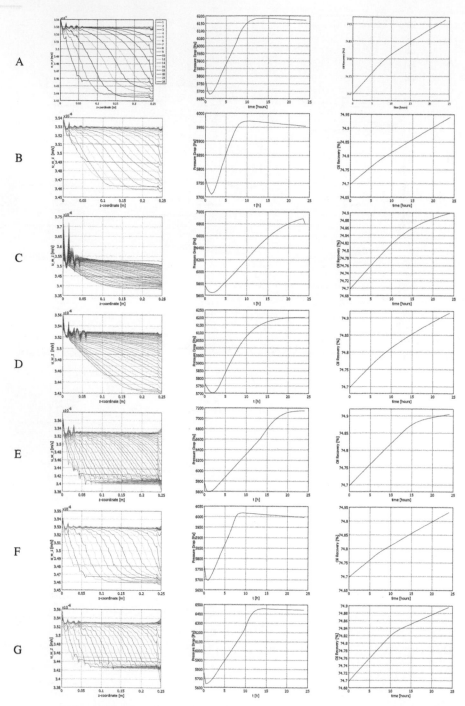

Figure 5.12   Profile modifications for the seven cases analyzed. The first column displays the water velocity (u^W) profiles along the path length, the second and the third columns show the corresponding oil pressure drop and oil recovery curves, respectively. Profiles are displayed for a 24 hours period of water flooding with microorganisms and nutrients for each case of Table 5.3.

case of oil displacement by the injection of water follows by the injection of water with microorganisms and nutrients, using data taken from the published literature, and then its performance was evaluated comparing variations of the reference study case.

In this paper, the porosity and permeability modification due to the variation in the biomass distribution along the core was investigated. The resemblance of the porosity (ϕ) and permeability (k) spatio-temporal distributions of the form of the sessile microorganisms concentration (σ), can be attributed to the simplicity of the porosity and permeability relationships applied, therefore the application of more realistic dependence functions is an open issue. In the cases with a significant difference between κ_a^m and κ_r^m values it can be clearly established distinguishable behaviors in terms of clogging intensity, pressure drop and amount of oil recovery.

The observed re-pressurization phenomenon associated with the modification of porosity and permeability due to the biomass growth could be used to plug highly permeable zones to redirect the flow and to increase the oil sweep efficiency.

Regarding the marginal additional recovery produced during the EOR microbial process, less than 1%, it can not be directly attributed to the mechanical effects such as the pressure increase due to microbial growth. It is pertinent to mention here, that the target of the study was the investigation of porous media microbial clogging phenomenon. A system to proper simulate oil recovery increase would require for example, that the porous column employed here would have a small high permeability flow channel. This channel will be clogged by a short pulse of microorganisms (i.e. no continuous microorganism injection), and then the subsequent water injected (not containing microorganisms) will swept the oil in the column region outside the original clogged flow channel.

To model other effects such as: rock wettability, viscosity ν, relative permeability k_r and capillary pressure p_c curves modification because of the bioproducts action over fluids and rock, further experimental investigation is required, to quantify in terms of constitutive relationships the interaction between petrophysical and fluid properties as a function of bioproducts concentrations.

ACKNOWLEDGEMENTS

This work was supported by the project D.00417 of the IMP Oil Recovery Program.

REFERENCES

Allen, M.B., Herrera, I. & Pinder, G.F. (1988) *Numerical modeling in science and engineering*. New York: John Wiley.

Bear, J. (1972) *Dynamics of fluids in porous media*. New York, Dover.

Brooks, R.J. & Corey, A.T. (1964) Hydraulic properties of porous media. *Hydrology Paper* 3, Fort Collins, Colorado State University.

Carman, P.C. (1956) Flow of gases through porous media. London, *Butterworth Scientific Publications*.

Chang, M.M., Chung, F.T-H., Bryant, R.S., Gao, H.W. & Burchfield, T.E. (1991) Modeling and laboratory investigation of microbial transport phenomena in porous media. *SPE Annual Technical Conference and Exhibition, 6–9 October 1991, Dallas, TX, USA*. SPE paper number 22845.

Chang, M.M., Bryant, R.S., Stepp, A.K. & Bertus, K.M. (1992) Modeling and Laboratory Investigations of Microbial Oil Recovery Mechanisms in Porous Media. *NIPER-629* Bartlesville, National Institute for Petroleum and Energy Research.

Chen, Z., Huan, G. & Ma, Y. (2006) *Computational methods for multiphase flows in porous media*. Philadelphia, SIAM.

Chen, Z. (2007) *Reservoir simulation: Mathematical techniques in oil recovery (CBMS-NSF Regional Conference Series in Applied Mathematics)*. Philadelphia, SIAM.

COMSOL Multiphysics (2007a) *Modeling Guide Version 3.4*. COMSOL AB.

COMSOL Multiphysics (2007b) *Earth Science Module, Users Guide Version 3.4*. COMSOL AB.

Corapcioglu, M.Y. & Haridas, A. (1984) Transport and fate of microorganisms in porous media: A theoretical investigation. *Journal of Hydrology*, 72: 149–169.

Diaz-Viera, M.A., Lopez-Falcon, D.A., Moctezuma-Berthier, A. & Ortiz-Tapia, A. (2008) COMSOL implementation of a multiphase fluid flow model in porous media. *COMSOL Conference, 9–11 October 2008, Boston MS, USA.*

Gaudy, A.F., Jr. & Gaudy, E.T. (1980) *Microbiology for Environmental Scientists and Engineers.* New York, McGraw-Hill.

Hoteit, H. & Firoozabadi, A. (2008) Numerical modeling of two-phase flow in heterogeneous permeable media with different capillarity pressures. *Advances in Water Resources,* 31, 56–73.

Lappan, E.R. & Fogler, H.S. (1996) Reduction of porous media permeability from in situ Leuconostoc mesenteroides growth and dextran production. *Biotechnology and Bioengineering,* 50(1): 6–15.

Lopez-Falcon, D.A., Diaz-Viera, M.A. & Ortiz-Tapia, A. (2008) Transport, growth, decay and sorption of microorganisms and nutrients through porous media: A simulation with COMSOL. *COMSOL Conference, 9–11 October 2008, Boston MS, USA.*

CHAPTER 6

Inter-well tracer test models for underground formations having conductive faults: development of a numerical model and comparison against analytical models

M. Coronado, J. Ramírez-Sabag & O. Valdiviezo-Mijangos

6.1 INTRODUCTION

Tracer test is an old technique originally developed to determine fluid flow connectivity in surface water paths and aquifers. In the last decades it has been greatly enhanced and is used in many earth science applications in hydrology, aquifers, mining, oil reservoirs, geothermal fields, pollutants dispersion, atmospheric circulation, oceanology, geology, etc. (Käss 1998). Presently, tracer tests constitute an important and decisive technique in determining communication channels in underground porous formations. It can provide fundamental information on the fluid motion, can determine formation properties and can offer solid elements when validating geological models. A traditional inter-well tracer test in underground formations consists in the release at a certain place (injection well) of a pulse of a chemical, biological or radioactive element, which is carried by the fluid following the flow trajectories and is later observed at other downstream sites (production or observation wells), normally at very low concentrations. The amount of tracer observed along the time (tracer breakthrough curve) provides therefore information on the flow pattern, communication channel characteristics and the porous media properties. In order to avoid the employment of huge amounts of a tracing product in tracer tests, those tracing elements that can be perfectly integrated in the fluid and can be detected at very low concentrations are preferred. Large amounts of a tracer are hard to manage and can modify the original system conditions. Therefore radioactive elements such as Tritium in tritiated water or in tritiated methane are frequently used. Many analytical and numerical models to describe tracer test in Geohydrology and oil reservoir research have been developed in the last decades (see for example Van Genuchten 1981, Maloszewski & Zuber 1990, Ramírez et al. 1993, Zemel 1995, Charbeneau 2000), which describe diverse circumstances and porous media types. A very important situtation is when geological faults are present, which are structural features frequently appearing in aquifers, oil reservoirs, and geothermal fields. Faults can work as a transversal fluid flow barrier or oppositely, as high permeability longitudinal conduit; therefore, their presence can seriously impact fluid flow dynamics. In oil reservoirs faults can dramatically reduce the efficiency of secondary and improved recovery procedures, and can importantly increase the undesired transport of pollutants in aquifers. The fault structure regularly consists of a central low permeability slab-like core, flanked by a high-permeability damaged zone at its both sides, which contains an intricate network of fractures with different sizes and characteristics. The fluid conduction properties of a fault depend on the composition and thickness of these damage zones (Caine et al. 1996, Fairley & Hinds 2004). For example, when conductive faults are present in a reservoir under secondary or improved oil recovery processes, conductive faults can seriously reduce the swept efficiency as a consequence of fluid oil-by-passing situations. The disposal of high radioactive waste in deep repositories located in fractured formations can represent a safety hazard. Also, contaminant spills on fields can rapidly pollute aquifers if underground conductive faults are present. Thus, knowing the presence of faults and their fluid conduction characteristics is of great relevance. Tracer tests can play a decisive role in determining those characteristics. To this purpose mathematical models to interpret field tracer data are necessary. Diverse models to describe tracer transport in a fracture network have been developed in order to treat tracer flow in faults (see for example Sudicky & Frind 1982 and Maloszewski & Zuber 1985) but they did not consider an important issue in oil reservoirs and geothermal fields, which is that injection and production wells can be located outside the fault. Recently some analytical tracer test models have been developed to treat this situation (Coronado & Ramírez-Sabag 2008, Coronado et al. 2010). An important point in these models is the introduction of three coupled tracer path regions, which represent the zone from the injection well to the fault, the path inside the fault self, and the region from the fault to the production well. In one of these works (Coronado et al. 2010) a successful application of the models to field data is made, through them the orientation of the fault could be calculated. Specifically, this is a very valuable information for geologists and reservoir engineers. In this work we present a complementary numerical model developed to describe a tracer test in geological formations with a conductive fault. The results from the analytical models are compared against the numerical model results, as a way to validate the structural model consistency. The numerical model has been set up by using the worldwide known simulator UTCHEM (Delshad et al. 1996).

6.2   DESCRIPTION OF THE ANALYTICAL MODELS

An illustrative representation of a working layer in an underground formation is displayed in Figure 6.1. Here, a fault, an injection well and two production wells in a 3D graph are shown.

The analytical models under analysis consider a simplified bi-dimensional porous media system in a X-Y plane, as illustrated in Figure 6.2. The injection fluid moves in the conventional path from Well A to the production Well C, but also through the fault to production Well B. Along this last path three clearly different regions are involved. Region 1, which covers the zone from the injection site in Well A to the fault, Region 2 that corresponds to the path along the fault, and finally Region 3, which is the zone from the fault to the production site in Well C. Each region might have its own average porosity and permeability. The permeability in the fault zone is expected to be notoriously higher than the permeability in the other areas. A tracer pulse introduced in the injector Well A would arrive faster to Well B than to Well C. Thus, the tracer breakthrough curve at those wells would take a shape as that exemplified in Figure 6.3. In order to appreciate the fault effects on the tracer breakthrough curve, this curve should be compared against the corresponding curve when the fault is not present. This has been done in the analytical models mentioned above. In one of these models (Coronado & Ramírez-Sabag 2008) a closed fault is treated, and in the second model (Coronado et al. 2010) an open fault is studied, which additionally includes the fracture-matrix interaction in the fault region, and no-dispersion-flow condition at the discharge site (production well). Closed fault means that there are not external fluid entering or leaving the fault along the fault line. This situation is schematically shown in

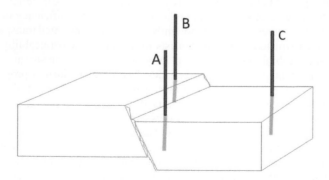

Figure 6.1   Pictorial representation of an underground layer with a fault and three wells.

Figure 6.4(A). On the other hand, open faults allow those external input or output flows. Two situations in an open fault are considered: injection-dominated-flow and fault-dominated-flow. These cases are displayed in Figures 6.4(B) and 6.4(C) respectively. In the first case the injection makes the fluid to flow outside the fault at the both fault sides, and in the second case the fault flow dominates and external fluid enters from the left fault side, as displayed in Figure 6.4(C).

6.2.1   *The closed fault model*

The model developed by Coronado and Ramírez-Sabag (2008) for a tracer test in a closed fault considers that the three regions form a aligned one-dimensional array as displayed in Figure 6.5. The tracer concentration $C(x,t)$ starts as a Dirac delta pulse injected at $x = 0$, and no tracer present in the rest of the system. In terms of the three regions it is

$$C_1(x, t = 0) = m\delta(x), \quad C_2(x,0) = 0, \quad C_3(x,0) = 0, \tag{6.1}$$

where the sub-index $i = 1, 2$ or 3 refers to the region, and m is the tracer mass injected by unit of transversal effective flow area. The tracer pulse flows along the x-direction through the three different regions subject to advection (at a constant velocity u_i) and hydrodynamic dispersion (at a constant dispersion coefficient D_i). The equation in each region i is

$$\frac{\partial C_i(x,t)}{\partial t} + u_i \frac{\partial C_i(x,t)}{\partial x} - D_i \frac{\partial^2 C_i(x,t)}{\partial x^2} = 0. \tag{6.2}$$

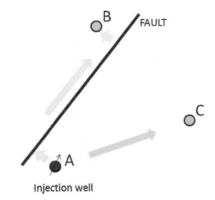

Figure 6.2   Plant view of the underground layer containing a fault, an injector and two production wells.

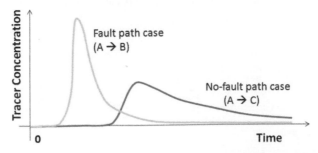

Figure 6.3   Illustrative tracer breakthrough curves for the inter-well path A → B, that goes through a conductive fault, and from A → C, not touching the fault.

The pulse reaches Region 2 in $x = L_1$. The border conditions on tracer concentration to be satisfied at that point are tracer concentration continuity and tracer mass conservation:

$$C_1(x = L_1, t) = C_2(x = L_1, t), \tag{6.3}$$

$$J_1(x = L_1, t) = J_2(x = L_1, t), \tag{6.4}$$

where $J_i = u_i C_i - D_i(\partial C_i / \partial x)$ is the tracer flux.

The complementary boundary condition is $C_1(x \to -\infty, t) = 0$. The pulse advances through Region 2 expectedly having a faster speed than in the previous region and a different dispersion rate. It arrives into the next region in $x = L_1 + L_2$. Again similar boundary conditions are set in this second inter-phase.

$$C_2(x = L_1 + L_2, t) = C_3(x = L_1 + L_2, t), \tag{6.5}$$

$$J_2(x = L_1 + L_2, t) = J_3(x = L_1 + L_2, t). \tag{6.6}$$

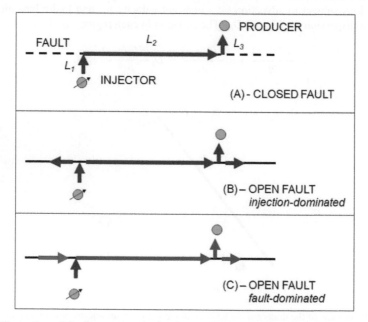

Figure 6.4   Illustration of the flow pattern in the three analytical models. (A) a closed fault, (B) an open fault with flow dominated by the injection conditions, and (C) an open fault with flow determined by the fault conditions.

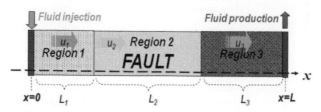

Figure 6.5   A one-dimensional three-coupled-region system representing the porous media between the injection well (at $x = 0$) and the production well (a $x = L$) in a closed fault.

Further, the tracer pulse moves through Region 3 presumable at a lower velocity than in Region 2 and at a different dispersion rate until it reaches the observation well in $x = L_1 + L_2 + L_3$. There is no boundary condition established locally on the production well since it is considered just as a monitoring (i.e. sampling) well, therefore the boundary condition is set in the right border of Region 3, which is $C_3(x \rightarrow \infty, t) = 0$. The analytical solution in Laplace space is presented in the above mentioned paper by Coronado and Ramírez-Sabag (2008). The analytical model involve multiple system parameters, which are: (i) the path length in each region (L_1, L_2 and L_3), (ii) the fluid velocity in each region (u_1, u_2 and u_3) and (iii) the corresponding region dispersivity (α_1, α_2 and α_3). Here the dispersion coefficients were written as $D_i = \alpha_i u_i$, been α_i the dispersivity in region i. In this paper diverse plot of the tracer pulse profile, i.e. tracer concentration as function of x, and the tracer breakthrough curve are shown for a short and a long fault. The curve sensitivity to diverse fluid velocities and dispersivities is there also analyzed. For simplicity the case when Region 1 and Region 3 have the same dispersivity and the same fluid velocity is treated. As expected, the effects of a short fault are weak (not remarkable) with respect to the no-fault case. Also, the dispersivity in the fault in comparison with the dispersivity in the regions outside the fault is not an important curve-modifying factor. However, for a long fault the effects on the breakthrough curve are clearly apparent. In this case the breakthrough curve is very sensitive to the velocity in the fault. In general, the tracer breakthrough curve features that manifest the presence of a long conductive fault are: a much shorter pulse transit time, a high curve peak, and a occasionally a long curve tail. These features make possible the development a tool for application on fault characterization using inter-well tracer tests. In this work, a validation of the model consistency will be made against a numerical model, in conjunction with other above mentioned analytical tracer test models that consider open conductive faults, which is the subject of the next section.

6.2.2   *The open fault model*

A new model has been more recently developed to consider tracer tests in reservoirs with conductive faults that are open to external flow (Coronado et al. 2010). In order to include this feature a slightly different basic flow system has been introduced. It consists in three independent one-dimensional fluid flow systems, one for each region. A graphical representation of the system is displayed in Figure 6.6. The coupling between the regions is made through a tracer source that captures the tracer output from the previous region. This source has a dependence on time, accordingly to how the tracer pulse reaches the region. The first system corresponds to the first region as previously described in the last section, it considers a one-dimensional streak using the space variable x_1, and assume tracer flow with uniform advection and constant hydrodynamic

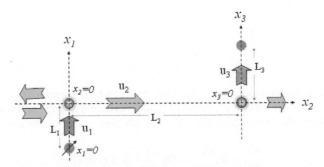

Figure 6.6   Three independent systems to describe the tracer pulse dynamics in open faults. Tracer pulse is injected in $x_1 = 0$ and flows to the fault located a $x_1 = L_1$. Here it appears as a tracer source in $x_2 = 0$ of Region 2. The tracer pulse continues in Region 2 and reaches Region 3 at $x_2 = L_2$. It merges as a tracer source at $x_3 = 0$ in Region 3 and flows to the production well located at $x_3 = L_3$.

dispersion. The equation to be solved in Region 1 is

$$\frac{\partial C_1}{\partial t} + u_1(x_1)\frac{\partial C_1}{\partial x_1} - D_1\frac{\partial^2 C_1}{\partial x^2} = 0. \tag{6.7}$$

Here $u_1(x_1)$ is a step function, equal to u_1 for $x_1 \geq 0$ and $-u_1$ for $x_1 < 0$. The dispersion coefficient is constant, written as $D_1 = \alpha_1 u_1$. The initial condition is

$$C_1(x_1, t = 0) = \frac{M_T}{\phi_1 A_1}\delta(x_1), \tag{6.8}$$

where M_T is the total tracer amount injected, and A_i is the macroscopic channel cross-section in Region i. The boundary conditions are $C_1(x \to \pm\infty, t) = 0$.

The second region, i.e. the fault zone, considers a two-dimensional system involving not only tracer advection and dispersion, but also tracer diffusion from the fault network to the porous rock matrix where fluid is stagnant, and vice versa. This double-porosity phenomenon is relevant in fractured systems, as it is the case in geological faults. The space variables in this region are two, x_2 and a perpendicular length z (see Coronado et al. 2010). There are two coupled mass balance equations, one for the flowing tracer concentration in the fracture network, $C_f(x_2, t)$, and another for the stagnant tracer concentration in the matrix system, $C_m(x_2, z, t)$, they are

$$\frac{\partial C_f}{\partial t} + u_2(x_2)\frac{\partial C_f}{\partial x_2} - D_f\frac{\partial^2 C_f}{\partial x_2^2} - \frac{\phi_m D_m}{w}\frac{\partial C_m}{\partial z}\bigg|_{z=w} = S_2(t)\delta(x_2). \tag{6.9}$$

$$\frac{\partial C_m}{\partial t} - \frac{D_m}{R}\frac{\partial^2 C_m}{\partial z^2} = 0. \tag{6.10}$$

Here the sub-script f refers to the fractured system and m to the matrix system, w is the average half-width of fractures, R is a retardation factor due to tracer adsorption-desorption, and $S_2(t)$ is the tracer source rate due to the pulse arrival from Region 1 at location $x_2 = 0$. The dispersion coefficient in the fracture network is written as $D_f = \alpha_2 u_2$. The coefficient D_m corresponds to diffusion of the tracer in the matrix. The initial conditions establish that no tracer is present in Region 2 at the initial time, it is $C_f(x_2, t = 0) = 0$ and $C_m(x_2, z, t = 0) = 0$. This source is given by

$$S_2(t) = \frac{A_1\phi_1}{A_2\phi_2}J_1(x_1 = L_1, t), \tag{6.11}$$

The velocity $u_2(x_2)$ is constant and given by u_2 for the fracture-dominated-flow case, and it is a step-function for the injection-dominated-flow case with u_2 for $x_2 \geq 0$ and $-u_2$ for $x_2 < 0$. The boundary conditions for the tracer population in the fracture network is $C_f(x \to \pm\infty, t) = 0$, and for the stagnant tracer population

$$C_m(x_2, z = w, t) = C_f(x_2, t), \tag{6.12}$$

and

$$\frac{C_m}{\partial z}\bigg|_{z=E/2} = 0 \tag{6.13}$$

with E the average matrix block size. The third region, which corresponds from the fault to the production well, is described by a one-dimensional system using the space variable x_3. The equation is similar to the first region but with a source present due to the tracer arriving from Region 3. it is

$$\frac{\partial C_3}{\partial t} + u_3\frac{\partial C_3}{\partial x_3} - D_3\frac{\partial^3 C_3}{\partial x_3^2} = S_3(t)\delta(x_3). \tag{6.14}$$

with u_3 constant, the dispersion coefficient written as $D_3 = \alpha_3 u_3$, and the source rate given by

$$S_3(t) = \frac{A_2\phi_2 F}{A_3\phi_3}J_2(x_2 = L_2, t), \tag{6.15}$$

where F is the fraction of the tracer amount in Region 2 that goes in Region 3. The initial condition requires that no tracer is present in Region 3, it is $C_3(x_3, t=0)=0$. A condition of no-dispersive-flow is set at the production well, $x_3 = L_3$ (this is the Danckwert's condition, which is a discharge condition from a porous medium interface into a non-porous medium), which is

$$\left. \frac{\partial C_3}{\partial x_3} \right|_{x_3=L_3} = 0 \qquad (6.16)$$

together with $C_3(x \to -\infty, t) = 0$. The amount of parameters in the model are the same as in the previous model for a closed fault plus two more parameters resulting from Region 2. These two parameter are β ($=\phi_m \sqrt{(RD_m t_T)/w^2}$), which is related to the diffusion process between the fracture network and matrix, and Θ ($=\sqrt{R/t_T D_m}(E/2 - w)$) which involves the matrix block size E. Here t_T is the average tracer pulse transit time. The case when Region 2 is not fractured is recovered when $\beta = 0$. As mentioned before, there are two flow sub-cases to be treated: the injection-dominated and the fault-dominated, which are graphically illustrated in Figures 6.4(B) and (C). In the first sub-case the fluid injected determines the flow pattern in the zone near the injection well, forcing the fluid to flow outside the fault, as displayed in the left fault side of Figure 6.4(B). In the second sub-case the flow in the fault dominates and therefore there is an external incoming fluid flow, as shown in the left side of the fault in Figure 6.4(C). In order to simplify the mathematical calculations the fluid velocity outside the left fault side is assumed to have the same speed value than the fluid speed inside the fault, only the direction is different in the two sub-cases. In the injection-dominated situation it points outwards, it is $-u_2$, and in the fault-dominated situation it points inwards, it is u_2. The analytical solution in Laplace space involves an additional model parameter which is the fraction of the originally injected total tracer amount that arrives the production well (see paper by Coronado et al. 2010). This parameter has to do with the fault total tracer losses. The amount of model parameters is large, however a sensitive analysis of the tracer breakthrough curve on the diverse parameters shows, as in the previous model for a closed fault, that long fault lengths and high fluid velocity in the fault display notorious fault signatures on the breakthrough curve. These features are short travel times and concentration high peaks. Low sensitivity of the breakthrough curve on the dispersivity is again found. The effect of the two new parameters (β and Θ) becomes relevant only when the fluid velocity in the fault is low. This can be understood by noticing that diffusion into the matrix is a slow process and therefore requires slow tracer pulse motion along the fracture network to become relevant. High β values (say $\beta \sim 3$) and high Θ values (say $\Theta \sim 10$) together with slow speeds make wide and long-tailed tracer breakthrough curves. In the mentioned paper, an application of the model to evaluate fault orientation by fitting field tracer breakthrough data is presented.

The three analytical models described in this section will be compared against the numerical results to be obtained by the tracer transport simulations described in the next sections.

6.3 THE NUMERICAL MODEL

One crucial issue in the previously presented analytical models concerns the assumption of uniform fluid velocity in each region. The velocity specific values are left as a externally given parameters. In reality the steady-state fluid velocity is not uniform, it should be determined by the injection and production well particular conditions (regularly fixed flow rate or given well pressure are considered) and the specific reservoir border conditions (commonly impermeable walls). Thus a comparison of these analytical models against a model not having this assumption results very relevant. An adequate way to perform this comparison is to set up a numerical code to describe the tracer transport in an equivalent bi-dimensional system, such as that displayed in Figure 6.2. There are diverse commercial and free numerical simulators that can be used for this purpose. We employ here the free code UTCHEM, which is a world wide recognized simulator been developed by the Center for Petroleum and Geosystems Engineering of

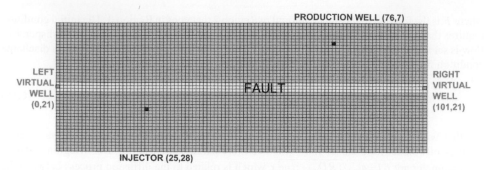

Figure 6.7   The cell grid used. It includes a horizontal high-conductivity fault, an injection well, a production well and a virtual well at the fault left and right border.

the University of Texas at Austin to describe improved oil recovery processes and tracer tests in reservoir and aquifers (see for example Delshad et al. 1996). The code considers a general tri-dimensional systems with four phases (water, oil, micro-emulsion and a passive gas phase for material balance purposes), multiple components, variable temperature and physical-chemistry reactions. Components considered by the UTCHEM that are important in enhanced oil recovery processes are polymers, gels, foams, alkali, surfactants and tracers. The numerical model can also treat double-porosity systems to simulated fractured rocks. The code solves the equations by a finite differences procedure. Among other applications of the simulator has been used in reservoir processes such as surfactant flooding, high pH chemical flooding, polymer flooding, profile control using gel, tracer tests, formation damage, soil remediation, microbial enhanced oil recovery, etc.

The situation we want to recreate is that for a steady-state single phase non-compressible fluid in an equivalent bi-dimensional system composed by an injection well, a production well and a conductive fault. The basic reasons for solving this numerical problem are: visualize the flow pattern and validate the analytical model consistency regrading the effective tracer transport. For this last purpose, specific system circumstances are simulated by considering certain parameters values, and the resulting tracer breakthrough data will be compared against the corresponding analytical model predictions. A remark should be made regarding the velocity and dispersivity values. The analytical models mentioned here assume a constant velocity inside each region. In the simulation however velocity is a space variable that depends on the local pressure gradient and the permeability. In order to introduce a meaningful velocity value in the analytical model, an effective tracer velocity in each region is estimated from the pulse dynamic observed in the simulation. On the other hand, the pulse dispersion coefficient is a product of dispersivity times velocity, thus this product changes point to point. Therefore an effective dispersivity for the analytical models can not be straightforwardly taken from the simulator input data and should be obtained from data fitting, as will be explained below.

The simulation system, as described in Figure 6.7, consists of a X-Y grid of 101×41 cells of size $\Delta x = \Delta y = 20$ m, an injection well located at cell $(25, 28)$, a production well in cell $(76, 7)$, and a high conductivity 3-cell-width channel extending from cell $(0, 21)$ to cell $(101, 21)$. At both fault borders an injection or production virtual well is introduced to emulate the fluid income or output of the open fault. In this system $L_1 = 7\Delta y = 140$ m, $L_2 = 51\Delta x = 1020$ m and $L_3 = 14\Delta y = 280$ m. The direct inter-well distance is $L = 1103.1$ m, total tracer path is $L_T = L_1 + L_2 + L_3 = 1440$ m. In reality, the simulator UTCHEM treats tri-dimensional systems, thus one cell in the overall system in z-direction having $\Delta z = 20$ m has been considered.

There are four simulation cases to be analyzed, which are graphically described in Figure 6.8. In the first case (A) there is no fault present. This situation is run for comparison purposes. The second situation (B) is when the fault is closed, this means that the virtual wells located at both fault horizontal borders have zero production or injection rate. The last two cases correspond to

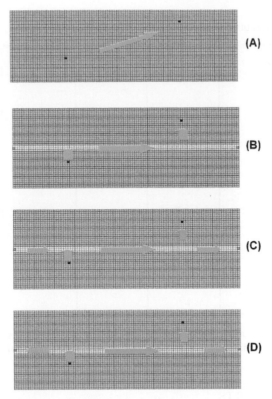

Figure 6.8 The four simulation cases analyzed: (A) no fault, (B) closed fault, (C) open fault with fault-dominated flow, and (D) open fault with injection-dominated flow.

open faults, the case (C) represents the fault-dominated situation, and the case (D) the injection-dominated situation. The left virtual well is an injector in case (C) and a producer in case (D). In both situations the virtual right well is a production well. The numerical tracer breakthrough results of cases (B), (C) and (D) will be compared against the results from the corresponding analytical models.

A single-porosity system with water as the single phase present is considered. The water injection rate in the injector is set to 2000 m^3/d. The production well works at a constant pressure of 5000 kPa ($\sim$50 kg/cm^2). A porosity of $\phi = 20\%$ and a permeability of $K_x = K_y = 100$ mD, $K_z = 0$ for the whole region outside the fault were taken. A constant overall dispersivity of 50 m was set (no space dependent dispersivity can be set in the simulator). In the fault it was assumed a porosity of $\phi = 8\%$, and a permeability of $K_x = 5000$ mD, $K_y = 100$ mD and $K_z = 0$. The fault is not assumed fractured, since the simulator treats double-porosity in the full system, not different values for diverse regions as we require for the fault region. A non-radioactive tracer was injected at a constant unit concentration during two days. In the fault-dominated case the virtual well at the left border is an injector with a water flow rate of 200 m^3/d and the virtual well at the right border is a producer at constant pressure of 60,000 kPa. The corresponding flow rate is 689.9 m^3/d for the production well and 1510.1 m^3/d for the virtual right well. In the injection-dominated case the same conditions as in the previous case are taken, but the left border virtual well is now a production well with a constant pressure of 75,000 kPa. The flow rate in this case is 475.6 m^3/d for the production well, 1013.8 m^3/d for the virtual left well and 510.6 m^3/d for the virtual right well. The results of the simulations are described in the next section.

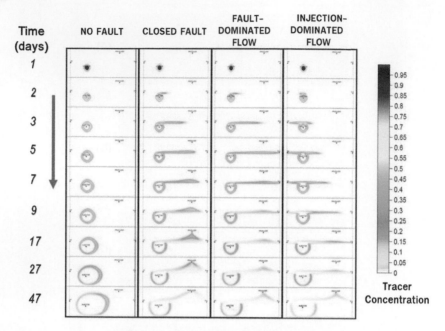

Figure 6.9   Flow pattern as a function of time for the four cases under study presented as columns. The colors correspond to tracer concentration accordingly to the bar at the right side of the figure.

6.4   NUMERICAL RESULTS

Diagrams of the tracer concentration dynamics for the four cases are shown in Figure 6.9. These columns represent from top to bottom the advance of the tracer pulse at nine times: $1, 2, 3, 5, 7, 9,$ $17, 27$ and 47 days respectively. The first column displays the case without the fault, in the second column the case of a closed fault (no external flow), and in the third and fourth columns the cases of fault-dominated (fluid incomes from fault left border and fluid outcomes at fault right border) and the injection-dominated (fluid outcomes at both left and right fault borders) respectively. To make pulse advance clearly apparent, tracer dispersivity in Figure 6.9 has been set to a very low value, it is 0.01 m. From these near dispersion-free figures an effective fluid velocity in Regions 1, 2 and 3 can be read, and by this mean the value of u_1, u_2 and u_3 can be estimated in each case. For this purpose, the tracer breakthrough curve at the virtual production wells was also examined.

An important feature to be observed in Figure 6.9 is that the presence of the conductive fault makes the tracer pulse to separate in two sections, above and below the fault. Since water is continuously pumped in the system, the lower tracer section becomes isolated and tracer can only hardly reach the production well located above the fault. This fact has been captured in the analytical models, since a divergent fluid flow in Region 1 has been taken. Further, at large times, the tracer arriving to the production well comes not in a straight path from the fault, but laterally from both sides. This phenomenon causes long tails in the tracer breakthrough curves. In real reservoir cases these long tails are not expected since reservoir borders are in general located far from the production well, not as in this simulation model. Therefore, numerical tracer breakthrough curve tails should be taken with care.

The tracer breakthrough curves for the four cases mentioned above are shown in Figure 6.10. There are some remarkable curve features. The average transit time (the first moment of the tracer breakthrough curve) is 87.3 days when no fault is present (black curve). The three synthetic situations with a fault, i.e. closed fault (green) (i.e. the curve with the highest peak), fault-dominated (red) (i.e. curve with the lowest peak) and injection-dominated (blue), show a tracer

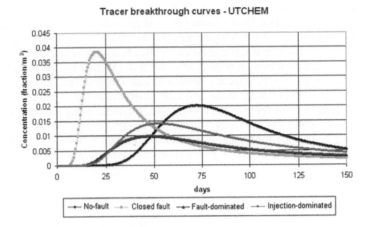

Figure 6.10   Tracer breakthrough curve of the four cases examined.

transit time shorter than the case without the fault, it is 46.2, 71.7 and 80.5 days respectively. The peak high and shape of the two last curves depend on the virtual well properties selected. They are smaller in height than the closed fault curve since external tracer losses through the border virtual wells are present. The total amount of tracer received at 150 days in the production well is 70.3% for the no-fault case, 75.9% for the closed fault situation, and 15.1% and 13.6% for the fault-dominated and injection-dominated cases respectively. (In the last case the tracer mass recovered is less than in the previous case since fluid production rate is lower.)

6.5   COMPARISON OF THE ANALYTICAL MODELS AGAINST NUMERICAL SIMULATIONS

In order to evaluate the potentiality of the analytical models to reproduce the breakthrough curves of the synthetic cases generated by the UTCHEM code, the simulation breakthrough data will be compared against the analytical model breakthrough predictions obtained by using the parameter values employed in the simulations. These fixed parameters are basically $L_1, L_2, L_3, u_1, u_2, u_3$, $\alpha_1, \alpha_2, \alpha_3, \beta = 0$ (single-porosity) and a multiplicative tracer concentration factor F_3 (which is related to the fraction of tracer recovered in the production well in Region 3). The velocity parameters are unknown, but they can be read from the graphs in Figure 6.9 (using its direct definition, tracer advance over time elapsed). The dispersivity is the same over the whole system (this restriction is imposed by some UTCHEM limitations), with a value designed as α. Accordingly to what was mentioned before, the pulse dispersion is proportional to velocity, thus an effective dispersivity is hard to be calculated a priori. Further, the breakthrough curve multiplicative factor F_3 will depend on this effective α value. For this reason, not a direct and meaningful comparison can be made, and therefore a simple data fitting procedure should be implemented as will be described below. The analytical models will be fitted to the corresponding simulation tracer breakthrough data using only two free parameters, they are α and F_3. For using the analytical models it is convenient to employ the dimensionless variables and parameters originally defined in the paper by Coronado et al. (2008) which are: $t_D = t/\langle t \rangle$, $x_D = x/L$, $d_i = L_i/L$, $b_i = u_i \langle t \rangle / L$ and $a = \alpha_i / L$ (here $\langle t \rangle$ means the average transit time). Also a dimensionless tracer concentration is used, which is defined as $C_D = C/C_{ref}$. Here L is the direct inter-well distance, which is $L = 1103.1$ m. This yields $d_1 = 0.127$, $d_2 = 0.925$ and $d_3 = 0.254$ (it is $d_1 + d_2 + d_3 = 1.306$). The reference concentration is defined as $C_{ref} = m_T / Q_1 \langle t \rangle$, with m_T the total tracer mass injected flowing through Region 1, and Q_1 is the volume fluid flow through Region 1.

The total injected tracer mass is $m_T = c_0 Q_{inj} \Delta t$, where c_0 is the tracer concentration in the injection fluid, Q_{inj} the injection fluid volume rate, and Δt is the tracer injection period. In the simulation it has been set, $\Delta t = 2d$, $c_0 = 1$ g/m^3 and $Q_{inj} = 2000$ m^3/d. Since $Q_1 = Q_{inj}/2$ it yields $C_{ref} = 4/\langle t \rangle$ in g/m^3. Further, to determine the parameters b_i the reference velocity $L/\langle t \rangle$ (direct inter-well distance over transit time) is needed. It is 23.9 m/d, 15.4 m/d and 14.8 m/d for the closed fault, the fault-dominated and the injection-dominated cases respectively shown in Figure 6.10. As a comparison value, the velocity in the non-fault case is 12.6 m/d. The three tracer models for faulted formations will be fitted to the corresponding data by using the two mentioned free parameters, a (dispersivity) and the factor F_3. The non-linear optimization procedure employed for the data fitting is Levenberg-Marquardt, which has been proven to be robust in tracer model applications (Ramírez-Sabag et al. 2005). For this purpose an standard algorithm of *Mathematica* was used (Wolfram 1999).

6.5.1   Injection-dominated flow case

In this case the analytical model for injection-dominated flow is fitted to the tracer breakthrough data displayed as black dots in Figure 6.11. Only 70% of the first data are employed to avoid final curve tail information distortion. As mentioned, the fitting parameters are the dispersivity, α, and the factor F_3. The velocity in each region is obtained by inspection of the pulse dynamics displayed in the last column of Figure 6.9. The velocity in Regions 1, Region 2 and Region 3 is 35 m/d, 120 m/d and 4.4 m/d respectively. To evaluate the specific reference velocity for this case, 13.7 m/d, should be used. It yields $b_1 = 2.55$, $b_2 = 8.76$ and $b_3 = 0.32$. These values together with $d_1 = 0.127$, $d_2 = 0.925$, $d_3 = 0.254$ and $\beta = 0$ are used. The optimization procedure yields the best parameters $a = 0.058$ and $F_3 = 0.152$. A plot of the corresponding breakthrough curve and data is displayed in Figure 6.11(A). The objective function (with 210 points) becomes 0.011. A tri-dimensional sensitivity plot of the objective function in terms of a and F_3 is displayed in Figure 6.11(B). This plot shows that the optimum value obtained (red dot) is indeed a global minimum. It is to be noticed that the model corresponding to fault-dominated flow yields a fitting that is as good as the injection-dominated flow treated here, but with a slightly different parameter values. The model for a closed fault fails in providing a good fitting; the objective function gives 0.082 in this case. The fluid transit time for each region, derived from the pulse dynamics, is 4.0d, 8.5d and 63d respectively. The total fluid transit time is then 75.5d, which as expected, is slighter smaller than but close to the tracer transit time of 80.5d. The time difference is small. This is probably one of the reasons why the model and data curve in Figure 6.11(A) are very similar.

Some information can be obtained from F_3. Its definition is $F_3 = (A_1 \phi_1 / A_3 \phi_3) F$, where A_i is effective macroscopic channel cross-section of Region i, and F the fraction of tracer mass that goes from Region 2 into Region 3. By employing the simulation tracer recovered amount in the virtual right well (17.0%) and the producer (13.6%) it follows $F = 0.44$. Therefore, since $\phi_1 = \phi_3$ it follows $A_3/A_1 = 2.9$. It means, the effective flow channel cross-section in Region 3 is around three times the cross-section in Region 1. This value seems to agree qualitatively with the illustrations in the 4th column of Figure 6.9.

6.5.2   Fault-dominated flow case

In this case the analytical model for fault-dominated flow is fitted with the tracer breakthrough data displayed in the red in Figure 6.12. Only 50% of the first data points are employed. Again, the fitting parameters are the dispersivity, a, and F_3. The velocity in each region is read from the pulse dynamics presented in the third column of Figure 6.9. The velocity in Region 1, Region 2 and Region 3 is 35 m/d, 300 m/d and 6.2 m/d respectively. The reference velocity is here 15.4 m/d and therefore it yields $b_1 = 2.27$, $b_2 = 19.48$ and $b_3 = 0.40$. The other parameters (d_i and β) are the same as in the previous case. The fitting procedure gives $a = 0.045$ and $F_3 = 0.059$. The objective function is 0.025 (150 data points). The model does not provide a data fit as good as in the previous case, as seen in Figure 6.12(A). In Figure 6.12(B) a 3D sensitivity plot of the

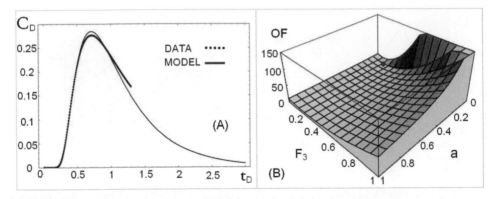

Figure 6.11   (A) Data fitting of tracer concentration VS time in the INJECTION DOMINATED flow case, and (B) objective function plot in terms of a (dimensionless dispersivity) and F_3 (multiplicative tracer-loss factor). The red dot indicates the objective function global minimum associated to the best fit (blue curve) in plot (A).

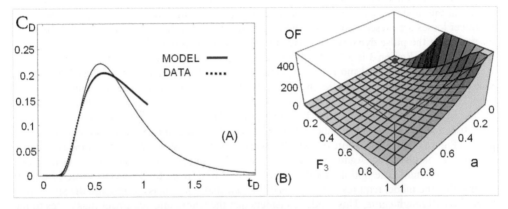

Figure 6.12   (A) Data fitting of tracer concentration VS time in the FAULT-DOMINATED flow case, and (B) Objective function as function of a (dimensionless dispersivity) and F_3 (tracer loss-factor). The best fit parameters are indicated by a red dot, and correspond to the red curve in plot (A).

objective function as function of a and F_3 is displayed. It shows that the minimum found is a global minimum (red dot). From the amount of tracer recovered in the virtual right well (62.5%) and in the producer (15.1%) it follows $F = 0.19$; thus, the effective channel cross-section ratio becomes $A_3/A_1 = F/F_3 = 3.2$. The fluid transit time in Regions 1, Region 2 and Region 3 is 4.0d, 3.4d and 45d respectively. Their sum yields 52.4d, which is much smaller than the tracer transit time of 71.7d. As a remark, it should be mentioned that the injection-dominated model gives a completely similar fitting than the fault-dominated model. However, the model corresponding to a closed fault yields a surprisingly good data fitting (the objective function is 0.005). No clear explanation for this result is found.

6.5.3   Closed fault case

The case of a closed fault is analyzed in order to explore the potentiality of the previous published model for tracer transport in a system with a closed fault (Coronado & Ramírez-Sabag 2008). To avoid the long tail simulation data, which involve border effects, only 30% of the initial tracer

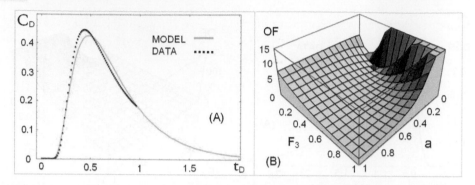

Figure 6.13   (A) Data fitting of tracer concentration VS time in the CLOSED fault case, and (B) Objective function as function of a (dimensionless dispersivity) and F_3 (tracer loss-factor).

breakthrough data points are here considered. Again, the two fitting parameters used are the dispersivity and the concentration multiplicative factor F_3. Strictly speaking the factor is equal to one in the model, since here the channel cross-section is the same in the three regions, and no tracer losses are present along the whole system path. The velocity in each region is obtained by exploring the pulse dynamics displayed in the second column of Figure 6.10. The velocity in Regions 1 to Region 3 is 35 m/d, 246 m/d and 17.5 m/d respectively. The fluid transit time in Regions 1 to 3 is 4.0d, 4.1d and 16.0d respectively. The total fluid transit time is 24.1d, which as a matter of fact is notoriously smaller than the tracer transit time of 46.2d. The reference velocity is here 23.9 m/d and therefore it follows $b_1 = 1.46$, $b_2 = 10.29$ and $b_3 = 0.73$. The other parameters are the same as in the previous case. The fitting procedure yields $a = 0.082$ and $F_3 = 0.311$. The objective function is 0.038 (90 data points). The model provides a relatively good data fitting as seen in Figure 6.13(A). In Figure 6.13(B) the sensitivity plot of the objective function as function of a and F_3 is displayed. Again, the red dot shows the minimum location, which results to be a global minimum. The optimum value notoriously differs from the expected value of one. It means, that the model requires only 31.1% of the injected tracer mass to provide the simulation tracer breakthrough curve. This model considers that the whole injected tracer mass goes to the fault, which is far from true, as can be seen from the simulation. Therefore, the obtained F_3 value can be understood. It is to be remarked that the new models presented in this paper consider that only 50% of the total injected tracer amount goes into the fault. Finally, it is also to be noticed that as might be expected, the injection-dominated and the fault-dominated models yield in this data case a bad fitting. Their corresponding objective function values are 0.146 and 0.117 respectively.

6.6   SUMMARY AND FINAL CONCLUSIONS

Numerical simulation is a valuable tool to graphically see the pulse dynamics and to provide important elements to explore the analytical model potential in describing tracer tests in reservoirs with conductive faults. Accordingly with the simulation model a global dispersivity $a(a_1 = a_2 = a_3)$ and a non-fractured fault media ($\beta = 0$) were set. Seven parameter values have been fixed using the simulation information, they are: $d_1, d_2, d_3, b_1, b_2, b_3$, and β, thus a minimum of only two optimizing parameters, a and F_3 were used to fit tracer breakthrough numerical data. Good results in the case of injection-dominated flow model, acceptable results in the case of fault-dominated flow model, and good results from a previous known closed fault model were obtained. The dispersivity value a obtained is 64.0 m, 49.5 m and 90.4 m for these three cases respectively. These values are relatively closed to the dispersivity employed in the simulation, which was 50 m. The obtained F_3 value seems to be also consistent with the effective channel cross-section ratio observed in the pulse dynamics illustrations for both the injection-dominated and the fault-dominated cases. Thus, it can be concluded that the UTCHEM simulations show that the analytical models have a good potential to describe tracer tests in reservoirs with a conductive fault. A final remark regarding the models and simulation cases treated here, is that for the same data set, the injection-dominated and fault-dominated cases give similar good tracer breakthrough data fittings (i.e. similar objective function value) with only slightly different parameter values. This results might have relevant practical applications. It means that the reservoir parameters obtained by data fitting are similar, not been importantly affected by which of both specific analytical models is used.

ACKNOWLEDGEMENTS

We thank Prof. Mojdeh Delshad for very valuable suggestions as we have set up our numerical model in UTCHEM.

REFERENCES

Caine J.S., Evans J.P. & Foster C.B. (1996) Fault zone architecture and permeability structure. *Geology*, 24, 1025–1028.

Charbeneau R.J. (2000) *Ground Water Hydraulics and Pollutant Transport*. Upper Saddle River, New Jersey, Prentice Hall.

Coronado M. & Ramírez-Sabag J. (2008) Analytical model for tracer transport in reservoirs having a conductive geological fault. *Journal of Petroleum Science and Engineering*, 62, 73–29.

Coronado M., Ramírez-Sabag J. & Valdiviezo-Mijangos O. (2011) Double-porosity model for tracer transport in reservoirs having an open conductive geological fault: determination of the fault orientation. *Journal of Petroleum Science and Engineering*, 78, 65–77.

Delshad M., Pope G.A. & Sepehrnoori K. (1996) A compositional simulator for modeling surfactant enhanced aquifer remediation, 1 Formulation. *Journal of Contaminant Hydrology*, 23, 303–327. See also the UTCHEM web site: www.cpge.utexas.edu/utchem.

Fairley J.P. & Hinds J.J. (2004) Field observation of fluid circulation patterns in a normal fault system. *Geophysical Research Letters*, 31, L19502 1–4. Available from: doi 10.1029/2004GL020812.

Käss W. (1998) *Tracing Technique in Geohydrology*: 1–15. Rotterdam, Balkema.

Maloszewski & Zuber, (1985) On the theory of tracer experiments in fissured rocks with a porous matrix. *Journal of Hydrology*, 79, 333–358.

Maloszewski P. & Zuber A. (1990) Mathematical modeling of tracer behavior in short-term experiments in fissured rocks. *Water Resources Research*, 26 (7), 1717–1528.

Ramírez J., Samaniego F., Rivera J. & Rodríguez F. (1993) Tracer flow in naturally fractured reservoirs. *Rocky Mountains Regional Low Permeability Reservoir Symposium*, 12–14 April 1993, Denver, CO, USA. Richardson TX, USA, Society of Petroleum Engineers. SPE data base paper number 25900.

Ramírez-Sabag J., Valdiviezo-Mijangos O. & Coronado M. (2005) Inter-well tracer tests in oil reservoirs using different optimization methods: A field case. *Geofisica Internacional*, 44 (1), 113.

Sudicky & Frind (1982) Contaminant transport in fractured porous media: Analytical solution for a system of parallel fractures. *Water Resources Research*, 18, 1634–1642.

Van Genuchten M.Th. (1981) Analytical solutions for chemical transport with simultaneous adsorption, zero-order production and first-order decay. *Journal of Hydrology*, 49, 213–233.

Wolfram S. (1999) *The Mathematica Book*, 4th ed. Cambridge, Cambridge University Press. See also Wolfram web site: www.wolfram.com

Zemel B. (1995) *Tracer in the oil field*. Amsterdam, Elsevier Science.

CHAPTER 7

Volume average transport equations for in-situ combustion

A.G. Vital-Ocampo & O. Cazarez-Candia

7.1 INTRODUCTION

The macro scale equations have received attention in recent years and thermo-hydrodynamic theories have been developed including interfacial effects (Hassanizadeh & Gray 1990, Soria & De Lasa 1991, Bousquet-Melou et al. 2002). Since averaged quantities are of engineering interest, one of the main approaches to multiphase flow modeling has been to average (in time, space, over an ensemble or in some combination of these) the original local instantaneous conservation equations. The volume averaging method has been widely used to develop the transport equation for multiphase flow (Gray 1975, Whitaker 1986, Nakayama et al. 2001, Bousquet-Melou et al. 2002, Andrew et al. 2003, Espinosa-Paredes & Cazarez-Candia 2004, Duval et al. 2004), and it can be used to solve an in-situ combustion problem. Predictive models for in-situ combustion have been formulated since early field applications of the process. Depending of their basic approach (analytical, numerical, scaled experimental, empirical, or correlative) and the specific objectives under which they were developed, they vary not only in complexity but also in applicability regarding process mode and parameters. For example, Genrich & Pope (1985) developed a simplified performance predictive model for in-situ combustion processes where they shown the calculations of production oil, gas, and water from forward in-situ combustion processes for given reservoir characteristics and injection rates. The model combines specific descriptions of fractional flow; combustion reactions, phase behavior, and heat transfer. Onyekonwu et al. (1986) discussed the effects of some parameters on in-situ combustion process derived from experimental and simulation studies of laboratory in-situ combustion recovery. On the other hand, Lu & Yortsos (2001) developed a pore-network model for in-situ combustion in porous media. They used dual pore networks (pores and solid sites) for modeling the effect of the microstructure on combustion processes. The model accounts for the transport of the gas phase in the pore space and for the heat transfer by conduction in solid phase. The development of sustained front propagation was studied as a function of various parameters, which include heat losses, instabilities, a correlated pore space and the distribution of fuel.

Also in literature there are works that focus on the simulation of in-situ combustion in laboratory tubes, e.g., Verma et al. (1978) presented a mathematical model to simulate the forward combustion in a laboratory combustion tube. The model incorporates the principal kinetic, thermodynamic and hydrodynamic aspects of the process. The authors made a detailed interpretation of the various computed profiles and identify some phenomena like the existence of an oil-vapor plateau and a peak on the water-vapor profiles. Soliman et al. (1981) developed and applied a multiphase mathematical model from mass and heat balance equations to simulate the process in a laboratory combustion tube. Gottfried (1965) proposed a mathematical model in which conduction-convection heat transfer, chemical reaction between oxygen and oil and aqueous phase change were included making the model applicable to a variety of thermal recovery processes. They used the model to simulate combustion tube experiments. On the other hand, Penberthy & Ramey (1965) developed an analytical model of movement of a burning front axially along a combustion tube with heat loss through an annular insulation. The model allows the identification of 1) the steady-state temperature, 2) distributions of temperature ahead and behind the burning front, with and without heat losses.

To date, the in-situ combustion process has been studied both experimentally and theoretically and it is clear that mathematical models to describe oil thermal recovery, take into account various aspects of the process such as, heat transfer with phase change, heat transfer with chemical reaction and so on. Even though, it is necessary to formulate and solve models that integrate the most important physical and chemical phenomena.

In this chapter, the volume average mass and energy equations for oil (o), water (w), gas (g), and coke (c) were obtained and used to simulate in-situ combustion. The equations developed taken into account: 1) generation of coke due to the chemical reaction of oil, 2) generation of gas due to the combustion reaction between coke and oxygen, and 3) mass transfer due to the phase change of oil and water. Closure equations for oil and water phase change were proposed.

The mathematical model can be used to predict the profiles along a combustion tube of: 1) temperature, 2) pressure, 3) oil, water and gas saturations, 4) oxygen, oil vapor and water steam mass fractions, and 5) coke concentration. The use of a mathematical model allows a wide range of operational conditions which involve substantial efforts to investigate and to complement laboratory experiments.

7.2   STUDY SYSTEM

To obtain local conservation equations for this study, where oil, water and gas interact among themselves and with the porous media and the coke (see Fig. 7.1), the next suppositions and considerations were made: 1) gas is ideal and compressible, 2) oil and water are incompressible and immiscible, 3) solid is impermeable, 4) Fluids are Newtonian, 5) oil and water phase can change, 6) mass can be generated (due to oil chemical reaction and reaction between coke and oxygen), 7) diffusive fluxes, 8) the porous media is homogeneous, isotropic and rigid, 9) coke does not affect to porosity, 10) rectangular coordinates are used, and 11) transitory state is assumed.

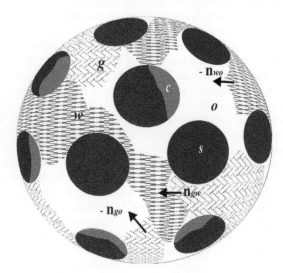

Figure 7.1   Averaging volume in the system composed by oil(o), gas (g), water (w), coke (c) and solid (s). The normal vectors between oil, water and gas are given by $n_{og} = -n_{go}$, $n_{wg} = -n_{gw}$ and $n_{wo} = -n_{ow}$.

7.2.1   *Local mass, momentum and energy equations*

The local mass equation for a homogeneous region inside the averaging volume is well established in literature (Soria & De Lasa 1991). It is given by

$$\underbrace{\frac{\partial \rho_k}{\partial t}}_{\text{Accumulation}} + \underbrace{\nabla \cdot (\rho_k \mathbf{v}_k)}_{\text{Inertial term}} + \underbrace{\nabla \cdot (\mathbf{j}_k)}_{\text{Diffusive Flux}} - \underbrace{\Phi_k}_{\text{Mass Generation}} = 0 \quad k = o, w, g, c \tag{7.1}$$

The local momentum equation was established by Bousquet-Melou et al. (2002) and it is given by:

$$\underbrace{\frac{\partial \rho_k \mathbf{v}_k}{\partial t}}_{\text{Accumulation}} + \underbrace{\nabla \cdot (\rho_k \mathbf{v}_k \mathbf{v}_k)}_{\text{Inertial term}} + \underbrace{\nabla (p_k)}_{\text{Pressure}} - \underbrace{\mu_k \nabla^2 \mathbf{v}_k}_{\text{ViscousStress}} - \underbrace{\rho_k \mathbf{g}}_{\text{Gravity}} = 0 \quad k = o, w, g \tag{7.2}$$

The energy equation proposed considers the fluid compressibility (β) and it is given by Espinosa-Parades & Cazarez-Candia (2004) as

$$\underbrace{\frac{\partial \rho_k Cp_k T_k}{\partial t}}_{\text{Accumulation}} + \underbrace{\nabla (\rho_k Cp_k \mathbf{v}_k T_k)}_{\text{Convective}} + \underbrace{\beta_k \frac{\partial p_k}{\partial t} + \beta_k \nabla \cdot (\mathbf{v}_k p_k)}_{\text{Compression}} = \underbrace{\nabla \cdot (K_k \nabla T_k)}_{\text{Conduction}} + \underbrace{\Phi_k}_{\text{Generation}} \quad k = o, w, g, c, s$$

$$\tag{7.3}$$

where:

$$\beta_k = \left[1 + \left(\frac{\partial \ln \rho}{\partial T} \right)_k \right] = \left[1 - \left(\rho T \frac{\partial \frac{1}{\rho}}{\partial T} \right) \right]_k \tag{7.4}$$

In Eqs. (7.1)–(7.4), ρ is mass density, $\mathbf{v}$ is flow velocity, $\mathbf{j}$ is diffusive flux, Φ is mass generation, p is pressure, μ is viscosity, $\mathbf{g}$ is the gravity acceleration, Cp is the heat capacity, T is temperature, K is thermal conductivity, the subscript k represents the phase o, w, g, c and s which represents oil, water, gas, coke and solid, respectively, and t is the time.

7.2.2   *Jump conditions*

Due to the presence of interfaces in the system, it is necessary to specify a set of local mass, momentum and energy balance equations that describe the transference processes at interfacial regions (see Fig. 7.2).

- Mass jump equations:

$$\{\rho_o(\mathbf{v}_o - \mathbf{w}_{og}) + \mathbf{j}_o\} \cdot \mathbf{n}_{og} = \{\rho_g(\mathbf{v}_g - \mathbf{w}_{go}) + \mathbf{j}_g\} \cdot \mathbf{n}_{go} \quad \text{at} \, A_{og} \tag{7.5}$$

$$\{\rho_w(\mathbf{v}_w - \mathbf{w}_{wg}) + \mathbf{j}_w\} \cdot \mathbf{n}_{wg} = \{\rho_g(\mathbf{v}_g - \mathbf{w}_{gw}) + \mathbf{j}_g\} \cdot \mathbf{n}_{gw} \quad \text{at} \, A_{wg} \tag{7.6}$$

- No-slip condition:

$$\mathbf{v}_k \cdot \mathbf{n}_{kc} = \mathbf{w}_{kc} \cdot \mathbf{n}_{kc} = 0 \quad \text{at} \, A_{kc} \quad k = o, w, g \tag{7.7}$$

$$\mathbf{v}_k \cdot \mathbf{n}_{ks} = \mathbf{w}_{ks} \cdot \mathbf{n}_{ks} = 0 \quad \text{at} \, A_{ks} \quad k = o, w, g \tag{7.8}$$

- Momentum jump equations:

$$-\left\{\rho_g \mathbf{v}_g(\mathbf{v}_g - \mathbf{w}_{go}) + p_g \mathbf{I} - \mu_g \nabla \mathbf{v}_g\right\} \cdot \mathbf{n}_{go}$$
$$= -\left\{\rho_o \mathbf{v}_o(\mathbf{v}_o - \mathbf{w}_{og}) + p_o \mathbf{I} - \mu_o \nabla \mathbf{v}_o\right\} \cdot \mathbf{n}_{og} + 2H_g \sigma \mathbf{n}_{go} \quad \text{at } A_{og} \tag{7.9}$$

$$-\left\{\rho_g \mathbf{v}_g(\mathbf{v}_g - \mathbf{w}_{gw}) + p_g \mathbf{I} - \mu_g \nabla \mathbf{v}_g\right\} \cdot \mathbf{n}_{gw}$$
$$= -\left\{\rho_w \mathbf{v}_w(\mathbf{v}_w - \mathbf{w}_{wg}) + p_w \mathbf{I} - \mu_w \nabla \mathbf{v}_w\right\} \cdot \mathbf{n}_{wg} + 2H_g \sigma \mathbf{n}_{gw} \quad \text{at } A_{wg} \tag{7.10}$$

$$\left\{-p_o \mathbf{I} + \mu_o \nabla \mathbf{v}_o\right\} \cdot \mathbf{n}_{ow} = \left\{-p_w \mathbf{I} + \mu_w \nabla \mathbf{v}_w\right\} \cdot \mathbf{n}_{wo} + 2H_o \sigma \mathbf{n}_{ow} \quad \text{at } A_{wo} \tag{7.11}$$

$$\left\{-p_k \mathbf{I} + \mu_k \nabla \mathbf{v}_k\right\} \cdot \mathbf{n}_{ks} + \left\{-p_s \mathbf{I} + \mu_s \nabla \mathbf{v}_s\right\} \cdot \mathbf{n}_{sk} = 0 \quad \text{at } A_{ks} \quad k = o, w, g \tag{7.12}$$

$$\left\{-p_k \mathbf{I} + \mu_k \nabla \mathbf{v}_k\right\} \cdot \mathbf{n}_{kc} + \left\{-p_c \mathbf{I} + \mu_c \nabla \mathbf{v}_c\right\} \cdot \mathbf{n}_{ck} = 0 \quad \text{at } A_{kc} \quad k = o, w, g \tag{7.13}$$

- Thermal equilibrium:

$$T_s = T_k \qquad \text{at } A_{sk} \quad k = o, w, g, c \tag{7.14}$$

$$T_k = T_m = T_{sat} \qquad \text{at } A_{km} \quad k = o, w, g; \; m = o, w, g; \; k \neq m \tag{7.15}$$

- Energy jump equations:

$$\mathbf{n}_{ks} \cdot K_k \nabla T_k = \mathbf{n}_{sk} \cdot K_s \nabla T_s \qquad \text{at } A_{ks} \quad k = o, w, g \tag{7.16}$$

$$\mathbf{n}_{ck} \cdot K_k \nabla T_k = \mathbf{n}_{ck} \cdot K_c \nabla T_c \qquad \text{at } A_{kc} \quad k = o, w, g \tag{7.17}$$

$$\mathbf{n}_{gw} \cdot \left\{Cp_g \rho_g T_g(\mathbf{v}_g - \mathbf{w}_{gw}) + K_g \nabla T_g\right\} = \mathbf{n}_{wg} \cdot \left\{Cp_w \rho_w T_w(\mathbf{v}_w - \mathbf{w}_{wg}) + K_w \nabla T_w\right\} \quad \text{at } A_{wg} \tag{7.18}$$

$$\mathbf{n}_{go} \cdot \left\{Cp_g \rho_g T_g(\mathbf{v}_g - \mathbf{w}_{go}) + K_g \nabla T_g\right\} = \mathbf{n}_{og} \cdot \left\{Cp_o \rho_o T_o(\mathbf{v}_o - \mathbf{w}_{og}) + K_o \nabla T_o\right\} \quad \text{at } A_{og} \tag{7.19}$$

where Eqs. (7.5)–(7.6) represent the interfacial mass transfer, Eqs. (7.7)–(7.8) are the no-slip conditions, Eqs. (7.9)–(7.13) represent the interfacial momentum transfer by mass, pressure and viscous stress, Eqs. (7.14)–(7.15) represent the interfacial thermal equilibrium and finally, Eqs. (7.16)–(7.19) represent the interfacial energy transfer by mass and heat conduction. In Eqs. (7.5)–(7.19), $\mathbf{w}_{og}$ and $\mathbf{w}_{gw}$ are the interfacial velocities between oil and gas, and gas and water respectively; $\mathbf{n}_{og}$ is the unit normal vector to the interface pointing outside of phase oil;

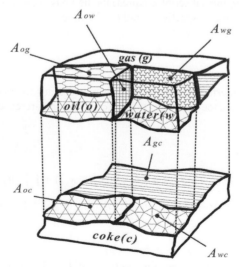

Figure 7.2   Interfacial areas in the averaging volume. Here $A_{og} = A_{go}$, $A_{wg} = A_{gw}$ and $A_{ow} = A_{wo}$ are the interfacial areas between oil, water and gas.

and $\mathbf{n}_{gw}$ is the unit normal vector to the interface pointing outside of phase gas; H is the mean interface curvature; σ is the superficial tension. It is important to note that: $\mathbf{w}_{og} = \mathbf{w}_{go}$, $\mathbf{w}_{wg} = \mathbf{w}_{gw}$, $\mathbf{w}_{ow} = \mathbf{w}_{wo}$, $A_{og} = A_{go}$, $A_{wg} = A_{gw}$, $A_{ow} = A_{wo}$, $\mathbf{n}_{og} = -\mathbf{n}_{go}$, $\mathbf{n}_{wg} = -\mathbf{n}_{gw}$ and $\mathbf{n}_{ow} = -\mathbf{n}_{wo}$.

7.3 AVERAGE VOLUME

In order to obtain an overall average description, volume and interface averaging theorems, which are defined in the works of Gray (1975) and Whitaker (1986) (see Table 7.1), must be applied to the local equations.

In Table 7.1 ψ_k is some function associated with the k-phase, $\langle \psi_k \rangle$ represents the averaged value of ψ_k for the k th-phase ($k = o, w, g, c$), $V_k(t)$ is the volume of the kth-phase contained within the averaging volume V, which is independent of space and time, $\varepsilon_k(= V_k/V)$ is the volume fraction, $A_{km}(= A_{mk})$ is the interfacial area formed by the k and m phases ($k \neq m$), $\sum_{km(k \neq m)}^{N}$ defines the summation of all the interfaces km that surround the k phase.

The averaging volume selected is constant (see Fig. 7.3), smaller than the characteristic macroscopic length of the porous media (L) larger than the characteristic lengths of phases (l_o, l_w, l_g, l_c). The characteristic length of the averaging volume (r_o) must satisfy the following inequality (Whitaker 1986):

$$\left(l_o, l_w, l_g, l_c \right) \ll r_o \ll L \tag{7.27}$$

Table 7.1 Averaging theorems.

Theorem	Equation	
Phase average	$$\langle \psi_k \rangle = \frac{1}{V} \int_{V_k(t)} \psi_k \, dV$$	(7.20)
Intrinsic phase average	$$\langle \psi_k \rangle^k = \frac{1}{V_k} \int_{V_k(t)} \psi_k \, dV$$	(7.21)
Related Eqs. 7.20–7.21	$$\langle \psi_k \rangle = \varepsilon_k \langle \psi_k \rangle^k$$	(7.22)
Transport	$$\left\langle \frac{\partial \psi_k}{\partial t} \right\rangle = \frac{\partial \langle \psi_k \rangle}{\partial t} - \sum_{km(k \neq m)}^{N} \frac{1}{V} \int_{A_{km}(t)} \psi_k \mathbf{w}_{km} \cdot \mathbf{n}_{km} \, dA$$	(7.23)
Spatial averaging	$$\langle \nabla \psi_k \rangle = \nabla \langle \psi_k \rangle + \sum_{km(k \neq m)}^{N} \frac{1}{V} \int_{A_{km}(t)} \psi_k \mathbf{n}_{km} \, dA$$	(7.24)
Volume fraction	$$\left(\frac{\partial \varepsilon_k}{\partial t} \right) = \sum_{km(k \neq m)}^{N} \frac{1}{V} \int_{A_{km}(t)} \mathbf{w}_{km} \cdot \mathbf{n}_{km} \, dA$$	(7.25)
	$$\nabla \varepsilon_k = - \sum_{km(k \neq m)}^{N} \frac{1}{V} \int_{A_{km}(t)} \mathbf{n}_{km} \, dA$$	(7.26)

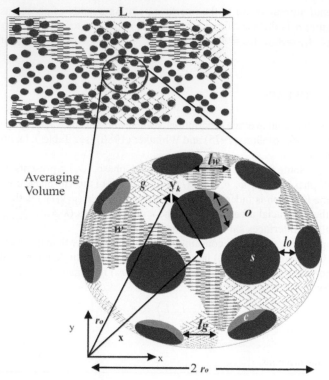

Figure 7.3   Averaging volume containing k phases. $\mathbf{x}$ is the position vector locating the centroid of the averaging volume, $\mathbf{y}_k$ is the position vector at any point in the k-phase relative to the centroid.

7.4   AVERAGE EQUATIONS

When the averaging theorems given by Eqs. (7.20)–(7.26) and the interface jump conditions given by Eqs. (7.5)–(7.8) are applied to Eq. (7.1) the next averaged mass equation is obtained:

$$\frac{\partial \langle \rho_k \rangle}{\partial t} + \nabla \cdot \langle \rho_k \mathbf{v}_k \rangle + \nabla \cdot \langle \mathbf{j}_k \rangle - \langle \Phi_k \rangle$$

$$= -\sum_{km(k \neq m)}^{N} \frac{1}{V} \int_{A_{km}(t)} \left\{ \rho_k (\mathbf{v}_k - \mathbf{w}_{km}) + \mathbf{j}_k \right\} \cdot \mathbf{n}_{km} \, dA \tag{7.28}$$

The average product between two local variables must be defined as the product of two averaged variables using the next expression (Espinoza-Paredes & Cazares-Candia 2004)

$$\langle \psi_{k_1} \psi_{k_2} \rangle = \langle \psi_{k_1} \rangle \langle \psi_{k_2} \rangle + \left\langle \tilde{\psi}_{k_1} \tilde{\psi}_{k_2} \right\rangle \tag{7.29}$$

where ψ is any local variable and $\tilde{\psi}$ is the spatial deviation of the property. The subscripts k_1 and k_2 represent two local variables of the same phase.

Applying Eq. (7.29) in the second term on the left side of Eq. (7.28)

$$\langle \rho_k \mathbf{v}_k \rangle = \langle \rho_k \rangle \langle \mathbf{v}_k \rangle + \langle \tilde{\rho}_k \tilde{\mathbf{v}}_k \rangle \tag{7.30}$$

it follows

$$\frac{\partial \varepsilon_k \langle \rho_k \rangle^k}{\partial t} + \nabla \cdot \varepsilon_k \langle \rho_k \rangle^k \langle \mathbf{v}_k \rangle^k + \nabla \cdot \varepsilon_k \langle \tilde{\rho}_k \tilde{\mathbf{v}}_k \rangle^k + \nabla \cdot \varepsilon_k \langle \mathbf{j}_k \rangle^k - \varepsilon_k \langle \Phi_k \rangle^k$$

$$= - \sum_{km(k \neq m)}^{N} \frac{1}{V} \int_{A_{km}(t)} \{ \rho_k(\mathbf{v}_k - \mathbf{w}_{km}) + \mathbf{j}_k \} \cdot \mathbf{n}_{km} dA \tag{7.31}$$

The second term on the right side of Eq. (7.30) was neglected assuming the pressure gradients are not large, since the density gradients at the microscopic level are very small compared with the velocity gradients. Then, all terms containing spatial variations of the density around its average value can be neglected ($\langle \tilde{\rho}_k \tilde{\mathbf{v}}_k \rangle^k = 0$) (Espinosa-Paredes & Cazarez-Candia 2004):

- Average Mass Equation:

$$\underbrace{\frac{\partial \varepsilon_k \langle \rho_k \rangle^k}{\partial t}}_{\text{Accumulation}} + \underbrace{\nabla \cdot \varepsilon_k \langle \rho_k \rangle^k \langle \mathbf{v}_k \rangle^k}_{\text{Inertial term}} + \underbrace{\nabla \cdot \varepsilon_k \langle \mathbf{j}_k \rangle^k}_{\text{Diffusive flux}} - \underbrace{\varepsilon_k \langle \Phi_k \rangle^k}_{\text{Mass generation}}$$

$$= \underbrace{- \sum_{km(k \neq m)}^{N} \frac{1}{V} \int_{A_{km}(t)} \{ \rho_k(\mathbf{v}_k - \mathbf{w}_{km}) + \mathbf{j}_k \} \cdot \mathbf{n}_{km} dA}_{\text{Mass transfer and diffusive flux on the interfaces}} \tag{7.32}$$

For the momentum equation (Eq. 7.2) a similar procedure is followed. However, in the integral area term, the Eq. (7.30) was also applied and finally the averaged momentum equation is defined by:

- Average Momentum Equation:

$$\underbrace{\frac{\partial \varepsilon_k \langle \rho_k \rangle^k \langle \mathbf{v}_k \rangle^k}{\partial t}}_{\text{Accumulation}} + \underbrace{\nabla \cdot (\varepsilon_k \langle \rho_k \rangle^k \langle \mathbf{v}_k \rangle^k \langle \mathbf{v}_k \rangle^k)}_{\text{Inertial term}} + \underbrace{\nabla \cdot (\varepsilon_k \langle \rho_k \rangle^k \langle \tilde{\mathbf{v}}_k \tilde{\mathbf{v}}_k \rangle^k)}_{\text{Dispersion}} + \underbrace{\varepsilon_k \nabla \langle p_k \rangle^k}_{\text{Pressure}} - \underbrace{\mu_k \varepsilon_k \nabla^2 \langle \mathbf{v}_k \rangle^k}_{\text{Viscous Stress}} - \underbrace{\varepsilon_k \langle \rho_k \rangle^k \mathbf{g}_k}_{\text{Gravity}}$$

$$= \underbrace{- \sum_{km(k \neq m)}^{N} \frac{1}{V} \int_{A_{km}(t)} \rho_k \mathbf{v}_k(\mathbf{v}_k - \mathbf{w}_{km}) \cdot \mathbf{n}_{km} dA}_{\text{Momentum by interfacial mass transfer}} + \underbrace{\sum_{km(k \neq m)}^{N} \frac{1}{V} \int_{A_{km}(t)} (-\bar{\bar{\mathbf{I}}} \tilde{p}_k + \mu_k \nabla \tilde{\mathbf{v}}_k) \cdot \mathbf{n}_{km} dA}_{\text{Interfacial forces}}$$

$$\underbrace{+ \mu_k \nabla \cdot \left(\sum_{km(k \neq m)}^{N} \frac{1}{V} \int_{A_{km}(t)} \tilde{\mathbf{v}}_k \cdot \mathbf{n}_{km} dA \right)}_{\text{Dispersion}} \tag{7.33}$$

A similar procedure was applied into the Eq. (7.3), so the averaged energy equation is:

- Average Energy Equation:

$$\underbrace{\frac{\partial Cp_k \langle \rho_k \rangle^k \varepsilon_k \langle T_k \rangle^k}{\partial t}}_{\text{Accumulacion}} + \underbrace{\nabla \cdot \varepsilon_k \langle Cp_k \rangle^k \langle \rho_k \rangle^k \langle \mathbf{v}_k \rangle^k \langle T_k \rangle^k}_{\text{Convection}} + \underbrace{\beta_k \varepsilon_k \frac{\partial \langle p_k \rangle^k}{\partial t}}_{\text{Compression}} + \underbrace{\beta_k \nabla \cdot \varepsilon_k \langle \mathbf{v}_k \rangle^k \langle p_k \rangle^k}_{\text{Compression}}$$

$$+ \underbrace{\nabla \cdot \left\langle \tilde{C} p_k \tilde{\mathbf{v}}_k \tilde{T}_k \right\rangle \langle \rho_k \rangle^k + \beta_k \langle \nabla \cdot \tilde{\mathbf{v}}_k \tilde{p}_k \rangle}_{\text{Dispersion}} + \underbrace{\sum_{km(k \neq m)}^{N} \frac{1}{V} \int_{A_{km}(t)} Cp_k \rho_k T_k (\mathbf{v}_k - \mathbf{w}_{km}) \cdot \mathbf{n}_{km} \, dA}_{\text{Phase change}}$$

$$+ \underbrace{\beta_k \left(\sum_{km(k \neq m)}^{N} \frac{1}{V} \int_{A_{km}(t)} p_k (\mathbf{v}_k - \mathbf{w}_{km}) \cdot \mathbf{n}_{km} dA \right)}_{\text{Interfacial compression}} \tag{7.34}$$

$$= \nabla \cdot \underbrace{\left[K_k \left(\varepsilon_k \nabla \langle T_k \rangle^k + \sum_{km(k \neq m)}^{N} \frac{1}{V} \int_{A_{km}(t)} \tilde{T}_k \cdot \mathbf{n}_{km} \, dA \right) \right]}_{\text{Conduction}}$$

$$+ \underbrace{\sum_{km(k \neq m)}^{N} \frac{1}{V} \int_{A_{km}(t)} K_k \nabla T_k \cdot \mathbf{n}_{km} dA}_{\text{Interfacial Flux}} + \underbrace{\varepsilon_k \langle \Phi_k \rangle^k}_{\text{Generation}}$$

In Eqs. (7.33) and (7.34), $\tilde{\mathbf{v}}_k$ is the spatial deviation velocity of the kth-phase, $\tilde{p}_k$ is the spatial deviation pressure of the kth-phase, $\tilde{T}_k$ is the spatial deviation temperature of the kth-phase.

The volume averaging method allows obtaining in a natural way, the mass, momentum and energy transfer terms on the interfaces.

7.5 PHYSICAL MODEL

Average equations obtained in Section 7.4 are used to simulate a combustion tube, then it is necessary to establish the physical model and write such equations in a convenient form. Physical model is described in this section and the corresponding equations are given in Section 7.6. Data from the work of Cazarez-Candia et al. (2010) (see Table 7.2) are used to test the mathematical model to be presented in Section 7.6. Therefore the physical model is established from such work, where two experiments were done in a combustion tube of stainless steel with an external diameter of 0.079375 m, a width of 0.015875 m and a length of 0.9906 m, which contains about 0.94996 m of an uniform mixture of sand, water and oil (see Fig. 7.4). In the experiments an insulation

Table 7.2 Experimental parameters.

Parameter	Experiment 1	Experiment 2
Water Saturation (sw)	0.23	0.275
Oil Saturation (so)	0.325	0.383
Porosity (ϕ)	0.41	0.41
Ignition Temperarure	427°C	462°C
Production Pressure	4.13×10^5 Pa	4.20×10^5 Pa
Air injection	3.166×10^{-5} m^3/s	3.166×10^{-5} m^3/s

band was place along the tube. The oil used had activation energy of 1.56×10^7 J/kg-mol and 27 API. The space formed from the sample to the top of the tube, was filled with clean sand, and an electric igniter was placed on this place. Electric current was gradually introduced into the igniter until the temperature in the combustion tube at the igniter level reached about 241°C and air injection was initiated. The production pressure was maintained constant. After the ignition, the combustion front moved from the igniter to bottom of the combustion tube. The combustion gases and liquids went out from the bottom of the tube. The end of the phenomenon occurred when the sand pack was burned until the bottom flange of the combustion tube.

7.6   EQUATIONS FOR IN-SITU COMBUSTION

To apply the average equations (Eqs. 7.32–7.34) on the simulation of the physical model presented in Section 7.4, the next considerations and assumptions are done: 1) fluids are Newtonian 2) oil and water phase changes are possible, 3) there is mass generation (due to oil chemical reaction and reaction between coke and oxygen), 4) the porous media is homogeneous, isotropic, and rigid, 5) coke does not affect to porosity 6) liquid phases are mono-component ($\mathbf{j}_k = 0$ in A_{km}), incompressible and immiscible, 7) gas is ideal and incompressible, 8) gas is a mixture composed by nitrogen, oxygen, oil vapor, steam, and combustion gases, 9) momentum equation is expressed using the Darcy velocity definition for every phase (the capillarity effects are neglected and the gravity segregation is not considered), 10) dispersive terms are neglected, 11) The local thermal equilibrium and the complete enthalpy definition (due to phase change) are considered in energy equations, so the energy balance can be expressed with only one equation, 12) no source is considered in the energy equation 13) heat is transferred from the system to the environment by convection, 14) a one-dimensional system is considered. Therefore Eqs. (7.32)–(7.34) can be written as:

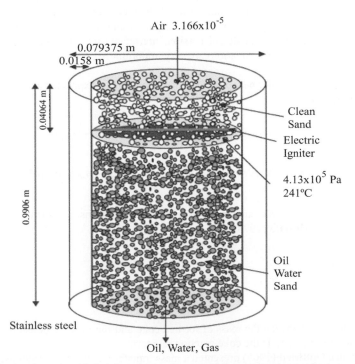

Figure 7.4   Combustion tube. Air is injected from the top and the combustion products are obtained from the bottom.

- Gas, oil, water and coke mass equations:

$$\frac{\partial \varepsilon_g \langle \rho_g \rangle^g}{\partial t} + \frac{\partial \varepsilon_g \langle \rho_g \rangle^g \langle \mathbf{v}_g \rangle^g}{\partial x} = -\frac{1}{V} \int_{A_{go}(t)} \rho_g (\mathbf{v}_g - \mathbf{w}_{go}) \cdot \mathbf{n}_{go} \, dA$$

$$-\frac{1}{V} \int_{A_{gw}(t)} \rho_g (\mathbf{v}_g - \mathbf{w}_{gw}) \cdot \mathbf{n}_{gw} \, dA - \varepsilon_g \langle \Phi_g \rangle^g \tag{7.35}$$

$$\frac{\partial \varepsilon_o \langle \rho_o \rangle^o}{\partial t} + \frac{\partial \varepsilon_o \langle \rho_o \rangle^o \langle \mathbf{v}_o \rangle^o}{\partial x} = -\frac{1}{V} \int_{A_{og}(t)} \rho_o (\mathbf{v}_o - \mathbf{w}_{og}) \cdot \mathbf{n}_{og} \, dA - \varepsilon_o \langle \Phi_o \rangle^o \tag{7.36}$$

$$\frac{\partial \varepsilon_w \langle \rho_w \rangle^w}{\partial t} + \frac{\partial \varepsilon_w \langle \rho_w \rangle^w \langle \mathbf{v}_w \rangle^w}{\partial x} = -\frac{1}{V} \int_{A_{wg}(t)} \rho_w (\mathbf{v}_w - \mathbf{w}_{wg}) \cdot \mathbf{n}_{wg} \, dA \tag{7.37}$$

$$\frac{\partial \varepsilon_c \langle \rho_c \rangle^c}{\partial t} = \varepsilon_c \langle \Phi_c \rangle^c \tag{7.38}$$

where $(\varepsilon_k \langle \Phi_k \rangle^k =) \langle \Phi_k \rangle$. To obtain Eqs. (7.35)–(7.38) the interfacial mass transfers oil-water, oil-coke, water-coke, and gas-coke, are not considered. On the other hand, interfacial mass transfers water-solid, oil-solid, gas-solid, and coke-solid are equal to zero due to the no-slip condition.

- Gas, oil and water momentum equations:

$$\langle \mathbf{v}_g \rangle = -\frac{\zeta_g}{\mu_g} \left(\frac{\partial p}{\partial x} \right); \quad \langle \mathbf{v}_o \rangle = -\frac{\zeta_o}{\mu_o} \left(\frac{\partial p}{\partial x} \right); \quad \langle \mathbf{v}_w \rangle = -\frac{\zeta_w}{\mu_w} \left(\frac{\partial p}{\partial x} \right) \tag{7.39}$$

where ζ is permeability.

In Eq. (7.39), the accumulation, inertial, dispersion, gravity, and interfacial momentum transfer by mass change were neglected, while the interfacial forces term was considered as the permeability definition (details about this term in Whitaker 1986, 1996).

As a consequence of averaging Eqs. (7.1)–(7.2), additional terms arose, which are known as closure terms: interfacial mass transfer, dispersion, interfacial flux, interfacial forces, etc. Closure terms together with constitutive equations are very important in order to simulate the phenomena properly.

The interfacial mass transfer between w and g and o and g is expressed according to the work of Duval et al. (2004) as:

$$-\frac{1}{V} \int_{A_{wg}(t)} \rho_w (\mathbf{v}_w - \mathbf{w}_{wg}) \cdot \mathbf{n}_{wg} \, dA = \dot{m}_{wg}; \quad -\frac{1}{V} \int_{A_{gw}(t)} \rho_g (\mathbf{v}_g - \mathbf{w}_{gw}) \cdot \mathbf{n}_{gw} \, dA = \dot{m}_{gw} \tag{7.40}$$

$$-\frac{1}{V} \int_{A_{go}(t)} \rho_g (\mathbf{v}_g - \mathbf{w}_{go}) \cdot \mathbf{n}_{go} \, dA = \dot{m}_{go}; \quad -\frac{1}{V} \int_{A_{og}(t)} \rho_o (\mathbf{v}_o - \mathbf{w}_{og}) \cdot \mathbf{n}_{og} \, dA = \dot{m}_{og} \tag{7.41}$$

The mass generation of coke and gas, due to chemical reactions, are written according to the work of Gottfried & Mustafa (1978) as:

$$\Phi_o = -S_l \tag{7.42}$$

$$\Phi_g = (1 - M_c) S_l + S_c \tag{7.43}$$

$$\Phi_c = M_c S_l - S_c \tag{7.44}$$

where M_c is the ratio between the mass of formed coke and the mass of consumed oil, S_l is the coke deposition term and S_c is the coke reaction rate term.

For Eq. (7.40), Gottfried (1965) presented a mass transfer equation for water as:

$$\dot{m}_{gw} = -\dot{m}_{wg} = ha(p_{lw} - p_{gw}) \tag{7.45}$$

where h is the evaporation or condensation coefficient, a is the gas-water interfacial area per unit of volume, p_{lw} is the vapor pressure of water, and p_{gw} is the partial pressure of the water vapor in the gas phase.

For Eq. (7.41), Chu (1964) presented a mass transfer equation as function of the total concentration. In this work the Chu's expression was modified as function of the vapor partial pressure of oil in the gas phase, p_{go} (pa) and vapor pressure of oil, p_{lo} (Pa) (Appendix 7.A). The modified expression is:

$$\dot{m}_{go} = -\dot{m}_{og} = \omega M_o(k_e p_{lo} - p_{go}),\tag{7.46}$$

where ω is defined by the Eq. (7.A4), M_o is the oil molecular weigh, and k_e is the vaporization equilibrium constant.

Regarding to the in-situ combustion phenomenon, it was assume that the following two reactions occur in the vicinity combustion zone:

$$\text{Oil} \longrightarrow \text{Coke} + \text{Hydrocarbon vapor}\tag{7.47}$$

$$\text{Coke} + \text{Oxygen} \longrightarrow \text{CO}, \text{CO}_2, \text{H}_2\text{O}\tag{7.48}$$

These two reactions represent an idealized simplification of the actual coking process, since effects such as low-temperature oxidation are not considered. However these mechanics offer a reasonable approximation to the true process without undue complexity.

If the coking reaction described by Eq. (7.47) depends only upon temperature, then the reaction rate can be expressed as a zero order Arrhenius reaction:

$$S_l = Z\,e^{-E/R(T+460)}\tag{7.49}$$

where Z is the Arrhenius rate coefficient for coke formation, E is the Arrhenius activation energy for coke formation, R is the ideal gas constant and T is the system temperature (°F).

The combustion reaction described by Eq. (7.48) is written as:

$$S_c = Z_c(\phi f_a \rho_g s_g)\,F\,e^{-E_c/R(T+460)}\tag{7.50}$$

where Z_c is the Arrhenius rate coefficient for combustion, f_a is the mass fraction of oxygen in the gas phase, ρ_g is the gas density, s_g is the gas saturation, F is the accumulated coke concentration and E_c is the Arrhenius activation energy for combustion.

Substituting the saturation ($s_k = V_k/V_H$), porosity ($\phi = V_H/V$), volume fraction ($\varepsilon_k = V_k/V$), and Darcy velocity definition ($\langle v \rangle = s_k \phi \langle v_k \rangle^k$) as well as the closure terms (Eqs. (7.45), (7.46), (7.49) and (7.50)) into the Eqs. (7.35)–(7.39) the next mass equations in terms of variables from the oil industry are obtained:

- Mass equations:

$$\frac{\partial s_g \phi \rho_g}{\partial t} + \frac{\partial}{\partial x}\left(-\frac{\rho_g \zeta_g}{\mu_g}\frac{\partial p}{\partial x}\right) = (1 - M_c)S_l + S_c + \dot{m}_{go} + \dot{m}_{gw}\tag{7.51}$$

$$\frac{\partial s_o \phi \rho_o}{\partial t} + \frac{\partial}{\partial x}\left(-\frac{\zeta_o \rho_o}{\mu_o}\frac{\partial p}{\partial x}\right) = -\dot{m}_{og} - S_l\tag{7.52}$$

$$\frac{\partial s_w \phi \rho_w}{\partial t} + \frac{\partial}{\partial x}\left(-\frac{\zeta_w \rho_w}{\mu_w}\frac{\partial p}{\partial x}\right) = -\dot{m}_{wg}\tag{7.53}$$

$$\frac{\partial F}{\partial t} = M_c S_l - S_c\tag{7.54}$$

where $(s_k \phi \langle \Phi_k \rangle^k =) \langle \Phi_k \rangle$, V_H is the hole volume, and $F(=\rho_c s_c \phi)$ is known as the accumulated coke concentration (Gottfried & Mustafa 1978). For in-situ combustion, using Eq. (7.51) the mass equation for oxygen, oil vapor, and steam can be written as:

$$\phi \frac{\partial}{\partial t}(f_a \rho_g s_g) - \frac{\partial}{\partial x}\left(\frac{f_a \rho_g \zeta_g}{\mu_g}\frac{\partial p}{\partial x}\right) = -M_a S_c \tag{7.55}$$

$$\phi \frac{\partial}{\partial t}(f_s \rho_g s_g) - \frac{\partial}{\partial x}\left(\frac{f_s \rho_g \zeta_g}{\mu_g}\frac{\partial p}{\partial x}\right) = M_s S_c + \dot{m}_{gw} \tag{7.56}$$

$$\phi \frac{\partial}{\partial t}(f_o \rho_g s_g) - \frac{\partial}{\partial x}\left(\frac{f_o \rho_g \zeta_g}{\mu_g}\frac{\partial p}{\partial x}\right) = M_{vo} S_l + \dot{m}_{go} \tag{7.57}$$

where f_s is the mass fraction of steam in the gas phase, f_o is the mass fraction of the oil vapor in the gas phase, M_a is the ratio between the mass of consumed oxygen and the mass of consumed coke, M_s is the ratio between the mass of produced steam and the mass of consumed coke, M_{vo} is the ratio between the mass of produced oil vapor and the mass of consumed oil, M_c is the ratio between the mass of formed coke and the mass of consumed oil, $\dot{m}_{gw} = -\dot{m}_{wg}$ is term of evaporation for the water, $\dot{m}_{go} = -\dot{m}_{og}$ is term of evaporation for the oil, S_l is the coke deposition term and S_c is the coke reaction rate term, the subscripts s, o, a, g, w represents steam oil, oxygen, gas and water, respectively.

To simplify the nomenclature in these equations are assumed that $\langle \psi_k \rangle^k = \psi_k$.

- Mixture energy equation: It was considered that there is thermal equilibrium among the phases then the energy equation can be written as:

$$\frac{\partial \varepsilon_g \langle Cp_g \rangle^g \langle \rho_g \rangle^g \langle T \rangle}{\partial t} + \frac{\partial \varepsilon_o \langle Cp_o \rangle^o \langle \rho_o \rangle^o \langle T \rangle}{\partial t} + \frac{\partial \varepsilon_w \langle Cp_w \rangle^w \langle \rho_w \rangle^w \langle T \rangle}{\partial t} + \frac{\partial \varepsilon_c \langle Cp_c \rangle^c \langle \rho_c \rangle^c \langle T \rangle}{\partial t}$$

$$+ \frac{\partial \varepsilon_s \langle Cp_s \rangle^s \langle \rho_s \rangle^s \langle T \rangle}{\partial t} + \frac{\partial \varepsilon_o \langle \rho_o \rangle^o \langle \lambda_o \rangle}{\partial t} + \frac{\partial \varepsilon_w \langle \rho_w \rangle^w \langle \lambda_w \rangle}{\partial t} + \frac{\partial \varepsilon_c \langle \rho_c \rangle^c \langle \lambda_c \rangle}{\partial t}$$

$$+ \frac{\partial \varepsilon_g \langle Cp_g \rangle^g \langle \rho_g \rangle^g \langle v_g \rangle^g \langle T \rangle}{\partial x} + \frac{\partial \varepsilon_o \langle Cp_o \rangle^o \langle \rho_o \rangle^o \langle v_o \rangle^o \langle T \rangle}{\partial x} + \frac{\partial \varepsilon_w \langle Cp_w \rangle^w \langle \rho_w \rangle^w \langle v_w \rangle^w \langle T \rangle}{\partial x}$$

$$+ \frac{\partial \varepsilon_o \langle \rho_o \rangle^o \langle v_o \rangle^o \langle \lambda_o \rangle}{\partial x} + \frac{\partial \varepsilon_w \langle \rho_w \rangle^w \langle v_w \rangle^w \langle \lambda_w \rangle}{\partial x} + h'a'(\langle T \rangle - T_r)$$

$$= \frac{\partial}{\partial x}\left[\left(K_g \varepsilon_g + K_o \varepsilon_o + K_w \varepsilon_w + K_c \varepsilon_c + K_s \varepsilon_s\right)\frac{\partial \langle T \rangle}{\partial x}\right] \tag{7.58}$$

where λ_w is the water heat of vaporization, λ_o is the latent heat associated with coke formation, λ_c is the heat of combustion, h' is the external heat loss coefficient, a' is the external heat loss area per unit of volume, T_r is the initial (ambient) temperature, and $\langle T \rangle$ is the space average temperature defined by:

$$\langle T \rangle = \frac{1}{V}\int_V T\, dV = \varepsilon_g \langle T_g \rangle^g + \varepsilon_o \langle T_o \rangle^o + \varepsilon_w \langle T_w \rangle^w + \varepsilon_c \langle T_c \rangle^c + \varepsilon_s \langle T_s \rangle^s \tag{7.59}$$

To simplify the nomenclature in these equations are assumed that $\langle \psi \rangle = \psi$ and $\langle \psi_k \rangle^k = \psi_k$.

For the energy equation, the Eqs. (7.52)–(7.54), (7.56) and (7.57), are substituted into the Eq. (7.58), so the energy equation can be written as:

$$\alpha \frac{\partial^2 T}{\partial x^2} + \beta \frac{\partial T}{\partial x} - \gamma T + \delta = \frac{\partial T}{\partial t} \tag{7.60}$$

where

$$\alpha = \frac{k_{ef}}{\theta} \tag{7.61}$$

$$\theta = \{(1 - \phi)Cp_s\rho_s + \phi[s_g\rho_gCp_wf_s + s_g\rho_gCp_of_o + s_g\rho_gCp_g(1 - f_s - f_o)$$
$$+ s_wCp_w\rho_w + s_oCp_o\rho_o] + FCp_c\} \tag{7.62}$$

$$\beta = \frac{1}{\theta}\left[\left(\frac{k_g\rho_g}{\mu_g}(f_sCp_w + f_oCp_o + Cp_g(1 - f_s - f_o))\right)\frac{\partial p}{\partial x}\right]$$
$$+ \frac{1}{\theta}\left[\left(\frac{k_w\rho_wCp_w}{\mu_w} + \frac{k_o\rho_oCp_o}{\mu_o}\right)\frac{\partial p}{\partial x}\right] \tag{7.63}$$

$$\gamma = \frac{1}{\theta}\{[(Cp_w - Cp_g)M_s + Cp_g - Cp_c]S_c + [(Cp_c - Cp_g)M_c]S_l\}$$
$$+ \frac{1}{\theta}\{[Cp_g - Cp_o + (Cp_o - Cp_g)M_{vo}]S_l + h'a'\} \tag{7.64}$$

$$\delta = \frac{1}{\theta}[(\lambda_c - \lambda_wM_s)S_c - \lambda_w\dot{m}_{gw} - \lambda_o\dot{m}_{go} - (\lambda_oM_{vo} + \lambda_cM_c)S_l + h'a'T_r] \tag{7.65}$$

where K_{ef} is the effective thermal conductivity given by:

$$K_{ef} = K_g\varepsilon_g + K_o\varepsilon_o + K_w\varepsilon_w + K_c\varepsilon_c + K_s\varepsilon_s \tag{7.66}$$

To solve the mass and energy equations presented in this section it is necessary to specify initial and boundary conditions.

- Initial Conditions: At time $t = 0$ the initial conditions in all the mesh as:

$$T(x,0) = TR \tag{7.67}$$

$$s_w(x,0) = swr \tag{7.68}$$

$$s_g(x,0) = sgr \tag{7.69}$$

$$fa(x,0) = far \tag{7.70}$$

$$fs(x,0) = fsr \tag{7.71}$$

where *TR* is the initial temperature, *swr* is the initial water saturation, *sgr* is the initial gas saturation, *far* is the initial oxygen mass fraction in the gas phase, *fsr* is the initial steam mass fraction in the gas phase.

- Boundary Conditions: The model is subjected to the following boundary conditions:

$$T(0,t) = TI \tag{7.72}$$

$$\left.\frac{\partial T}{\partial x}\right|_{x=L} = 0 \tag{7.73}$$

$$p(L,t) = p_o \tag{7.74}$$

$$s_o(0,t) = 0 \tag{7.75}$$

$$fa(0,t) = fai \tag{7.76}$$

$$fs(0,t) = fsi \tag{7.77}$$

where *TI* is the input temperature of the system.

7.7 NUMERICAL SOLUTION

The numerical technique, applied to solve Eqs. (7.51)–(7.57), and (7.60) was the finite differences. An explicit scheme was used to solve mass equations. The time and spatial derivatives were approximated using the next equations:

$$\frac{\partial \Psi}{\partial t} = \frac{\Psi_i^{t+\Delta t} - \Psi_i^t}{\Delta t} \tag{7.78}$$

$$\frac{\partial \Psi}{\partial x} = \frac{\Psi_i^{t+\frac{\Delta t}{2}} - \Psi_{i-1}^{t+\frac{\Delta t}{2}}}{\Delta x} \tag{7.79}$$

Energy equation was solved using the Crank-Nicolson approximation given by:

$$\frac{\partial \Psi}{\partial t} = \frac{\Psi_i^{t+\Delta t} - \Psi_i^t}{\Delta t} \tag{7.80}$$

$$\frac{\partial \Psi}{\partial x} = \frac{\Psi_i^{t+\Delta t} + \Psi_i^t - \Psi_{i-1}^{t+\Delta t} - \Psi_{i-1}^t}{2\Delta x} \tag{7.81}$$

$$\frac{\partial^2 \Psi}{\partial x^2} = \frac{\Psi_{i+1}^{t+\Delta t} + \Psi_{i+1}^t + \Psi_{i-1}^{t+\Delta t} + \Psi_{i-1}^t - 2\Psi_i^{t+\Delta t} - 2\Psi_i^t}{2\Delta x^2} \tag{7.82}$$

where Ψ is some dependent variable, Δx is the space step size, Δt is the time step size, the subscripts $i-1$, i, and $i+1$ represent backward, current and forward nodes, respectively. The superscripts t and $t + \Delta t$ represent current and later times, respectively.

7.8 SOLUTION

The next procedure was used to solve the average mass and energy equations obtained in Section 7.6.

Input Parameters
The oil (ρ_o), gas (ρ_g) and water (ρ_w) mass densities, oil (μ_o), water (μ_w) and gas (μ_g) viscosities and oil (k_o), water (k_w) and gas (k_g) effective permeabilities are obtained from the correlations defined in the literature (Gotffired 1965).

On the other hand, the oil (Cp_o), water (Cp_w), gas (Cp_g), and coke (Cp_c) heat capacities, the Arrhenius activation energy for coke formation (E), the Arrhenius activation energy for combustion (E_c), the Arrhenius rate coefficient for coke formation (Z), the Arrhenius rate coefficient for combustion (Z_c), absolute permeability (ζ), thermal conductivity, (K_{ef}), the ratio between the mass of consumed oxygen and the mass of consumed coke (M_a), the ratio between the mass of produced steam and the mass of consumed coke (M_s), the ratio between the mass of produced oil vapor and the mass of consumed oil (M_{vo}), the ratio between the mass of formed coke and the mass of consumed oil (M_c), the water heat of vaporization (λ_w), the latent heat associated with coke formation (λ_o), the heat of combustion (λ_c), the external heat loss coefficient (h'), the external heat loss area per unit of volume (a') all of them are parameter necessaries for the simulation.

Closure Relations
The closure equations are determinate by Eqs. (7.45), (7.46), (7.49), (7.50) and describe the oil and gas interfacial mass transfer as well as the mass generation due to chemical reactions between oil and coke and oxygen.

Solution Procedure:

Step 1: Input data: mix length, gas injection rate, time simulation, initial temperature and pressure, time step size, mesh size

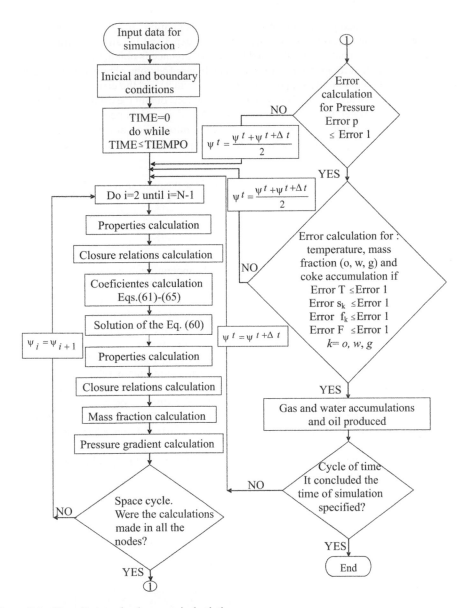

Figure 7.5   Flow diagram for the numerical solution.

Step 2: Initial condition in $t = 0$
Step 3: Boundary Conditions
Step 4: Calculate fluid properties
Step 5: Calculate the closure equations
Step 6: Evaluate each coefficient in the energy equation ($\alpha, \beta, \gamma, \delta, \theta$)
Step 7: The energy equation is solved by using the Newton-Raphson-Bisection method
Step 8: Fluid properties are again calculated
Step 9: The closure equations are again calculated
Step 10: The oxygen, steam and oil vapor mass fraction are calculated
Step 11: A new pressure distribution is obtained

Step 12: The oil, water and gas saturations and coke accumulation are again calculated

Step 13: The distribution pressure is compared (between step 3 and 11) and the error is evaluated. The steps 5–13 are repeated until the established convergence is obtained

Step 14: The distribution temperatures, saturation and mass fractions obtained in the steps 7, 10 and 12 are compared with the supposed data (steps 2 and 3) and the error is evaluated

Step 15: The steps 2–14 are repeated until all the distributions obtain the established convergence

Step 16: The gas and water accumulations and oil produced are calculated

Step 17: The simulated time is advanced by Δt; all the information at time $t + \Delta t$ is analyzed to time t, and the computation of a new time step begins. The computation ceases when the desired time has been simulated

All the steps above mentioned can be seen into the flow diagram shown in Figure 7.5.

7.9   RESULTS

Figure 7.6 shows the comparison between experimental temperature from Experiment 1 (Cazarez-Candia et al., 2010) and the results from the model form by Eqs. (7.51)–(7.57), and (7.60). The front velocity and the temperature profiles obtained with the model are in agreement with experimental data. The temperature profiles behind the combustion front show a little different behaviour until about 250°C and a bigger difference for smaller temperatures. This because the effect of the electric igniter was not taken into account in the model.

In Figure 7.6, it can be seen that due to the injected air temperature (150°C), the temperature profiles grow until the combustion front temperature (maximum temperatures in the profiles), ahead of it temperature falls quickly (coke formation zone) and finally it remains almost constant.

On the other hand, in Figure 7.7 it can be seen that the front velocity and the temperature profiles obtained with the model are in agreement with the data from experiment 2. However the temperatures profiles ahead of the combustion front show a little difference behaviour until about 200°C.

The data obtained from a combustion tube test are: 1) temperature, 2) volume of injected air, 3) oil, gas and water yield, 4) chromatographic data, etc.

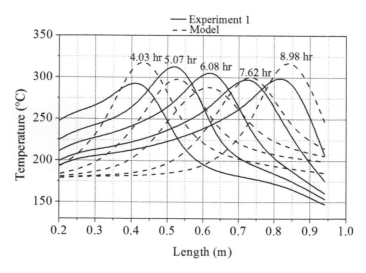

Figure 7.6   Temperature profiles for diverse times in Experiment 1.

However, pressure, saturation,and mass fraction profiles along the tube can not be measured. To know this information the mathematical model (Eqs. (7.51)–(7.57) and (7.60).) and data given for experiment 1 (see Table 7.2) were used.

In the experiments pressure, at the outlet end and the air flux at the time inlet end of the tube were kept constant. These caused that pressure presents a fluctuating behaviour at the inlet out of the tube just as it can be seen in Figure 7.8.

The total amount of water into the tube is due to the original water and the water produced from combustion. The water is displaced along the tube forming banks (Fig. 7.9). These banks travel behind oil banks causing the oil production.

On the other hand, there is only air behind the combustion front as it travels along the tube. Then the gas phase, in that zone, is formed only by air, and the gas saturation takes the value of 1.

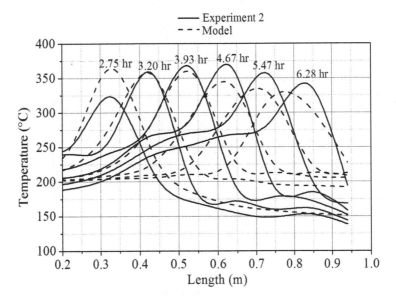

Figure 7.7   Temperature profiles for various times in the case of Experiment 2.

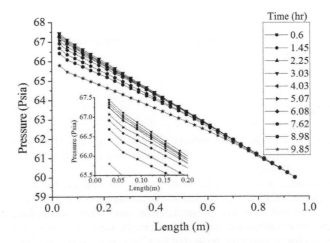

Figure 7.8   Pressure gradient along the tube for various times.

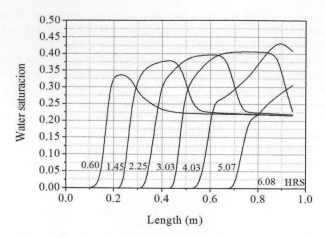

Figure 7.9   Water saturation along the tube for variuos times.

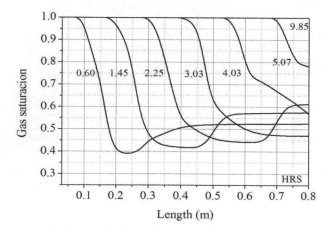

Figure 7.10   Gas saturation along the tube for diverse times (in hours).

When the air touches the coke occur combustion and then the gas saturation falls until values near to 0.38. This can be seen in Figure 7.10.

7.10   CONCLUSIONS

In this chapter a set of averaged mass, momentum and energy equations for a multiphase flow of gas, two immiscible liquids (water and oil) and coke into a homogeneous, isotropic and rigid porous media was developed.

Such set of equations was written for water, coke, oil and gas for an in-situ combustion application. The closure relationships needed for the mass generation of coke and gas phases, and the phase change of oil and water were presented. Equations for the water phase change, the coke formation and the coke combustion, were taken from the work by Gottfried & Mustafa (1978), while the oil phase change equation, presented by Chu (1964), was modified as function of the partial pressure of the oil vapor in the gas phase and the oil vapor pressure.

The simulation of a combustion tube was done using the averaged mass, momentum and energy equations. It was found that: 1) predictions for temperature are in agreement with experimental data when the oil evaporation term was consider, 2) the profiles along the tube of other parameters present an expected behaviour, 3) the effect of the electric igniter must be include into the model in order to improve the predictions.

ACKNOWLEDGMENTS

The authors wish to thank the financial support provided by the Consejo Nacional de Ciencia y Tecnología (CONACyT).

7.A   APPENDIX

7.A.1   *Oil vaporization*

The evaporation term as function of concentrations can be written as Chu (1964):

$$\chi = k_{mo}(k_e C_{lo} - C_{go}) \tag{7.A1}$$

where k_{mo} is the oil mass transfer coefficient (1/hr), k_e is the vaporization equilibrium constant (dimensionless), C_{lo} is the concentration of the vaporizable oil fraction in the liquid oil phase (lb mol/ft^3) and C_{go} is the concentration of the vaporizable oil fraction in the gas phase (lb mol/ft^3) which can be given by:

$$C_{lo} = \frac{p_{lo}}{RT}; \quad C_{go} = \frac{p_{go}}{RT} \tag{7.A2}$$

then,

$$\chi = \frac{k_{mo}}{RT}(k_e p_{lo} - p_{go}) \tag{7.A3}$$

In Eq. (7.A2), p_{lo} is the vapor pressure of oil (psia), p_{go} is the partial pressure of the oil vapor in the gas phase (psia), R is the ideal gas constant (psia-ft^3/°F-lb mol) and T is the system temperature (°F). If one assumed that,

$$\omega = \frac{k_{mo}}{RT} \tag{7.A4}$$

then

$$\chi = \omega (k_e p_{lo} - p_{go}) \tag{7.A5}$$

Multiplying Eq. (7.A5) for the oil molecular weigh M_o, we have:

$$\dot{m}_{go} = \chi M_o = \omega M_o(k_e p_{lo} - p_{go}) \tag{7.A6}$$

where $\dot{m}_{go}$ is the evaporation term as function of pressure. To calculate ω, it is necessary to calculate k_{mo} and k_e, which are given by Chu (1964) as:

$$k_{mo} = 2253 + 6.33T; \quad k_e = e^{6.74-(8930/T_a)} \tag{7.A7}$$

where Ta is temperature (R).

It is necessary to clarify that the parameters $(C_{lo}, C_{go}, k_{mo}, p_{lo}, p_{go}, R, T, T_a)$, in Eqs. (7.A1)–(7.A7), should be used in the units mentioned in the text.

REFERENCES

Andrew, S.A., Dennis, E.R. & Whitaker, S. (2003) New equations for binary gas transport in porous media, Part I: Equations development. *Advances in Water Resources*, 26, 695–715.

Bousquet-Melou, P., Goyeau, B., Quintard, M., Fichot, F. & Gobin, D. (2002) Average momentum equation for interdentritic flow in a solidifying columnar mushy zone. *International Journal of Heat and Mass Transfer*, 45, 3651–3665.

Cazarez-Candia, O., Cruz-Hernández, J., Islas-Juárez, R. & Márquez, R.E. (2010) Theoretical and experimental study of combustion tubes. *Petroleum Science and Technology*, 28, 1186–1196.

Chu, C. (1964) The vaporization-condensation phenomenon in a linear heat wave. *Society of Petroleum Engineers Journal*, 680, 85–95.

Duval, F., Fichot, F. & Quintard, M. (2004) A local thermal non-equilibrium model for two-phase flow with phase change in porous media. *International Journal of Heat and Mass Transfer*, 47, 613–639.

Gray, W.G. (1975) A derivation of the equations for multiphase transport. *Chemical Engineering Science*, 30, 229–233.

Espinosa-Paredes, G. & Cazarez-Candia, O. (2004) Modelo fenomenológico para la determinación de temperaturas en un yacimiento petrolero: desarrollo matemático. *Internal technical report for the Programa de Yacimientos Naturalmente Fracturados of the Instituto Mexicano del Petróleo*, 27–41.

Genrich, J.F. & Pope, G.A. (1985) A simplified performance-predictive model for in-situ combustion processes. *SPE Annual Technical Conference and Exhibition*. 22–25 September. SPE paper number 14242.

Gottfried, B.S. (1965) A mathematical model of thermal oil recovery in linear systems. *SPE Production Research Symposium*, 3–4 May, Tulsa Okla. 196–210. SPE paper number 1117.

Gottfried, B.S. & Mustafa, K. (1978) Computer simulation of thermal oil recovery processes. *Society of Petroleum Engineers Journal*, 7824, 1–42.

Hassanizadeh, S.M. & Gray, W.G. (1990) Mechanics and thermodynamics of multiphase flow in porous media including interface boundaries. *Advances in Water Resources*, 13(4), 169–185.

Lu, C. & Yortsos, Y.C. (2001) A pore-network model of in-situ combustion in porous media. *SPE International Thermal Operations and Heavy Oil Symposium*. 12–14 March, Margarita, Venezuela. SPE paper number 69705.

Nakayama, A., Kuwahara, F., Sugiyama, M. & Xu, G. (2001) A two energy equations model for conduction and convection in porous media. *International Journal of Heat and Mass Transfer*, 44, 4375–4379.

Onyekonwu, M.O., Pande, K., Ramey, H.J. & Brigham, W.E. (1986) Experimental and Simulation Studies of Laboratory In-Situ Combustion Recovery. *SPE 56th California Regional Meeting, 2–4 April, Oakland, CA*. SPE paper number 15090.

Penberthy, W.L. & Ramey, H.J. (1965) Design and operation of laboratory combustion tubes. *SPE Annual Fall Meeting, 3–6 October, Denver, Colo.* SPE paper number 1290.

Soliman, M.Y., Brigham, W.E. & Raghavan, R. (1981) Numerical simulation of thermal recovery processes. *SPE California Regional Meeting, 25–26 March, Bakersfield*. SPE paper number 9942.

Soria, A. & De Lasa, H.I. (1991) Averaged transport equations for multiphase systems with interfacial effects. *Chemical Engineering Science*, 46 (8), 2093–2111.

Verma, V.B., Reynolds, A.C. & Thomas, G.W. (1978) A theoretical investigation of forward combustion in a one-dimensional system. *SPE-AIME 53rd Annual Fall Technical Conference and Exhibition, 1–3 October, Houston*. SPE paper number 7526.

Whitaker, S.A. (1986) Flow in porous media II: The governing equations for immiscible, two-phase flow. *Transport Porous Media*, 1, 105–125.

Whitaker, S.A. (1996) The Forchheimer equation: A theoretical development. *Transport Porous Media*, 25, 27–61.

CHAPTER 8

Biphasic isothermal tricomponent model to simulate advection-diffusion in 2D porous media

A. Moctezuma-Berthier

8.1 INTRODUCTION

Dynamic and static characterization of reservoirs is of great interest in the oil industry. A powerful tool to describe fluid flow in porous media is the numerical modeling of the involved physical processes. Through it, the behavior of the reservoir pressure and the in-situ oil saturation distribution can be determined. A biphasic isothermal tri-component model to analyze advection-diffusion in 2D porous media can be used to simulate oil and gas flow (in presence of static water) together with a third component called tracer. This model can describe flowing conditions in the area of well influence (radial model), 2D inter-well sections, and reservoir sectors (areal view), in which tracer injection studies are conducted. Models considering only the advective term in the fluid balance equation can be solved for pressure and saturation. However, the use of the diffusive term in the equation allows the dynamic characterization of the porous formation as fluids move towards producing wells. Within this subject, the use of tracers (chemical agents) has been seen as a very exciting technology (Zemel 1995). Tracers are transported through the diverse fluid phases in motion and can be detected at their arrival in the producing wells. Tracer breakthrough curves can be used therefore to identify flow paths in the reservoir, which manifest the specific heterogeneity in the test area.

To evaluate the tracer concentration profile in a porous media, a multi-component model that contains both convective and diffusive terms in the tracer balance equation is required. In order to develop a 2D tool with these characteristics, we present here a biphasic tri-component isothermal model that simulates the advection-diffusion phenomena in 2D porous media for the specific case of oil and gas flow in the presence of static water, and a tracer in the gas phase. Here, the mathematical and the numerical models are described, as well as some possible applications of them. The equation solution scheme is shown, with a detailed illustration of how derivatives are generated by the Newton's method. The flow model validation and some preliminary findings are also discussed.

8.2 MODEL DESCRIPTION

In this section the mathematical and numerical models as well as the solution scheme are described. It is to be mentioned that the technical description of this section covers from the beginner up to the specialist level in reservoir dynamics.

8.2.1 *General considerations*

The model has been developed based in the following general considerations.

The fluids flow is isothermal, at the average reservoir temperature. The moving phases are oil and gas. Water in the porous media is considered incompressible and static. Oil and gas phases are composed of a mixture of three components: oil, gas and tracer (as illustrated in Figure 8.1).

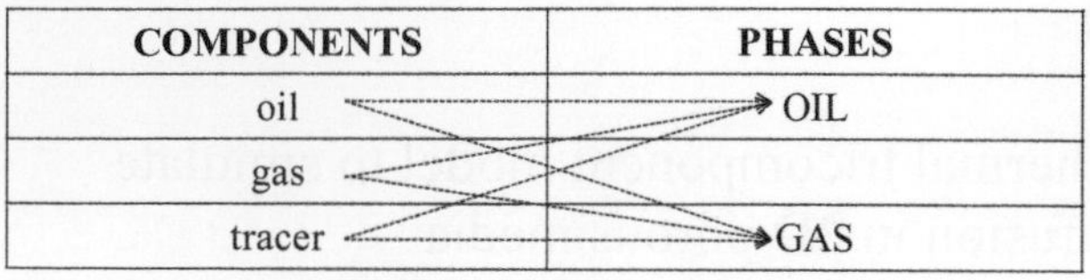

COMPONENTS	PHASES
oil	OIL
gas	
tracer	GAS

Figure 8.1   Scheme of components and phases.

After a pressure change, the thermodynamic equilibrium is reached instantaneously. Diffusion processes are included, additionally to convection.

The composition of the hydrocarbon phase mixture is a function of reservoir pressure and temperature. It comprehends the mole fraction of each component z_1, z_2 and z_3, which correspond to oil, gas and tracer respectively,

$$z_n = \frac{moles_{nM}}{moles_M} \tag{8.1}$$

where $n = 1, 2$ and 3 identifies the component (oil, gas and tracer), and M corresponds to the total mole amount in the mixture. It is therefore established that

$$z_n = \frac{moles_{no} + moles_{ng}}{moles_M} \tag{8.2}$$

The molar fractions of each component will be defined by x and y (for oil and gas phase respectively), as well as L and V for the mole fractions of the oil and gas phase in the mixture respectively. They are given by

$$x_n = \frac{moles_{no}}{moles_o}, \quad y_n = \frac{moles_{ng}}{moles_g}, \quad L = \frac{moles_o}{moles_M} \quad \text{and} \quad V = \frac{moles_g}{moles_M} \tag{8.3}$$

where $moles_p = \sum_{n=1}^{3} moles_{np}$, being $p = o, g$, for the oil and gas phase respectively.

The following relationships hold

$$\sum_{n=1}^{3} x_n = 1, \quad \sum_{n=1}^{3} y_n = 1, \quad L + V = 1 \quad \text{and} \quad \sum_{n=1}^{3} z_n = 1 \tag{8.4}$$

together with

$$z_n = x_n L + y_n V \tag{8.5}$$

The total molar flow in terms of convection and diffusion is represented by

$$\mathbf{N}_{np} = \mathbf{J}_{np} + x_{np}\hat{\rho}_p \mathbf{v}_p; \quad \text{for } p = o, g \quad \text{and} \quad n = 1, 2, 3 \tag{8.6}$$

Here $\mathbf{N}_{np}$ represents the molar flow of component n in the phase p. $\mathbf{J}_{np}$ represents the molar diffusive flux of component n in the phase p. $x_{np}\hat{\rho}_p$ corresponds to the molar concentracion of component n in the phase p, and $\mathbf{v}_p$ the average velocity of phase p.

8.2.2   Mathematical model

For the development of a two-dimensional model, it is possible to treat three different cases, as shown in Figure 8.2:

- Radial flow in a well, for the analysis of the area of drainage (well scale),
- Sectional flow between two wells, for the analysis of a vertical section (inter-well scale), and
- Areal flow in a reservoir, for the analysis of any area or an entire field (field scale).

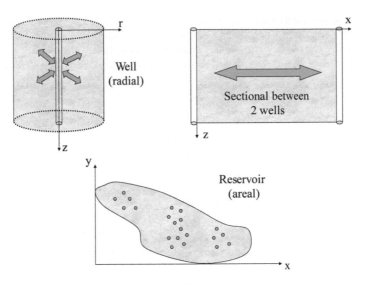

Figure 8.2   Three different 2D cases that can be simulated.

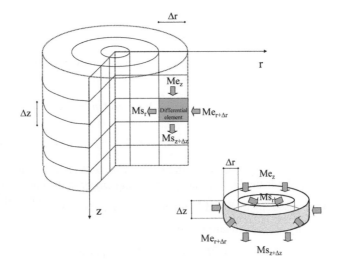

Figure 8.3   Elementary 2D cell in cilindrical coordiantes with axial simmetry.

In Figure 8.3 the two-dimensional radial flow pattern in r and z coordinates is shown. Performing the mass balance analysis in a differential element in coordinates r and z we have:

$$M_e - M_s + M_{PI} = M_{ac} \qquad (8.7)$$

Here M_e and M_s are the input and output mass rates, M_{ac} the cumulative rate in time and M_{PI} the source rate.

By considering areas perpendicular to flow in the volume of the differential element, and taking limit when the differential element approaches to zero, we get the mass balance equation for component n, where $n = 1, 2, 3$ (oil, gas and tracer):

$$-\nabla \cdot \left(x_n \hat{\rho}_o \mathbf{v}_o + \mathbf{J}_{no} + y_n \hat{\rho}_g \mathbf{v}_g + \mathbf{J}_{ng} \right) + \left(x_n \hat{\rho}_o \tilde{q}_o^* + y_n \hat{\rho}_g \tilde{q}_g^* \right)$$
$$= \frac{\partial}{\partial t} \left[\phi \left(x_n \hat{\rho}_o S_o + y_n \hat{\rho}_g S_g \right) \right] \qquad (8.8)$$

This expression is known as the "diffusion equation" representing the molar flow in two-phase (oil-gas) of each component (for oil, gas and tracer) in a porous media.

The divergence operator appearing in the balance equations takes different forms depending of the geometry of the three cases mentioned above. The divergence of a vector field $\mathbf{A}$ in the radial case is given by

$$\nabla \cdot \mathbf{A} = \frac{1}{r}\frac{\partial}{\partial r}(rA_r) + \frac{\partial}{\partial z}A_z \tag{8.9}$$

for the 2D section between the wells by

$$\nabla \cdot \mathbf{A} = \frac{\partial}{\partial x}A_x + \frac{\partial}{\partial z}A_z \tag{8.10}$$

and for the 2D reservoir areal case by

$$\nabla \cdot \mathbf{A} = \frac{\partial}{\partial x}A_x + \frac{\partial}{\partial y}A_y \tag{8.11}$$

in which the velocity vectors and diffusive flux in phase p (for $p = o, g$) are, for the radial flow

$$\mathbf{v}_p = v_{pr}\delta_r + v_{pz}\delta_z \quad \text{and} \quad \mathbf{J}_p = J_{pr}\delta_r + J_{pz}\delta_z \tag{8.12}$$

for the sectional

$$\mathbf{v}_p = v_{px}\delta_x + v_{pz}\delta_z \quad \text{and} \quad \mathbf{J}_p = J_{px}\delta_x + J_{pz}\delta_z \tag{8.13}$$

for the reservoir areal

$$\mathbf{v}_p = v_{px}\delta_x + v_{py}\delta_y \quad \text{and} \quad \mathbf{J}_p = J_{px}\delta_x + J_{py}\delta_y \tag{8.14}$$

For the mechanism of convection, velocity, $\mathbf{v}_p$, is represented by the consideration of laminar flow in an homogeneous porous media, defined by Darcy's law

$$\mathbf{v}_p = -\frac{\mathbf{k}\,k_{rp}}{\mu_p} \cdot (\nabla p_p - \gamma_p \nabla Z) \quad \text{for } p = o, g \tag{8.15}$$

where γ_p, is specific weight, given by $\gamma_p = \rho_p(g/g_c)$; where g and g_c are gravity and the gravitational constant.

μ_p is viscosity and k_{rp} are relative permeability of phase p. Z is the depth with respect to a "plane of reference" and $\mathbf{k}$ is the 2D absolute permeability tensor in the porous medium, which in the X-Y case with anisotropy in X and Y direction respectively, is $\mathbf{k} = \begin{bmatrix} k_x & 0 \\ 0 & k_y \end{bmatrix}$.

The diffusive flow, $\mathbf{J}_p$, is given in terms of the Fick's law

$$\mathbf{J}_{no} = -\hat{\rho}_o \mathbf{D}_{no} \cdot \nabla x_n \quad \text{fo the oil phase} \tag{8.16}$$

$$\mathbf{J}_{ng} = -\hat{\rho}_g \mathbf{D}_{ng} \cdot \nabla y_n \quad \text{for the gas phase} \tag{8.17}$$

where $\mathbf{D}_{np}$ is an effective diffusion coefficient of component n in the phase p.

Substituting Darcy's law for velocity and Fick's law for the diffusive flow then we have

$$\nabla \cdot \left(x_n \hat{\rho}_o \frac{\mathbf{k}\,k_{ro}}{\mu_o} \cdot (\nabla p_o - \gamma_o \nabla Z) + (\hat{\rho}_o \mathbf{D}_{no} \cdot \nabla x_n) \right)$$

$$+ \nabla \cdot \left(y_n \hat{\rho}_g \frac{\mathbf{k}\,k_{rg}}{\mu_g} \cdot (\nabla p_g - \gamma_g \nabla Z) + (\hat{\rho}_g \mathbf{D}_{ng} \cdot \nabla y_n) \right) + \left(x_n \hat{\rho}_o \tilde{q}_o^* + y_n \hat{\rho}_g \tilde{q}_g^* \right) \tag{8.18}$$

$$= \frac{\partial}{\partial t} \left[\phi \left(x_n \hat{\rho}_o S_o + y_n \hat{\rho}_g S_g \right) \right]; \quad \text{for } n = 1, 2, 3$$

Given the condition of thermodynamic equilibrium between oil and gas phase, the fugacity of both phases should be the same

$$f_{no} = f_{ng} \tag{8.19}$$

The fugacity is function of pressure, reservoir temperature (T) and composition

$$f_{np} = f(p_p, T, x_n, y_n) \quad \text{for } p = o, g \quad \text{and} \quad n = 1, 2, 3 \tag{8.20}$$

Additionally, we have the relations of consistency regarding capillary pressure, saturation and molar fraction given by

$$P_{cgo} = p_g - p_o, \quad S_o + S_g = 1, \quad \sum_{n=1}^{3} x_n = 1 \quad \text{and} \quad \sum_{n=1}^{3} y_n = 1 \tag{8.21}$$

In this way, we can verify that Eqs. (8.18), (8.19) and (8.21) form an equation set with: three molar flow balance equations, three thermo dynamical equilibrium equations, one capillary pressure equation, one saturation equation, and two relations involving the sum of molar fractions. This gives a system of 10 equations with the 10 followings unknowns

$$p_o, p_g, S_o, S_g, x_o, y_o, x_g, y_g, x_T, y_T \tag{8.22}$$

Considerations for the tracer component

Because tracer concentrations are very small, compared to the main gas and oil components, it is possible to consider that the oil and gas components determine completely the phase behavior, and therefore the tracer concentration does not influence the phase distribution or the PVT (Zemel 1995). However, the tracer will be partitioning between the oil and gas.

Under this consideration, the flow system will be associated to a bi-component oil-gas in which the thermodynamic equilibrium of the phases will be given by the PVT properties obtained in the laboratory

$$x_o = f(p_o, B_o), \quad x_g = f(p_o, B_o, R_s), \quad y_o = f(p_g, B_g, r_s) \quad \text{and} \quad y_g = f(p_g, B_g) \tag{8.23}$$

where B_p are the volumetric factor of each phase, $p = o, g$; R_s is the solubility of gas in the oil phase; and r_s is the vaporization of oil in the gas phase.

This consideration implies that the saturation and velocity of the phases are not influenced by the tracer component; therefore can be solved them independently. We then first solve the equation system for pressure and saturation, and later the system for tracer concentration. It means, once the distribution of phases and their velocities are determined, the tracer distribution can be solved only as a function of the effects of diffusion and partition coefficient between tracer and phases, defined by K_T.

That is, the balance of components for the oil phase and the gas meet

$$x_o + x_g = 1 \quad \text{and} \quad y_o + y_g = 1 \tag{8.24}$$

Thus, using the definition of the mole fractions, and using the relationship between moles and mass, we have for the oil component,

$$\nabla \cdot \left(\frac{1}{B_o} \lambda_o (\nabla p_o - \gamma_o \nabla Z) + \left(D_{oo} \nabla \frac{1}{B_o} \right) + \frac{r_s}{B_g} \lambda_g (\nabla p_g - \gamma_g \nabla Z) + \left(r_s D_{og} \nabla \frac{1}{B_g} \right) \right)$$
$$+ \left(q_o^* + r_s q_g^* \right) = \frac{\partial}{\partial t} \left[\phi \left(\frac{S_o}{B_o} + \frac{r_s S_g}{B_g} \right) \right] \tag{8.25}$$

for the gas component,

$$\nabla \cdot \left(\frac{R_s}{B_o} \lambda_o (\nabla p_o - \gamma_o \nabla Z) + \left(R_s D_{go} \nabla \frac{1}{B_o} \right) + \frac{1}{B_g} \lambda_g (\nabla p_g - \gamma_g \nabla Z) + \left(D_{gg} \nabla \frac{1}{B_g} \right) \right)$$
$$+ \left(R_s q_o^* + q_g^* \right) = \frac{\partial}{\partial t} \left[\phi \left(\frac{R_s S_o}{B_o} + \frac{S_g}{B_g} \right) \right] \tag{8.26}$$

where q_o^* and q_g^* represent the surface flow rates, (measured or referenced at standard conditions, denoted by @SC) by phase per unit volume of rock. The flow rates can be either production or injection, depending on the case.

The following constraint equations hold

$$S_o + S_g = 1 \quad \text{and} \quad Pc_{go} = p_g - p_o \tag{8.27}$$

The amount of equations and independent variables is in this case four: p_o, p_g, S_o, S_g. If there is no oil component to vaporize into the gas phase, knowing as "Black Oil" type model, this implies that $r_s = 0$, however, this is not valid for the case of simulating gas and condensate crudes.

In these expressions it is also considered that the density gradients, molar solubility and vaporization are small, implying that they are "constant position."

For the component tracer, T, molar flow equation will be only function of the carrying fluid velocity (due to pressure potential) and tracer concentration in each phase, thus, replacing the molar densities of the phases we have

$$\nabla \cdot \left(x_T \frac{\lambda_o}{B_o} (\nabla p_o - \gamma_o \nabla Z) + \left(\frac{D_{To}}{B_o} \nabla x_T \right) \right)$$

$$+ \nabla \cdot \left(y_T \frac{\lambda_g}{B_g} R_{ro} (\nabla p_g - \gamma_g \nabla Z) + \left(\frac{D_{Tg}}{B_g} R_{ro} \nabla y_T \right) \right) \tag{8.28}$$

$$+ (x_T q_o^* + y_T R_{ro} q_g^*) = \frac{\partial}{\partial t} \left[\phi \left(x_T \frac{S_o}{B_o} + y_T R_{ro} \frac{S_g}{B_g} \right) \right]$$

where R_{ro} is a constant relationship between the densities @SC, and K_T partition coefficient of the tracer, given by

$$R_{ro} = \frac{r_s \hat{\rho}_o + \hat{\rho}_g}{\hat{\rho}_o + R_s \hat{\rho}_g} \quad K_T(p_o, T_y, x_T, y_T) = \frac{y_T}{x_T} \tag{8.29}$$

The molar balance of the tracer applies both for the fractions in liquid and vapor phases

$$x_T + y_T = 1 \tag{8.30}$$

In this way, the tracer concentrations are given by solving Eqs. (8.28) and (8.30), which is a system 2 equations and 2 unknowns,

$$x_T, y_T \tag{8.31}$$

It should be noted that since capillary pressure relates the pressure phases, and that both the saturation and the mole fraction of tracer are defined with a balance, we get the following dependence for the different properties for the phases and rock:

$$\mu_o = f(p_o), \ \mu_g = f(p_o), \ B_o = f(p_o), \ B_g = f(p_o), \ \gamma_o = f(p_o), \ \gamma_g = f(p_o)$$

$$r_s = f(p_o), \ R_s = f(p_o), \ S_g = f(S_o), \ K_T = f(x_T), \tag{8.32}$$

$$\phi = f(p_o), \ Pc_{go} = f(S_o), \ k_{ro} = f(S_o), \ k_{rg} = f(S_o)$$

In this way the tri-component phase flow model is a system of 3 equations and 3 unknowns, p_o, S_o, x_T and the equations are, for the oil component,

$$\nabla \cdot \left(\frac{\lambda_o}{B_o} (\nabla p_o - \gamma_o \nabla Z) + \left(D_{oo} \nabla \frac{1}{B_o} \right) \right)$$

$$+ \nabla \cdot \left(\frac{r_s \lambda_g}{B_g} (\nabla (p_o + Pc_{go}) - \gamma_g \nabla Z) + \left(r_s D_{og} \nabla \frac{1}{B_g} \right) \right) \tag{8.33}$$

$$+ (q_o^* + r_s q_g^*) = \frac{\partial}{\partial t} \left[\phi \left(S_o \left(\frac{1}{B_o} - \frac{r_s}{B_g} \right) + \frac{r_s}{B_g} \right) \right]$$

for the gas component,

$$\nabla \cdot \left(\frac{R_s \lambda_o}{B_o} (\nabla p_o - \gamma_o \nabla Z) + \left(R_s D_{go} \nabla \frac{1}{B_o} \right) \right)$$

$$+ \nabla \cdot \left(\frac{\lambda_g}{B_g} (\nabla (p_o + Pc_{go}) - \gamma_g \nabla Z) + \left(D_{gg} \nabla \frac{1}{B_g} \right) \right) \tag{8.34}$$

$$+ \left(R_s q_o^* + q_g^* \right) = \frac{\partial}{\partial t} \left[\phi \left(S_o \left(\frac{R_s}{B_o} - \frac{1}{B_g} \right) + \frac{1}{B_g} \right) \right]$$

for the tracer component,

$$\nabla \cdot \left(x_T \frac{\lambda_o}{B_o} (\nabla p_o - \gamma_o \nabla Z) + \left(\frac{D_{To}}{B_o} \nabla x_T \right) \right)$$

$$+ \nabla \cdot \left(K_T x_T \frac{\lambda_g}{B_g} R_{ro} (\nabla (p_o + Pc_{go}) - \gamma_g \nabla Z) + \left(\frac{D_{Tg}}{B_g} R_{ro} \nabla (K_T x_T) \right) \right) \tag{8.35}$$

$$+ \left(x_T (q_o^* + K_T R_{ro} q_g^*) \right) = \frac{\partial}{\partial t} \left[\phi \left(x_T \left(S_o \left(\frac{1}{B_o} - \frac{K_T R_{ro}}{B_g} \right) + \frac{K_T R_{ro}}{B_g} \right) \right) \right]$$

As was mentioned, knowing pressure and saturations, the tracer distribution is solved only as a function of the effects of velocity, diffusion and partition coefficient K_T.

8.2.3 *Numerical model*

The numerical model will be described for the three components and analyzed considering the two systems that were mentioned in the previous Section 2.2. The numerical system is based on differential operators in the expansion of the terms of the derivatives calculated by finite difference approximations (Hildebrand 1968, Levy 1992).

In this way the equations to be solved expressed in finite difference operators are for the oil component,

$$\Delta \left[R_s T_o (\Delta p_o - \gamma_o \Delta Z) + R_s \Upsilon_{go} \Delta b_o \right]_{i,j}^{n+1}$$

$$+ \Delta \left[T_g (\Delta p_o + \Delta Pc_{go} - \gamma_g \Delta Z) + \Upsilon_{gg} \Delta b_g \right]_{i,j}^{n+1} \tag{8.36}$$

$$+ \left[R_s q_o + q_g \right]_{i,j}^{n+1} = \frac{Vr_{i,j}}{\Delta t} \Delta_t \left[\phi \left(S_o b_o R_s + (1 - S_o) b_g \right) \right]_{i,j}$$

for the gas component,

$$\Delta \left[R_s T_o (\Delta p_o - \gamma_o \Delta Z) + R_s \Upsilon_{go} \Delta b_o \right]_{i,j}^{n+1}$$

$$+ \Delta \left[T_g (\Delta p_o + \Delta Pc_{go} - \gamma_g \Delta Z) + \Upsilon_{gg} \Delta b_g \right]_{i,j}^{n+1} + \left[R_s q_o + q_g \right]_{i,j}^{n+1} \tag{8.37}$$

$$= \frac{Vr_{i,j}}{\Delta t} \Delta_t \left[\phi \left(S_o b_o R_s + (1 - S_o) b_g \right) \right]_{i,j}$$

for the tracer component,

$$\Delta \left[x_T T_o (\Delta p_o - \gamma_o \Delta Z) + \Upsilon_{To} b_o \Delta x_T \right]_{i,j}^{n+1}$$

$$+ \Delta \left[(K_T x_T) T_g R_{ro} (\Delta p_o + \Delta Pc_{go} - \gamma_g \Delta Z) + \Upsilon_{Tg} b_g R_{ro} \Delta (K_T x_T) \right]_{i,j}^{n+1} \tag{8.38}$$

$$+ \left[x_T q_o + x_T K_T R_{ro} q_g \right]_{i,j}^{n+1} = \frac{Vr_{i,j}}{\Delta t} \Delta_t \left[\phi \left(x_T S_o b_o + (1 - S_o) x_T K_T R_{ro} b_g \right) \right]_{i,j}$$

where

Δ corresponds to the central difference operator (associated to the indices i, j for the directions: i for r or x, j for y or z)

Δ_t corresponds to the forward difference operator (associated to the time level, n and $n + 1$).
Δt corresponds to the differential operator in time (associated to the period between the time level n and $n + 1$).
n corresponds to the level in time, and
$Vr_{i,j}$ corresponds to the pore volume of rock in cell i, j, depending on the case.

For radial geometry is given by:

$$Vr_{i,j} = \pi \left(r_{i+\frac{1}{2}}^2 - r_{i-\frac{1}{2}}^2 \right) \left(z_{j+\frac{1}{2}} - z_{j-\frac{1}{2}} \right) \tag{8.39}$$

For the linear geometry is given by

$$Vr_{i,j} = h \left(x_{i+\frac{1}{2}} - x_{i-\frac{1}{2}} \right) \left(y_{j+\frac{1}{2}} - y_{j-\frac{1}{2}} \right) \tag{8.40}$$

for areal reservoir consideration (h is the net porous formation thickness)

$$Vr_{i,j} = w \left(x_{i+\frac{1}{2}} - x_{i-\frac{1}{2}} \right) \left(z_{j+\frac{1}{2}} - z_{j-\frac{1}{2}} \right) \tag{8.41}$$

for sectional consideration (w is the width of the section)
$q_p|_{i,j}$ represents the volumetric flow rate of the phase @SC $p = o, g$ in cell i, j.
b_p represents the inverse of the volume factors, $1/B_p$, for each p phase at level in time n.

The terms T_{np}, represent the advective transmissivity given by the permeability properties and flow area between cells. The terms Υ_{np}, represent the diffusive transmissivity between cells given by the diffusion coefficients and flow areas. Both terms will be defined after expansion of the operators.

The expansion for the oil component of both the finite differences and differential operators gives

$$
\begin{aligned}
&T_o|_{i+\frac{1}{2}}^{n+1} \left[\left(p_o|_{i+1} - p_o|_i \right) - \gamma_o|_{i+\frac{1}{2}}^{n+1} (Z_{i+1} - Z_i) \right] + \Upsilon_{oo}|_{i+\frac{1}{2}}^{n+1} \left[b_o|_{i+1} - b_o|_i \right] \\
&- T_o|_{i-\frac{1}{2}}^{n+1} \left[\left(p_o|_i - p_o|_{i-1} \right) - \gamma_o|_{i-\frac{1}{2}}^{n+1} (Z_i - Z_{i-1}) \right] - \Upsilon_{oo}|_{i-\frac{1}{2}}^{n+1} \left[b_o|_i - b_o|_{i-1} \right] \\
&+ T_o|_{j+\frac{1}{2}}^{n+1} \left[\left(p_o|_{j+1} - p_o|_j \right) - \gamma_o|_{j+\frac{1}{2}}^{n+1} (Z_{j+1} - Z_j) \right] + \Upsilon_{oo}|_{j+\frac{1}{2}}^{n+1} \left[b_o|_{j+1} - b_o|_j \right] \\
&- T_o|_{j-\frac{1}{2}}^{n+1} \left[\left(p_o|_j - p_o|_{j-1} \right) - \gamma_o|_{j-\frac{1}{2}}^{n+1} (Z_j - Z_{j-1}) \right] - \Upsilon_{oo}|_{j-\frac{1}{2}}^{n+1} \left[b_o|_j - b_o|_{j-1} \right] \\
&+ r_s|_{i+\frac{1}{2}}^{n+1} \left[T_g|_{i+\frac{1}{2}}^{n+1} \left[\left(p_o|_{i+1} - p_o|_i \right) + \left(Pc_{go}|_{i+1} - Pc_{go}|_i \right) \right] \right] \\
&- r_s|_{i+\frac{1}{2}}^{n+1} \left[T_g|_{i+\frac{1}{2}}^{n+1} \left[\gamma_g|_{i+\frac{1}{2}}^{n+1} (Z_{i+1} - Z_i) \right] \right] + r_s|_{i+\frac{1}{2}}^{n+1} \left[\Upsilon_{og}|_{i+\frac{1}{2}}^{n+1} \left(b_g|_{i+1} - b_g|_i \right) \right] \\
&- r_s|_{i-\frac{1}{2}}^{n+1} \left[T_g|_{i-\frac{1}{2}}^{n+1} \left[\left(p_o|_i - p_o|_{i-1} \right) + \left(Pc_{go}|_i - Pc_{go}|_{i-1} \right) \right] \right] \\
&+ r_s|_{i-\frac{1}{2}}^{n+1} \left[T_g|_{i-\frac{1}{2}}^{n+1} \left[\gamma_g|_{i-\frac{1}{2}}^{n+1} (Z_i - Z_{i-1}) \right] \right] - r_s|_{i-\frac{1}{2}}^{n+1} \left[\Upsilon_{og}|_{i-\frac{1}{2}}^{n+1} \left(b_g|_i - b_g|_{i-1} \right) \right] \\
&+ r_s|_{j+\frac{1}{2}}^{n+1} \left[T_g|_{j+\frac{1}{2}}^{n+1} \left[\left(p_o|_{j+1} - p_o|_j \right) + \left(Pc_{go}|_{j+1} - Pc_{go}|_j \right) \right] \right] \\
&- r_s|_{j+\frac{1}{2}}^{n+1} \left[T_g|_{j+\frac{1}{2}}^{n+1} \left[\gamma_g|_{j+\frac{1}{2}}^{n+1} (Z_{j+1} - Z_j) \right] \right] + r_s|_{j+\frac{1}{2}}^{n+1} \left[\Upsilon_{og}|_{j+\frac{1}{2}}^{n+1} \left(b_g|_{j+1} - b_g|_j \right) \right] \\
&- r_s|_{j-\frac{1}{2}}^{n+1} \left[T_g|_{j-\frac{1}{2}}^{n+1} \left[\left(p_o|_j - p_o|_{j-1} \right) + \left(Pc_{go}|_j - Pc_{go}|_{j-1} \right) \right] \right] \\
&+ r_s|_{j-\frac{1}{2}}^{n+1} \left[T_g|_{j-\frac{1}{2}}^{n+1} \left[\gamma_g|_{j-\frac{1}{2}}^{n+1} (Z_j - Z_{j-1}) \right] \right] - r_s|_{j-\frac{1}{2}}^{n+1} \left[\Upsilon_{og}|_{j-\frac{1}{2}}^{n+1} \left(b_g|_j - b_g|_{j-1} \right) \right] \\
&+ \left[q_o + r_s q_g \right]_{i,j}^{n+1} \\
&= \frac{Vr_{i,j}}{\Delta t} \left[\phi \left(S_o b_o + (1 - S_o) b_g r_s \right) \right]_{i,j}^{n+1} - \frac{Vr_{i,j}}{\Delta t} \left[\phi \left(S_o b_o + (1 - S_o) b_g r_s \right) \right]_{i,j}^n
\end{aligned}
\tag{8.42}
$$

For the gas component,

$$
\begin{aligned}
&R_s\big|_{i+\frac{1}{2}}^{n+1} T_o\big|_{i+\frac{1}{2}}^{n+1}\left[\left(p_o\big|_{i+1}-p_o\big|_i\right)-\gamma_o\big|_{i+\frac{1}{2}}^{n+1}(Z_{i+1}-Z_i)\right]+R_s\big|_{i+\frac{1}{2}}^{n+1}\Upsilon_{go}\big|_{i+\frac{1}{2}}^{n+1}\left(b_o\big|_{i+1}-b_o\big|_i\right)\\
&-R_s\big|_{i-\frac{1}{2}}^{n+1} T_o\big|_{i-\frac{1}{2}}^{n+1}\left[\left(p_o\big|_i-p_o\big|_{i-1}\right)-\gamma_o\big|_{i-\frac{1}{2}}^{n+1}(Z_i-Z_{i-1})\right]-R_s\big|_{i-\frac{1}{2}}^{n+1}\Upsilon_{go}\big|_{i-\frac{1}{2}}^{n+1}\left(b_o\big|_i-b_o\big|_{i-1}\right)\\
&+R_s\big|_{j+\frac{1}{2}}^{n+1} T_o\big|_{j+\frac{1}{2}}^{n+1}\left[\left(p_o\big|_{j+1}-p_o\big|_j\right)-\gamma_o\big|_{j+\frac{1}{2}}^{n+1}(Z_{j+1}-Z_j)\right]+R_s\big|_{j+\frac{1}{2}}^{n+1}\Upsilon_{go}\big|_{j+\frac{1}{2}}^{n+1}\left(b_o\big|_{j+1}-b_o\big|_j\right)\\
&-R_s\big|_{j-\frac{1}{2}}^{n+1} T_o\big|_{j-\frac{1}{2}}^{n+1}\left[\left(p_o\big|_j-p_o\big|_{j-1}\right)-\gamma_o\big|_{j-\frac{1}{2}}^{n+1}(Z_j-Z_{j-1})\right]-R_s\big|_{j-\frac{1}{2}}^{n+1}\Upsilon_{go}\big|_{j-\frac{1}{2}}^{n+1}\left(b_o\big|_j-b_o\big|_{j-1}\right)\\
&+T_g\big|_{i+\frac{1}{2}}^{n+1}\left[\left(p_o\big|_{i+1}-p_o\big|_i\right)+\left(Pc_{go}\big|_{i+1}-Pc_{go}\big|_i\right)-\gamma_g\big|_{i+\frac{1}{2}}^{n+1}(Z_{i+1}-Z_i)\right]\\
&+\Upsilon_{gg}\big|_{i+\frac{1}{2}}^{n+1}\left(b_g\big|_{i+1}-b_g\big|_i\right)\\
&-T_g\big|_{i-\frac{1}{2}}^{n+1}\left[\left(p_o\big|_i-p_o\big|_{i-1}\right)+\left(Pc_{go}\big|_i-Pc_{go}\big|_{i-1}\right)-\gamma_g\big|_{i-\frac{1}{2}}^{n+1}(Z_i-Z_{i-1})\right]\\
&-\Upsilon_{gg}\big|_{i-\frac{1}{2}}^{n+1}\left(b_g\big|_i-b_g\big|_{i-1}\right)\\
&+T_g\big|_{j+\frac{1}{2}}^{n+1}\left[\left(p_o\big|_{j+1}-p_o\big|_j\right)+\left(Pc_{go}\big|_{j+1}-Pc_{go}\big|_j\right)-\gamma_g\big|_{j+\frac{1}{2}}^{n+1}(Z_{j+1}-Z_j)\right]\\
&+\Upsilon_{gg}\big|_{j+\frac{1}{2}}^{n+1}\left(b_g\big|_{j+1}-b_g\big|_j\right)\\
&-T_g\big|_{j-\frac{1}{2}}^{n+1}\left[\left(p_o\big|_j-p_o\big|_{j-1}\right)+\left(Pc_{go}\big|_j-Pc_{go}\big|_{j-1}\right)-\gamma_g\big|_{j-\frac{1}{2}}^{n+1}(Z_j-Z_{j-1})\right]\\
&-\Upsilon_{gg}\big|_{j-\frac{1}{2}}^{n+1}\left(b_g\big|_j-b_g\big|_{j-1}\right)+[R_sq_o+q_g]_{i,j}^{n+1}\\
&=\frac{Vr_{i,j}}{\Delta t}\left[\left[\phi\left(S_ob_oR_s+(1-S_o)b_g\right)\right]_{i,j}^{n+1}-\left[\phi\left(S_ob_oR_s+(1-S_o)b_g\right)\right]_{i,j}^n\right]
\end{aligned}
\tag{8.43}
$$

For the tracer component,

$$
\begin{aligned}
&[x_TT_o]_{i+\frac{1}{2}}^{n+1}\left[\left(p_o\big|_{i+1}-p_o\big|_i\right)-\gamma_o\big|_{i+\frac{1}{2}}^{n+1}(Z_{i+1}-Z_i)\right]+[\Upsilon_{To}b_o]_{i+\frac{1}{2}}^{n+1}\left(x_T\big|_{i+1}-x_T\big|_i\right)\\[4pt]
&-[x_TT_o]_{i-\frac{1}{2}}^{n+1}\left[\left(p_o\big|_i-p_o\big|_{i-1}\right)-\gamma_o\big|_{i-\frac{1}{2}}^{n+1}(Z_i-Z_{i-1})\right]-[\Upsilon_{To}b_o]_{i-\frac{1}{2}}^{n+1}\left(x_T\big|_i-x_T\big|_{i-1}\right)\\[4pt]
&+[x_TT_o]_{j+\frac{1}{2}}^{n+1}\left[\left(p_o\big|_{j+1}-p_o\big|_j\right)-\gamma_o\big|_{j+\frac{1}{2}}^{n+1}(Z_{j+1}-Z_j)\right]+[\Upsilon_{To}b_o]_{j+\frac{1}{2}}^{n+1}\left(x_T\big|_{j+1}-x_T\big|_j\right)\\[4pt]
&-[x_TT_o]_{j-\frac{1}{2}}^{n+1}\left[\left(p_o\big|_j-p_o\big|_{j-1}\right)-\gamma_o\big|_{j-\frac{1}{2}}^{n+1}(Z_j-Z_{j-1})\right]-[\Upsilon_{To}b_o]_{j-\frac{1}{2}}^{n+1}\left(x_T\big|_j-x_T\big|_{j-1}\right)\\[4pt]
&+R_{ro}\left[K_Tx_TT_g\right]_{i+\frac{1}{2}}^{n+1}\left[\left(p_o\big|_{i+1}-p_o\big|_i\right)+\left(Pc_{go}\big|_{i+1}-Pc_{go}\big|_i\right)-\gamma_g\big|_{i+\frac{1}{2}}^{n+1}(Z_{i+1}-Z_i)\right]\\[4pt]
&+R_{ro}\left[\Upsilon_{Tg}b_g\right]_{i+\frac{1}{2}}^{n+1}\left[(K_Tx_T)_{i+1}-(K_Tx_T)_i\right]\\[4pt]
&-R_{ro}\left[K_Tx_TT_g\right]_{i-\frac{1}{2}}^{n+1}\left[\left(p_o\big|_i-p_o\big|_{i-1}\right)+\left(Pc_{go}\big|_i-Pc_{go}\big|_{i-1}\right)-\gamma_g\big|_{i-\frac{1}{2}}^{n+1}(Z_i-Z_{i-1})\right]\\[4pt]
&-R_{ro}\left[\Upsilon_{Tg}b_g\right]_{i-\frac{1}{2}}^{n+1}\left[(K_Tx_T)_i-(K_Tx_T)_{i-1}\right]\\[4pt]
&+R_{ro}\left[K_Tx_TT_g\right]_{j+\frac{1}{2}}^{n+1}\left[\left(p_o\big|_{j+1}-p_o\big|_j\right)+\left(Pc_{go}\big|_{j+1}-Pc_{go}\big|_j\right)-\gamma_g\big|_{j+\frac{1}{2}}^{n+1}(Z_{j+1}-Z_j)\right]\\[4pt]
&+R_{ro}\left[\Upsilon_{Tg}b_g\right]_{j+\frac{1}{2}}^{n+1}\left[(K_Tx_T)_{j+1}-(K_Tx_T)_j\right]\\[4pt]
&-R_{ro}\left[K_Tx_TT_g\right]_{j-\frac{1}{2}}^{n+1}\left[\left(p_o\big|_j-p_o\big|_{j-1}\right)+\left(Pc_{go}\big|_j-Pc_{go}\big|_{j-1}\right)-\gamma_g\big|_{j-\frac{1}{2}}^{n+1}(Z_j-Z_{j-1})\right]
\end{aligned}
\tag{8.44}
$$

$$- R_{ro} \left[\Upsilon_{Tg} b_g\right]_{j-\frac{1}{2}}^{n+1} \left[(K_T x_T)_j - (K_T x_T)_{j-1}\right] + \left[x_T q_o + x_T K_T R_{ro} q_g\right]_{i,j}^{n+1}$$

$$= \frac{Vr_{i,j}}{\Delta t} \left[\phi \left(x_T S_o b_o + (1 - S_o) x_T K_T R_{ro} b_g\right)\right]_{i,j}^{n+1}$$

$$- \frac{Vr_{i,j}}{\Delta t} \left[\phi \left(x_T S_o b_o + (1 - S_o) x_T K_T R_{ro} b_g\right)\right]_{i,j}^{n}$$

which are the finite difference expressions of the balance equations of the components involved in the flow.

The advective transmissivity terms are defined by the following considerations $T_p\big|_{i\pm\frac{1}{2}} = (A\lambda_{px}/B_p\Delta d_x)_{i\pm\frac{1}{2}}$ for the horizontal direction, $T_p\big|_{j\pm\frac{1}{2}} = (A\lambda_{py}/B_p\Delta d_y)_{j\pm\frac{1}{2}}$ for the vertical direction for phase p, oil and gas.

Diffusive transmissivity terms will be given by $\Upsilon_{np}\big|_{i\pm\frac{1}{2}} = \left(AD_{np}/\Delta d_x\right)_{i\pm\frac{1}{2}}$ for component n (oil, gas and tracer) in the p phase (oil and gas) and the horizontal direction; $\Upsilon_{np}\big|_{j\pm\frac{1}{2}} = \left(AD_{np}/\Delta d_y\right)_{j\pm\frac{1}{2}}$ for component n (oil, gas and tracer) in the p phase (oil and gas) and the vertical direction.

In these expressions $A\big|_{i\pm\frac{1}{2}}$ represents the area perpendicular to flow direction r or x and $A\big|_{j\pm\frac{1}{2}}$ the area perpendicular to the direction or flow, which will be given by

$$A_{i\pm\frac{1}{2}} = 2\pi r_{i\pm\frac{1}{2}} \cdot \Delta z_j; \; A_{j\pm\frac{1}{2}} = 2\pi \left(r_{i+\frac{1}{2}}^2 - r_{i-\frac{1}{2}}^2\right), \text{ if the case is for radial flow.}$$

$$A_{i\pm\frac{1}{2}} = w \cdot \Delta z_j; \; A_{j\pm\frac{1}{2}} = w \cdot \Delta x_i, \text{ if the case is for sectional areal between 2 wells.}$$

$$A_{i\pm\frac{1}{2}} = h \cdot \Delta y_j; \; A_{j\pm\frac{1}{2}} = h \cdot \Delta x_i, \text{ if the case is for a reservoir.}$$

The term Δd defines the cell size in the flow direction d, and will be for one direction $\Delta d_{xi\pm\frac{1}{2}} = \left(d_{xi+\frac{1}{2}} - d_{xi-\frac{1}{2}}\right)$, where dx corresponds to the r or x direction, and depends on the case, r for radial flow, x for linear flow.

For the other direction it holds $\Delta d_{yj\pm\frac{1}{2}} = \left(d_{yj+\frac{1}{2}} - d_{yj-\frac{1}{2}}\right)$ where dy corresponds to the y or z direction, and depends on the case, z for radial flow, y for linear flow.

The determination of the mobility terms in the advective transmissivity will be given by

$$\frac{\lambda_p}{B_p}\bigg|_{i\pm\frac{1}{2}} = \left(\frac{k\,k_{rp}}{B_p\mu_p}\right)_{i\pm\frac{1}{2}} \tag{8.45}$$

Their definition for the i, j-position is defined by the "upstream" criterium where the properties that depend on pressure, and saturation are evaluated on the node with the greatest potential:

for the term in $i + \frac{1}{2}$ we have

$$\frac{k_{rp}}{B_p\mu_p}\bigg|_{i+\frac{1}{2}} = \delta_p\left(\frac{k_{rp}}{B_p\mu_p}\right)\bigg|_i + (1 - \delta_p)\left(\frac{k_{rp}}{B_p\mu_p}\right)\bigg|_{i+1} \tag{8.46}$$

where δ_p is a weight factor applying the "upstream" criterium in the calculation of the term for each phase p is given by the following condition

$$\text{if } \left(\Delta p_p - \gamma_p \Delta Z\right)\big|_{i+\frac{1}{2}} \geq 0 \quad \text{then } \delta_p = 0; \quad \text{otherwise } \delta_p = 1, \quad \text{for } p = o, g \tag{8.47}$$

for terms in $i - \frac{1}{2}$

$$\frac{k_{rp}}{B_p\mu_p}\bigg|_{i-\frac{1}{2}} = \delta_p\left(\frac{k_{rp}}{B_p\mu_p}\right)\bigg|_i + (1 - \delta_p)\left(\frac{k_{rp}}{B_p\mu_p}\right)\bigg|_{i-1} \tag{8.48}$$

and the weighting factor is given by the following

$$\text{if } \left(\Delta p_p - \gamma_p \Delta Z\right)\big|_{i-\frac{1}{2}} \leq 0 \quad \text{then } \delta_p = 0; \quad \text{otherwise } \delta_p = 1, \quad \text{for } p = o, g \tag{8.49}$$

Similar expressions apply for the other direction by replacing the index i by j.
The absolute permeability values k depends on direction, and is calculated by

$$k_{i+\frac{1}{2}} = \left(\frac{k_i k_{i+1} \ln\left(\frac{r_{i+1}}{r_i}\right)}{k_{i+1} \ln\left(\frac{r_{i+1/2}}{r_i}\right) + k_i \ln\left(\frac{r_{i+1}}{r_{i+1/2}}\right)} \right) \quad \text{for radial flow;}$$

(8.50)

$$k_{i+\frac{1}{2}} = \left(\frac{k_i k_{i+1} \Delta x_{i+1/2}}{k_{i+1}\Delta x_i + k_i \Delta x_{i+1}} \right) \quad \text{for linear flow}$$

For the z or y directions we have

$$k|_{j+\frac{1}{2}} = \left(\frac{k_j k_{j+1} \Delta d_{yj+\frac{1}{2}}}{k_{j+1}\Delta d_{yj} + k_j \Delta d_{yj+1}} \right)$$

(8.51)

The properties that depend only on the pressure, γ_p, R_s, r_s, and K_T, are calculated with the interpolated value between cells considering the cell size, in that way
As a reference on x and y directions, we have,

$$PROP_{i+\frac{1}{2}} = 1 \left/ \left(1 + \left(\frac{\Delta d_x|_{i+1}}{\Delta d_x|_i}\right)\right) PROP_{i+1} + \left(1 - 1 \left/ \left(1 + \left(\frac{\Delta d_x|_{i+1}}{\Delta d_x|_i}\right)\right)\right)\right) PROP_i \right.$$

(8.52)

$$PROP_{j+\frac{1}{2}} = 1 \left/ \left(1 + \left(\frac{\Delta d_y|_{j+1}}{\Delta d_y|_j}\right)\right) PROP_{j+1} + \left(1 - 1 \left/ \left(1 + \left(\frac{\Delta d_y|_{j+1}}{\Delta d_y|_j}\right)\right)\right)\right) PROP_j \right.$$

(8.53)

where $PROP$ is the property that is calculated, γ_p, R_s, r_s, or K_T.
For the determination of the terms of molecular diffusivity in x direction will be given by

$$\Upsilon_{np}|_{i\pm\frac{1}{2}} = \left(\frac{AD_{np}}{\Delta d_x}\right)_{i\pm\frac{1}{2}}$$

(8.54)

As for mobility terms they are defined by the "upstream" criteria, and applied to the diffusion coefficient that depends on the pressure, so for the term in $i + \frac{1}{2}$ we have

$$D_{np}|_{i+\frac{1}{2}} = \delta_p \left(D_{np}\right)|_i + \left(1 - \delta_p\right)\left(D_{np}\right)|_{i+1}$$

(8.55)

where δ_p is a weighting factor applying the "upstream" criterium in the calculation of the term for each phase p is given by the following condition

$$\text{if } \Delta x_T|_{i+\frac{1}{2}} \geq 0 \quad \delta_p = 0, \quad \text{otherwise } \delta_p = 1, \quad \text{for } p = o, g$$

(8.56)

for the terms in $i - \frac{1}{2}$

$$D_{np}|_{i-\frac{1}{2}} = \delta_p (D_{np})|_i + (1 - \delta_p)(D_{np})|_{i-1}$$

(8.57)

and the weighting factor is given by the following

$$\text{if } \Delta x_T|_{i-\frac{1}{2}} \leq 0 \quad \delta_p = 0, \quad \text{otherwise } \delta_p = 1, \quad \text{for } p = o, g$$

(8.58)

Similar expressions apply for the other direction by replacing the index i by j.
The concentration x_T in, $i \pm \frac{1}{2}, j \pm \frac{1}{2}$ will be calculated with the interpolated value between cells considering their size

$$x_T|_{i+\frac{1}{2}} = 1 \left/ \left(1 + \left(\frac{\Delta d_x|_{i+1}}{\Delta d_x|_i}\right)\right) x_T|_{i+1} + \left(1 - 1 \left/ \left(1 + \left(\frac{\Delta d_x|_{i+1}}{\Delta d_x|_i}\right)\right)\right)\right) x_T|_i \right.$$

(8.59)

$$x_T|_{j+\frac{1}{2}} = 1 \left/ \left(1 + \left(\frac{\Delta d_y|_{j+1}}{\Delta d_y|_j}\right)\right) x_T|_{j+1} + \left(1 - 1 \left/ \left(1 + \left(\frac{\Delta d_y|_{j+1}}{\Delta d_y|_j}\right)\right)\right)\right) x_T|_j \right.$$

(8.60)

8.2.4 *Solution of the system*

To obtain the solution of the equation system we use the Newton-Raphson method.

Thus, the system to be solved expressed in terms of residual form functions, F, is given for the oil component by

$$
F_o|_{i,j}^{n+1} \left(p_o|_{i-1,j}, S_o|_{i-1,j}, p_o|_{i,j}, S_o|_{i,j}, p_o|_{i+1,j}, S_o|_{i+1,j}, p_o|_{i,j-1}, S_o|_{i,j-1}, p_o|_{i,j+1}, S_o|_{i,j+1}\right)
$$

$$
= \Delta\left[T_o(\Delta p_o - \gamma_o \Delta Z) + \Upsilon_{oo}\Delta b_o\right]_{i,j}^{n+1} + \Delta\left[r_s T_g(\Delta p_o + \Delta Pc_{go} - \gamma_g \Delta Z) + r_s \Upsilon_{og}\Delta b_g\right]_{i,j}^{n+1}
$$

$$
+ \left[q_o + r_s q_g\right]_{i,j}^{n+1} - \frac{Vr_{i,j}}{\Delta t}\Delta_t\left[\phi\left(S_o b_o + (1 - S_o)b_g r_s\right)\right]_{i,j} \tag{8.61}
$$

$$
= 0
$$

for the gas component,

$$
F_g|_{i,j}^{n+1} \left(p_o|_{i-1,j}, S_o|_{i-1,j}, p_o|_{i,j}, S_o|_{i,j}, p_o|_{i+1,j}, S_o|_{i+1,j}, p_o|_{i,j-1}, S_o|_{i,j-1}, p_o|_{i,j+1}, S_o|_{i,j+1}\right)
$$

$$
= \Delta\left[R_s T_o(\Delta p_o - \gamma_o \Delta Z) + R_s \Upsilon_{go}\Delta b_o\right]_{i,j}^{n+1} + \Delta\left[T_g(\Delta p_o + \Delta Pc_{go} - \gamma_g \Delta Z) + \Upsilon_{gg}\Delta b_g\right]_{i,j}^{n+1}
$$

$$
+ \left[R_s q_o + q_g\right]_{i,j}^{n+1} - \frac{Vr_{i,j}}{\Delta t}\Delta_t\left[\phi\left(S_o b_o R_s + (1 - S_o)b_g\right)\right]_{i,j} \tag{8.62}
$$

$$
= 0
$$

for the tracer component,

$$
F_T|_{i,j}^{n+1} \left(p_o|_{i-1,j}, S_o|_{i-1,j}, x_T|_{i-1,j}, p_o|_{i,j}, S_o|_{i,j}, x_T|_{i,j}, p_o|_{i+1,j},\right.
$$

$$
\left. S_o|_{i+1,j}, x_T|_{i+1,j}, p_o|_{i,j-1}, S_o|_{i,j-1}, x_T|_{i,j-1}, p_o|_{i,j+1}, S_o|_{i,j+1}, x_T|_{i,j+1}\right)
$$

$$
= \Delta\left[x_T T_o(\Delta p_o - \gamma_o \Delta Z) + \Upsilon_{To}b_o\Delta x_T\right]_{i,j}^{n+1} \tag{8.63}
$$

$$
+ \Delta\left[(K_T x_T)T_g R_{ro}(\Delta p_o + \Delta Pc_{go} - \gamma_g \Delta Z) + \Upsilon_{Tg}b_g R_{ro}\Delta(K_T x_T)\right]_{i,j}^{n+1}
$$

$$
+ \left[x_T(q_o + K_T R_{ro}q_g)\right]_{i,j}^{n+1} - \frac{Vr_{i,j}}{\Delta t}\Delta_t\left[\phi\left(x_T S_o b_o + (1 - S_o)x_T K_T R_{ro}b_g\right)\right]_{i,j}
$$

$$
= 0
$$

for $i = 1, 2, 3, \ldots, I$, $j = 1, 2, 3, \ldots, J$ and $n = 0, 1, 2, 3, \ldots$

The Newton-Raphson method solves the unknowns $p_o|_{i,j}$, $S_o|_{i,j}$, $x_T|_{i,j}$, in an iterative procedure (Deuflhard 2004). The iterative process is based on the expansion of functions, $F|_{i,j}^{n+1}$, of each component, in Taylor series around the iteration level, v. The expression for the oil component is,

$$
F_o|_{i,j}^{(v+1)} = F_o|_{i,j}^{(v)} + \left(\frac{\partial F_{oij}}{\partial p_{oi-1}}\right)^{(v)} \delta p_o|_{i-1}^{(v+1)} + \left(\frac{\partial F_{oij}}{\partial S_{oi-1}}\right)^{(v)} \delta S_o|_{i-1}^{(v+1)} + \left(\frac{\partial F_{oij}}{\partial p_{oi}}\right)^{(v)} \delta p_o|_i^{(v+1)}
$$

$$
+ \left(\frac{\partial F_{oij}}{\partial S_{oi}}\right)^{(v)} \delta S_o|_i^{(v+1)} + \left(\frac{\partial F_{oij}}{\partial p_{oi+1}}\right)^{(v)} \delta p_o|_{i+1}^{(v+1)} + \left(\frac{\partial F_{oij}}{\partial S_{oi+1}}\right)^{(v)} \delta S_o|_{i+1}^{(v+1)}
$$

$$
+ \left(\frac{\partial F_{oij}}{\partial p_{oj-1}}\right)^{(v)} \delta p_o|_{j-1}^{(v+1)} + \left(\frac{\partial F_{oij}}{\partial S_{oj-1}}\right)^{(v)} \delta S_o|_{j-1}^{(v+1)} + \left(\frac{\partial F_{oij}}{\partial p_{oj+1}}\right)^{(v)} \delta p_o|_{j+1}^{(v+1)} \tag{8.64}
$$

$$
+ \left(\frac{\partial F_{oij}}{\partial S_{oj+1}}\right)^{(v)} \delta S_o|_{j+1}^{(v+1)}
$$

$$
= 0
$$

for the gas component we get similar expressions replacing F_o by F_g.

For the tracer component we have,

$$
\begin{aligned}
F_T|_{i,j}^{(v+1)} &= F_T|_{i,j}^{(v)} + \left(\frac{\partial F_{Tij}}{\partial p_{oi-1}}\right)^{(v)} \delta p_o|_{i-1}^{(v+1)} + \left(\frac{\partial F_{Tij}}{\partial S_{oi-1}}\right)^{(v)} \delta S_o|_{i-1}^{(v+1)} + \left(\frac{\partial F_{Tij}}{\partial x_{Ti-1}}\right)^{(v)} \delta x_T|_{i-1}^{(v+1)} \\
&+ \left(\frac{\partial F_{Tij}}{\partial p_{oi}}\right)^{(v)} \delta p_o|_{i}^{(v+1)} + \left(\frac{\partial F_{Tij}}{\partial S_{oi}}\right)^{(v)} \delta S_o|_{i}^{(v+1)} + \left(\frac{\partial F_{Tij}}{\partial x_{Ti}}\right)^{(v)} \delta x_T|_{i}^{(v+1)} \\
&+ \left(\frac{\partial F_{Tij}}{\partial p_{oi+1}}\right)^{(v)} \delta p_o|_{i+1}^{(v+1)} + \left(\frac{\partial F_{Tij}}{\partial S_{oi+1}}\right)^{(v)} \delta S_o|_{i+1}^{(v+1)} + \left(\frac{\partial F_{Tij}}{\partial x_{Ti+1}}\right)^{(v)} \delta x_T|_{i+1}^{(v+1)} \\
&+ \left(\frac{\partial F_{Tij}}{\partial p_{oj-1}}\right)^{(v)} \delta p_o|_{j-1}^{(v+1)} + \left(\frac{\partial F_{Tij}}{\partial S_{oj-1}}\right)^{(v)} \delta S_o|_{j-1}^{(v+1)} + \left(\frac{\partial F_{Tij}}{\partial x_{Tj-1}}\right)^{(v)} \delta x_T|_{j-1}^{(v+1)} \\
&+ \left(\frac{\partial F_{Tij}}{\partial p_{oj+1}}\right)^{(v)} \delta p_o|_{j+1}^{(v+1)} + \left(\frac{\partial F_{Tij}}{\partial S_{oj+1}}\right)^{(v)} \delta S_o|_{j+1}^{(v+1)} + \left(\frac{\partial F_{Tij}}{\partial x_{Tj+1}}\right)^{(v)} \delta x_T|_{j+1}^{(v+1)} \\
&= 0
\end{aligned}
\tag{8.65}
$$

where

$$
\delta p_o|_{i,j}^{(v+1)} = p_o|_{i,j}^{(v+1)} - p_o|_{i,j}^{(v)}; \quad \delta S_o|_{i,j}^{(v+1)} = S_o|_{i,j}^{(v+1)} - S_o|_{i,j}^{(v)}; \quad \delta x_T|_{i,j}^{(v+1)} = x_T|_{i,j}^{(v+1)} - x_T|_{i,j}^{(v)}
\tag{8.66}
$$

As we mentioned, these expressions represent the iterative changes for the unknowns, in the level of iteration v.

The iterative process starts with

$$
p_o|_{i,j}^{(0)} = p_o|_{i,j}^{(n)}; \quad S_o|_{i,j}^{(0)} = S_o|_{i,j}^{(n)}; \quad x_T|_{i,j}^{(0)} = x_T|_{i,j}^{(n)}
\tag{8.67}
$$

Once the iterative changes are solved $\delta p_o|_{i,j}^{(v+1)}, \delta S_o|_{i,j}^{(v+1)}, \delta x_T|_{i,j}^{(v+1)}$ the unknowns are re-evaluated at iteration $(v+1)$ by

$$
p_o|_{i,j}^{(v+1)} = p_o|_{i,j}^{(v)} + \delta p_o|_{i,j}^{(v+1)}; \quad S_o|_{i,j}^{(v+1)} = S_o|_{i,j}^{(v)} + \delta S_o|_{i,j}^{(v+1)}; \quad x_T|_{i,j}^{(v+1)} = x_T|_{i,j}^{(v)} + \delta x_T|_{i,j}^{(v+1)}
\tag{8.68}
$$

The iterative process ends (converges) after the iterative changes are below a preset tolerance for the variables

$$
\left| \delta p_o|_{i,j}^{(v+1)} \right| \le \varepsilon_p; \quad \left| \delta S_o|_{i,j}^{(v+1)} \right| \le \varepsilon_S; \quad \left| \delta x_T|_{i,j}^{(v+1)} \right| \le \varepsilon_x
\tag{8.69}
$$

Expressing the system in matrix form the equation to be solved is

$$
\mathbf{M}^{(v)} \mathbf{d}^{(v+1)} = -\mathbf{F}^{(v)}
\tag{8.70}
$$

where $\mathbf{M}^{(v)}$ is the matrix, or Jacobian, evaluated at iteration level v, $\mathbf{d}^{(v+1)}$ is the vector of unknowns of iterative changes (level iteration $v + 1$), and $\mathbf{F}^{(v)}$ the vector of the residual functions at iteration level v.

To analyze the matrix structure and vector terms, consider coefficients a, b, c, e, f, associated to the 2D space position to keep the cells in reference to a unit element (unit element of the finite difference expansion is the central node i, j) given by the diagram shown in Figure 8.4.

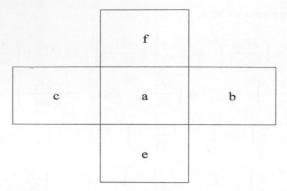

Figure 8.4 The scheme of the unit cell with node identification. a analyzed cell (i, j) which is in the center of the unit element, b is the cell in position $(i + 1, j)$, c the cell $(i - 1, j)$, e the cell $(i, j + 1)$ and f the cell $(i, j - 1)$.

Thus, the matrix structure $\mathbf{M}^{(\nu)}$ for a system of I, J cells is given by

$$
\begin{bmatrix}
a_{1,1} & b_{1,1} & \dots & & e_{1,1} & & & & & \\
c_{2,1} & a_{2,1} & b_{2,1} & \dots & & e_{2,1} & & & & \\
\dots & c_{3,1} & a_{3,1} & b_{3,1} & \dots & & e_{3,1} & & & \\
\dots & \dots & c_{4,1} & a_{4,1} & \dots & & \dots & e_{4,1} & & \\
& \dots & \dots & \dots & \dots & \dots & \dots & \dots & & \\
& & \dots & \dots & \dots & \dots & \dots & \dots & \dots & \\
& & f_{I-3,J} & \dots & & c_{I-3,J} & a_{I-3,J} & \dots & \dots & \dots \\
& & & f_{I-2,J} & \dots & & c_{I-2,J} & a_{I-2,J} & b_{I-2,J} & \dots \\
& & & & f_{I-1,J} & \dots & & c_{I-1,J} & a_{I-1,J} & b_{I-1,J} \\
& & & & & f_{I,J} & \dots & & c_{I,J} & a_{I,J}
\end{bmatrix}
\tag{8.71}
$$

The structure of bands in the system is called pentadiagonal and the order of the matrix is $IJ \times IJ$. Each element of the matrix is a 3×3 sub-matrix and, in the case of the elements $a_{i,j}$, are defined by

$$
a_{i,j} =
\begin{bmatrix}
\dfrac{\partial F_{oij}}{\partial p_{oij}} & \dfrac{\partial F_{oij}}{\partial S_{oij}} & \dfrac{\partial F_{oij}}{\partial x_{Tij}} \\[2ex]
\dfrac{\partial F_{gij}}{\partial p_{oij}} & \dfrac{\partial F_{gij}}{\partial S_{oij}} & \dfrac{\partial F_{gij}}{\partial x_{Tij}} \\[2ex]
\dfrac{\partial F_{Tij}}{\partial p_{oij}} & \dfrac{\partial F_{Tij}}{\partial S_{oij}} & \dfrac{\partial F_{Tij}}{\partial x_{Tij}}
\end{bmatrix}
\tag{8.72}
$$

for the other elements $b_{i,j}, c_{i,j}, e_{i,j}, f_{i,j}$ we have similar expressions.

Analyzing the consideration that the tracer does not affect the flow properties and phase behavior, the partial derivation of the residual functions are independent of molar fraction of the tracer, the sub-matrix elements are then

$$
a_{i,j} =
\begin{bmatrix}
\dfrac{\partial F_{oij}}{\partial p_{oij}} & \dfrac{\partial F_{oij}}{\partial S_{oij}} & 0 \\[2ex]
\dfrac{\partial F_{gij}}{\partial p_{oij}} & \dfrac{\partial F_{gij}}{\partial S_{oij}} & 0 \\[2ex]
0 & 0 & \dfrac{\partial F_{Tij}}{\partial x_{Tij}}
\end{bmatrix}
\tag{8.73}
$$

for the other elements $b_{i,j}, c_{i,j}, e_{i,j}, f_{i,j}$ we have similar expressions.

The tracer system can be "decoupled" from the oil-gas phase system, and can be solved given the oil pressure and oil saturation. For the tracer concentration, pressure and saturation are already solved in terms of iteration, so its derivatives regarding pressure and saturation are zero.

By matrix algebraic manipulation, the vector of unknowns is given by

$$\mathbf{d} = \begin{bmatrix} d_{1,1} \\ d_{2,1} \\ d_{3,1} \\ \dots \\ d_{I-2,J} \\ d_{I-1,J} \\ d_{I,J} \end{bmatrix}_{IJ \times 1} \quad \text{where } d_{i,j} = \begin{bmatrix} \delta p_{oi,j} \\ \delta S_{oi,j} \\ \delta x_{Ti,j} \end{bmatrix} \tag{8.74}$$

which is an IJ column vector elements. And the vector of residual functions of

$$\mathbf{F} = \begin{bmatrix} F_{1,1} \\ F_{2,1} \\ F_{3,1} \\ \dots \\ F_{I-2,J} \\ F_{I-1,J} \\ F_{I,J} \end{bmatrix}_{IJ \times 1} \quad \text{where } F_{i,j} = \begin{bmatrix} F_{oi,j} \\ F_{gi,j} \\ F_{Ti,j} \end{bmatrix} \tag{8.75}$$

which is an IJ column vector elements in which each element is a column vector of 3 elements, so restating the matrix equation for the decoupling of the tracer component is

$$\mathbf{M} = \begin{bmatrix} [\mathbf{M}_{bf}] & [0] \\ [0] & [\mathbf{M}_T] \end{bmatrix}_{2IJ \times IJ} ; \quad \mathbf{d} = \begin{bmatrix} [\mathbf{d}_{bf}] \\ [\mathbf{d}_T] \end{bmatrix}_{2IJ \times 1} ; \quad \mathbf{F} = \begin{bmatrix} [\mathbf{F}_{bf}] \\ [\mathbf{F}_T] \end{bmatrix}_{2IJ \times 1} \tag{8.76}$$

where, for the matrix $\mathbf{M}_{bf}$ of the system of oil and gas components is given by

$$\begin{bmatrix} a^b_{1,1} & b^b_{1,1} & \dots & & e^b_{1,1} & & & & & \\ c^b_{2,1} & a^b_{2,1} & b^b_{2,1} & \dots & & e^b_{2,1} & & & & \\ \dots & c^b_{3,1} & a^b_{3,1} & b^b_{3,1} & \dots & & e^b_{3,1} & & & \\ \dots & \dots & c^b_{4,1} & a^b_{4,1} & \dots & & \dots & e^b_{4,1} & & \\ & \dots & \dots & \dots & \dots & \dots & \dots & \dots & & \\ & & \dots & \dots & \dots & \dots & \dots & \dots & \dots & \\ & & & f^b_{I-3,J} & \dots & & c^b_{I-3,J} & a^b_{I-3,J} & \dots & \dots & \dots \\ & & & & f^b_{I-2,J} & \dots & & c^b_{I-2,J} & a^b_{I-2,J} & b^b_{I-2,J} & \dots \\ & & & & & f^b_{I-1,J} & \dots & & c^b_{I-1,J} & a^b_{I-1,J} & b^b_{I-1,J} \\ & & & & & & f^b_{I,J} & \dots & & c^b_{I,J} & b^b_{I,J} \end{bmatrix} \tag{8.77}$$

with $\mathbf{d}_{bf}\big|_{i,j} = \begin{bmatrix} \delta p_o|_{i,j} \\ \delta S_o|_{i,j} \end{bmatrix}$ and $\mathbf{F}_{bf}\big|_{i,j} = \begin{bmatrix} F_o|_{i,j} \\ F_g|_{i,j} \end{bmatrix}$.

And for the component tracer

$$
\begin{bmatrix}
a^T_{1,1} & b^T_{1,1} & \cdots & & e^T_{1,1} & & & & \\
c^T_{2,1} & a^T_{2,1} & b^T_{2,1} & \cdots & & e^T_{2,1} & & & \\
\cdots & c^T_{3,1} & a^T_{3,1} & b^T_{3,1} & \cdots & & e^T_{3,1} & & \\
\cdots & \cdots & c^T_{4,1} & a^T_{4,1} & \cdots & & & e^T_{4,1} & \\
& \cdots & \cdots & \cdots & \cdots & \cdots & \cdots & & \\
& & \cdots & \cdots & \cdots & \cdots & \cdots & \cdots & \\
& & f^T_{I-3,J} & \cdots & c^T_{I-3,J} & a^T_{I-3,J} & \cdots & \cdots & \cdots \\
& & & f^T_{I-2,J} & \cdots & c^T_{I-2,J} & a^T_{I-2,J} & b^T_{I-2,J} & \cdots \\
& & & & f^T_{I-1,J} & \cdots & c^T_{I-1,J} & a^T_{I-1,J} & b^T_{I-1,J} \\
& & & & & f^T_{I,J} & \cdots & c^T_{I,J} & b^T_{I,J}
\end{bmatrix}
\tag{8.78}
$$

with $\mathbf{d}_T|_{i,j} = \delta x_T|_{i,j}$ and the residual function for the tracer, $\mathbf{F}_T|_{i,j}$.

The system maintains its penta-diagonal structure of $IJ \times IJ$ order, and the elements of the matrix of the oil and gas, $\mathbf{M}_{bf}$ components are elements of 2×2 matrix.

The elements are then defined by the partial derivatives as follows matrix components for oil and gas

$$
a^b_{i,j} = \begin{bmatrix} \dfrac{\partial F_{oij}}{\partial p_{oij}} & \dfrac{\partial F_{oij}}{\partial S_{oij}} \\[2mm] \dfrac{\partial F_{gij}}{\partial p_{oij}} & \dfrac{\partial F_{gij}}{\partial S_{oij}} \end{bmatrix}; \quad
b^b_{i,j} = \begin{bmatrix} \dfrac{\partial F_{oij}}{\partial p_{oi+1j}} & \dfrac{\partial F_{oij}}{\partial S_{oi+1j}} \\[2mm] \dfrac{\partial F_{gij}}{\partial p_{oi+1j}} & \dfrac{\partial F_{gij}}{\partial S_{oi+1j}} \end{bmatrix}; \quad
c^b_{i,j} = \begin{bmatrix} \dfrac{\partial F_{oij}}{\partial p_{oi-1j}} & \dfrac{\partial F_{oij}}{\partial S_{oi-1j}} \\[2mm] \dfrac{\partial F_{gij}}{\partial p_{oi-1j}} & \dfrac{\partial F_{gij}}{\partial S_{oi-1j}} \end{bmatrix}
\tag{8.79}
$$

$$
e^b_{i,j} = \begin{bmatrix} \dfrac{\partial F_{oij}}{\partial p_{oij+1}} & \dfrac{\partial F_{oij}}{\partial S_{oij+1}} \\[2mm] \dfrac{\partial F_{gij}}{\partial p_{oij+1}} & \dfrac{\partial F_{gij}}{\partial S_{oij+1}} \end{bmatrix}; \quad
f^b_{i,j} = \begin{bmatrix} \dfrac{\partial F_{oij}}{\partial p_{oij-1}} & \dfrac{\partial F_{oij}}{\partial S_{oij-1}} \\[2mm] \dfrac{\partial F_{gij}}{\partial p_{oij-1}} & \dfrac{\partial F_{gij}}{\partial S_{oij-1}} \end{bmatrix}
\tag{8.80}
$$

and the tracer component

$$
a^T_{i,j} = \left(\dfrac{\partial F_{Tij}}{\partial x_{Tij}}\right); \quad b^T_{i,j} = \left(\dfrac{\partial F_{Tij}}{\partial x_{Ti+1j}}\right); \quad c^T_{i,j} = \left(\dfrac{\partial F_{Tij}}{\partial x_{Ti-1j}}\right); \quad e^T_{i,j} = \left(\dfrac{\partial F_{Tij}}{\partial x_{Tij+1}}\right); \quad f^T_{i,j} = \left(\dfrac{\partial F_{Tij}}{\partial x_{Tij-1}}\right)
\tag{8.81}
$$

Thus, the solution procedure consists first in solving the system of oil and gas components for values $p_o^{(n+1)}$, $S_o^{(n+1)}$ given by

$$
\mathbf{M}_{bf}^{(v)} \mathbf{d}_{bf}^{(v+1)} = -\mathbf{F}_{bf}^{(v)}
\tag{8.82}
$$

And second, solving the tracer mole fraction in the system

$$
\mathbf{M}_T^{(n+1)} \mathbf{d}_T^{(n+1)} = -\mathbf{F}_T^{(n)}
\tag{8.83}
$$

8.2.5　Management of the partials derivatives

8.2.5.1　Matrix for oil-gas phases $\mathbf{M}_{bf}$ (oil and gas components)

By analyzing the system for oil and gas components we can see that the nonlinear flux terms are of the form

$T(p, S)^{n+1} [\Delta p + \Delta Pc(S) - \gamma(p, S) \Delta Z]^{n+1}$ for advection terms,
$q(p, S)^{n+1}$ for the source terms, and
$(\phi b S)^{n+1}$ for terms of accumulation.

Thus, the standard form of nonlinear functions are summarized in the scheme known as the "fully implicit" (TI), which considers all the system parameters evaluated at time level $n+1$,

$$F|_{i,j}^{n+1} \rightarrow \{T(p,S)^{n+1}[\Delta p + \Delta Pc(S) - \gamma(p,S)\Delta Z]^{n+1}, q(p,S)^{n+1}, (\phi bS)^{n+1}\} \qquad (8.84)$$

For oil and gas components the dependency is

$$T_o(p_o, S_o), \quad \gamma_o(p_o), \quad R_s(p_o), \quad b_o(p_o), \quad T_g(p_o, S_o), \quad \gamma_g(p_o, S_o), \quad r_s(p_o, S_o), \quad b_g(p_o, S_o) \qquad (8.85)$$

In this way the elements of the Jacobian IT for the biphasic oil and gas components will be configured for the following derivatives respect to pressure for oil

$$
\begin{aligned}
\frac{\partial F_{oi,j}}{\partial p_{oi,j-1}} =\ & T_o|_{j-\frac{1}{2}} + \left(T_o|_{j-\frac{1}{2}}\,\Delta Z|_{j-\frac{1}{2}}\right)\frac{\partial \gamma_{o\,j-\frac{1}{2}}}{\partial p_{oi,j-1}} - \left(\Delta p_o|_{j-\frac{1}{2}} - \gamma_o|_{j-\frac{1}{2}}\,\Delta Z|_{j-\frac{1}{2}}\right)\frac{\partial T_{o\,j-\frac{1}{2}}}{\partial p_{oi,j-1}} \\
& + r_s|_{j-\frac{1}{2}}\,T_g\big|_{j-\frac{1}{2}} + r_s|_{j-\frac{1}{2}}\left(T_g\big|_{j-\frac{1}{2}}\,\Delta Z|_{j-\frac{1}{2}}\right)\frac{\partial \gamma_{g\,j-\frac{1}{2}}}{\partial p_{oi,j-1}} \\
& - r_s|_{j-\frac{1}{2}}\left(\Delta p_o|_{j-\frac{1}{2}} + \Delta Pc_{go}\big|_{j-\frac{1}{2}} - \gamma_g|_{j-\frac{1}{2}}\,\Delta Z|_{j-\frac{1}{2}}\right)\frac{\partial T_{g\,j-\frac{1}{2}}}{\partial p_{oi,j-1}} \\
& - T_g\big|_{j-\frac{1}{2}}\left(\Delta p_o|_{j-\frac{1}{2}} + \Delta Pc_{go}\big|_{j-\frac{1}{2}} - \gamma_g|_{j-\frac{1}{2}}\,\Delta Z|_{j-\frac{1}{2}}\right)\frac{\partial r_{s\,j-\frac{1}{2}}}{\partial p_{oi,j-1}}
\end{aligned}
\qquad (8.86)
$$

$$
\begin{aligned}
\frac{\partial F_{oi,j}}{\partial p_{oi-1,j}} =\ & T_o|_{i-\frac{1}{2}} + \left(T_o|_{i-\frac{1}{2}}\,\Delta Z|_{i-\frac{1}{2}}\right)\frac{\partial \gamma_{oi-\frac{1}{2}}}{\partial p_{oi-1,j}} - \left(\Delta p_o|_{i-\frac{1}{2}} - \gamma_o|_{i-\frac{1}{2}}\,\Delta Z|_{i-\frac{1}{2}}\right)\frac{\partial T_{oi-\frac{1}{2}}}{\partial p_{oi-1,j}} \\
& + r_s|_{i-\frac{1}{2}}\,T_g\big|_{i-\frac{1}{2}} + r_s|_{i-\frac{1}{2}}\left(T_g\big|_{i-\frac{1}{2}}\,\Delta Z|_{i-\frac{1}{2}}\right)\frac{\partial \gamma_{gi-\frac{1}{2}}}{\partial p_{oi-1,j}} \\
& - r_s|_{i-\frac{1}{2}}\left(\Delta p_o|_{i-\frac{1}{2}} + \Delta Pc_{go}\big|_{i-\frac{1}{2}} - \gamma_g|_{i-\frac{1}{2}}\,\Delta Z|_{i-\frac{1}{2}}\right)\frac{\partial T_{gi-\frac{1}{2}}}{\partial p_{oi-1,j}} \\
& - T_g\big|_{i-\frac{1}{2}}\left(\Delta p_o|_{i-\frac{1}{2}} + \Delta Pc_{go}\big|_{i-\frac{1}{2}} - \gamma_g|_{i-\frac{1}{2}}\,\Delta Z|_{i-\frac{1}{2}}\right)\frac{\partial r_{si-\frac{1}{2}}}{\partial p_{oi-1,j}}
\end{aligned}
\qquad (8.87)
$$

$$
\begin{aligned}
\frac{\partial F_{oi,j}}{\partial p_{oi,j}} =\ & - T_o|_{i+\frac{1}{2}} - \left(T_o|_{i+\frac{1}{2}}\,\Delta Z|_{i+\frac{1}{2}}\right)\frac{\partial \gamma_{oi+\frac{1}{2}}}{\partial p_{oi,j}} + \left(\Delta p_o|_{i+\frac{1}{2}} - \gamma_o|_{i+\frac{1}{2}}\,\Delta Z|_{i+\frac{1}{2}}\right)\frac{\partial T_{oi+\frac{1}{2}}}{\partial p_{oi,j}} \\
& - T_o|_{i-\frac{1}{2}} + \left(T_o|_{i-\frac{1}{2}}\,\Delta Z|_{i-\frac{1}{2}}\right)\frac{\partial \gamma_{oi-\frac{1}{2}}}{\partial p_{oi,j}} - \left(\Delta p_o|_{i-\frac{1}{2}} - \gamma_o|_{i-\frac{1}{2}}\,\Delta Z|_{i-\frac{1}{2}}\right)\frac{\partial T_{oi-\frac{1}{2}}}{\partial p_{oi,j}} \\
& - T_o|_{j+\frac{1}{2}} - \left(T_o|_{j+\frac{1}{2}}\,\Delta Z|_{j+\frac{1}{2}}\right)\frac{\partial \gamma_{o\,j+\frac{1}{2}}}{\partial p_{oi,j}} + \left(\Delta p_o|_{j+\frac{1}{2}} - \gamma_o|_{j+\frac{1}{2}}\,\Delta Z|_{j+\frac{1}{2}}\right)\frac{\partial T_{o\,j+\frac{1}{2}}}{\partial p_{oi,j}} \\
& - T_o|_{j-\frac{1}{2}} + \left(T_o|_{j-\frac{1}{2}}\,\Delta Z|_{j-\frac{1}{2}}\right)\frac{\partial \gamma_{o\,j-\frac{1}{2}}}{\partial p_{oi,j}} - \left(\Delta p_o|_{j-\frac{1}{2}} - \gamma_o|_{j-\frac{1}{2}}\,\Delta Z|_{j-\frac{1}{2}}\right)\frac{\partial T_{o\,j-\frac{1}{2}}}{\partial p_{oi,j}} \\
& - r_s|_{i+\frac{1}{2}}\,T_g\big|_{i+\frac{1}{2}} - r_s|_{i+\frac{1}{2}}\left(T_g\big|_{i+\frac{1}{2}}\,\Delta Z|_{i+\frac{1}{2}}\right)\frac{\partial \gamma_{gi+\frac{1}{2}}}{\partial p_{oi,j}} \\
& + r_s|_{i+\frac{1}{2}}\left(\Delta p_o|_{i+\frac{1}{2}} + \Delta Pc_{go}\big|_{i+\frac{1}{2}} - \gamma_g|_{i+\frac{1}{2}}\,\Delta Z|_{i+\frac{1}{2}}\right)\frac{\partial T_{gi+\frac{1}{2}}}{\partial p_{oi,j}} \\
& + T_g\big|_{i+\frac{1}{2}}\left(\Delta p_o|_{i+\frac{1}{2}} + \Delta Pc_{go}\big|_{i+\frac{1}{2}} - \gamma_g|_{i+\frac{1}{2}}\,\Delta Z|_{i+\frac{1}{2}}\right)\frac{\partial r_{si+\frac{1}{2}}}{\partial p_{oi,j}}
\end{aligned}
$$

$$- r_s|_{i-\frac{1}{2}} \, T_g|_{i-\frac{1}{2}} + r_s|_{i-\frac{1}{2}} \left(T_g|_{i-\frac{1}{2}} \, \Delta Z|_{i-\frac{1}{2}} \right) \frac{\partial \gamma_{g\,i-\frac{1}{2}}}{\partial p_{oi,j}}$$

$$- r_s|_{i-\frac{1}{2}} \left(\Delta p_o|_{i-\frac{1}{2}} + \Delta Pc_{go}|_{i-\frac{1}{2}} - \gamma_g|_{i-\frac{1}{2}} \, \Delta Z|_{i-\frac{1}{2}} \right) \frac{\partial T_{g\,i-\frac{1}{2}}}{\partial p_{oi,j}}$$

$$- T_g|_{i-\frac{1}{2}} \left(\Delta p_o|_{i-\frac{1}{2}} + \Delta Pc_{go}|_{i-\frac{1}{2}} - \gamma_g|_{i-\frac{1}{2}} \, \Delta Z|_{i-\frac{1}{2}} \right) \frac{\partial r_{s\,i-\frac{1}{2}}}{\partial p_{oi,j}}$$

$$- r_s|_{j+\frac{1}{2}} \, T_g|_{j+\frac{1}{2}} - r_s|_{j+\frac{1}{2}} \left(T_g|_{j+\frac{1}{2}} \, \Delta Z|_{j+\frac{1}{2}} \right) \frac{\partial \gamma_{g\,j+\frac{1}{2}}}{\partial p_{oi,j}}$$

$$+ r_s|_{j+\frac{1}{2}} \left(\Delta p_o|_{j+\frac{1}{2}} + \Delta Pc_{go}|_{j+\frac{1}{2}} - \gamma_g|_{j+\frac{1}{2}} \, \Delta Z|_{j+\frac{1}{2}} \right) \frac{\partial T_{g\,j+\frac{1}{2}}}{\partial p_{oi,j}}$$

$$+ T_g|_{j+\frac{1}{2}} \left(\Delta p_o|_{j+\frac{1}{2}} + \Delta Pc_{go}|_{j+\frac{1}{2}} - \gamma_g|_{j+\frac{1}{2}} \, \Delta Z|_{j+\frac{1}{2}} \right) \frac{\partial r_{s\,j+\frac{1}{2}}}{\partial p_{oi,j}}$$

$$- r_s|_{j-\frac{1}{2}} \, T_g|_{j-\frac{1}{2}} + r_s|_{j-\frac{1}{2}} \left(T_g|_{j-\frac{1}{2}} \, \Delta Z|_{j-\frac{1}{2}} \right) \frac{\partial \gamma_{g\,j-\frac{1}{2}}}{\partial p_{oi,j}}$$

$$- r_s|_{j-\frac{1}{2}} \left(\Delta p_o|_{j-\frac{1}{2}} + \Delta Pc_{go}|_{j-\frac{1}{2}} - \gamma_g|_{j-\frac{1}{2}} \, \Delta Z|_{j-\frac{1}{2}} \right) \frac{\partial T_{g\,j-\frac{1}{2}}}{\partial p_{oi,j}}$$

$$- T_g|_{j-\frac{1}{2}} \left(\Delta p_o|_{j-\frac{1}{2}} + \Delta Pc_{go}|_{j-\frac{1}{2}} - \gamma_g|_{j-\frac{1}{2}} \, \Delta Z|_{j-\frac{1}{2}} \right) \frac{\partial r_{s\,j-\frac{1}{2}}}{\partial p_{oi,j}}$$

$$+ \frac{\partial q_{oi,j}}{\partial p_{oi,j}} + r_s|_{i,j} \frac{\partial q_{gi,j}}{\partial p_{oi,j}} + q_g|_{i,j} \frac{\partial r_{si,j}}{\partial p_{oi,j}} - \frac{Vr|_{i,j}}{\Delta t} \left\{ \left(S_o b_o + (1 - S_o) \, b_g r_s \right) \frac{\partial \phi}{\partial p_{oi,j}} \right\}$$

$$- \frac{Vr|_{i,j}}{\Delta t} \left\{ \phi \left(S_o \frac{\partial b_o}{\partial p_{oi,j}} + (1 - S_o) \left(b_g \frac{\partial r_s}{\partial p_{oi,j}} + r_s \frac{\partial b_g}{\partial p_{oi,j}} \right) \right) \right\} \tag{8.88}$$

$$\frac{\partial F_{oi,j}}{\partial p_{oi+1,j}} = T_o|_{i+\frac{1}{2}} - \left(T_o|_{i+\frac{1}{2}} \, \Delta Z|_{i+\frac{1}{2}} \right) \frac{\partial \gamma_{oi+\frac{1}{2}}}{\partial p_{oi+1,j}} + \left(\Delta p_o|_{i+\frac{1}{2}} - \gamma_o|_{i+\frac{1}{2}} \, \Delta Z|_{i+\frac{1}{2}} \right) \frac{\partial T_{oi+\frac{1}{2}}}{\partial p_{oi+1,j}}$$

$$+ r_s|_{i+\frac{1}{2}} \, T_g|_{i+\frac{1}{2}} - r_s|_{i+\frac{1}{2}} \left(T_g|_{i+\frac{1}{2}} \, \Delta Z|_{i+\frac{1}{2}} \right) \frac{\partial \gamma_{g\,i+\frac{1}{2}}}{\partial p_{oi+1,j}}$$

$$+ r_s|_{i+\frac{1}{2}} \left(\Delta p_o|_{i+\frac{1}{2}} + \Delta Pc_{go}|_{i+\frac{1}{2}} - \gamma_g|_{i+\frac{1}{2}} \, \Delta Z|_{i+\frac{1}{2}} \right) \frac{\partial T_{g\,i+\frac{1}{2}}}{\partial p_{oi+1,j}}$$

$$+ T_g|_{i+\frac{1}{2}} \left(\Delta p_o|_{i+\frac{1}{2}} + \Delta Pc_{go}|_{i+\frac{1}{2}} - \gamma_g|_{i+\frac{1}{2}} \, \Delta Z|_{i+\frac{1}{2}} \right) \frac{\partial r_{si+\frac{1}{2}}}{\partial p_{oi+1,j}} \tag{8.89}$$

$$\frac{\partial F_{oi,j}}{\partial p_{oi,j+1}} = T_o|_{j+\frac{1}{2}} - \left(T_o|_{j+\frac{1}{2}} \, \Delta Z|_{j+\frac{1}{2}} \right) \frac{\partial \gamma_{oj+\frac{1}{2}}}{\partial p_{oi,j+1}} + \left(\Delta p_o|_{j+\frac{1}{2}} - \gamma_o|_{j+\frac{1}{2}} \, \Delta Z|_{j+\frac{1}{2}} \right) \frac{\partial T_{oj+\frac{1}{2}}}{\partial p_{oi,j+1}}$$

$$+ r_s|_{j+\frac{1}{2}} \, T_g|_{j+\frac{1}{2}} - r_s|_{j+\frac{1}{2}} \left(T_g|_{j+\frac{1}{2}} \, \Delta Z|_{j+\frac{1}{2}} \right) \frac{\partial \gamma_{g\,j+\frac{1}{2}}}{\partial p_{oi,j+1}}$$

$$+ r_s|_{j+\frac{1}{2}} \left(\Delta p_o|_{j+\frac{1}{2}} + \Delta Pc_{go}|_{j+\frac{1}{2}} - \gamma_g|_{j+\frac{1}{2}} \, \Delta Z|_{j+\frac{1}{2}} \right) \frac{\partial T_{g\,j+\frac{1}{2}}}{\partial p_{oi,j+1}}$$

$$+ T_g|_{j+\frac{1}{2}} \left(\Delta p_o|_{j+\frac{1}{2}} + \Delta Pc_{go}|_{j+\frac{1}{2}} - \gamma_g|_{j+\frac{1}{2}} \, \Delta Z|_{j+\frac{1}{2}} \right) \frac{\partial r_{s\,j+\frac{1}{2}}}{\partial p_{oi,j+1}} \tag{8.90}$$

And for the derivative terms with respect to saturation, for oil

$$
\begin{aligned}
\frac{\partial F_{oi,j}}{\partial S_{oi,j-1}} = {} & -\left(\Delta p_o|_{j-\frac{1}{2}} - \gamma_o|_{j-\frac{1}{2}}\,\Delta Z|_{j-\frac{1}{2}}\right)\frac{\partial T_{o\,j-\frac{1}{2}}}{\partial S_{oi,j-1}} \\
& + r_s|_{j-\frac{1}{2}}\,T_g|_{j-\frac{1}{2}}\frac{\partial Pc_{go\,i,j-1}}{\partial S_{oi,j-1}} + r_s|_{j-\frac{1}{2}}\left(T_g|_{j-\frac{1}{2}}\,\Delta Z|_{j-\frac{1}{2}}\right)\frac{\partial \gamma_{g\,j-\frac{1}{2}}}{\partial S_{oi,j-1}} \\
& - r_s|_{j-\frac{1}{2}}\left(\Delta p_o|_{j-\frac{1}{2}} + \Delta Pc_{go}|_{j-\frac{1}{2}} - \gamma_g|_{j-\frac{1}{2}}\,\Delta Z|_{j-\frac{1}{2}}\right)\frac{\partial T_{g\,j-\frac{1}{2}}}{\partial S_{oi,j-1}} \\
& - T_g|_{j-\frac{1}{2}}\left(\Delta p_o|_{j-\frac{1}{2}} + \Delta Pc_{go}|_{j-\frac{1}{2}} - \gamma_g|_{j-\frac{1}{2}}\,\Delta Z|_{j-\frac{1}{2}}\right)\frac{\partial r_{s\,j-\frac{1}{2}}}{\partial p_{g\,i,j-1}}\frac{\partial Pc_{go\,i,j-1}}{\partial S_{oi,j-1}}
\end{aligned}
\tag{8.91}
$$

$$
\begin{aligned}
\frac{\partial F_{oi,j}}{\partial S_{oi-1,j}} = {} & -\left(\Delta p_o|_{i-\frac{1}{2}} - \gamma_o|_{i-\frac{1}{2}}\,\Delta Z|_{i-\frac{1}{2}}\right)\frac{\partial T_{oi-\frac{1}{2}}}{\partial S_{oi-1,j}} \\
& + r_s|_{i-\frac{1}{2}}\,T_g|_{i-\frac{1}{2}}\frac{\partial Pc_{go\,i-1,j}}{\partial S_{oi-1,j}} + r_s|_{i-\frac{1}{2}}\left(T_g|_{i-\frac{1}{2}}\,\Delta Z|_{i-\frac{1}{2}}\right)\frac{\partial \gamma_{g\,i-\frac{1}{2}}}{\partial S_{oi-1,j}} \\
& - r_s|_{i-\frac{1}{2}}\left(\Delta p_o|_{i-\frac{1}{2}} + \Delta Pc_{go}|_{i-\frac{1}{2}} - \gamma_g|_{i-\frac{1}{2}}\,\Delta Z|_{i-\frac{1}{2}}\right)\frac{\partial T_{g\,i-\frac{1}{2}}}{\partial S_{oi-1,j}} \\
& - T_g|_{i-\frac{1}{2}}\left(\Delta p_o|_{i-\frac{1}{2}} + \Delta Pc_{go}|_{i-\frac{1}{2}} - \gamma_g|_{i-\frac{1}{2}}\,\Delta Z|_{i-\frac{1}{2}}\right)\frac{\partial r_{s\,i-\frac{1}{2}}}{\partial p_{g\,i-1,j}}\frac{\partial Pc_{go\,i-1,j}}{\partial S_{oi-1,j}}
\end{aligned}
\tag{8.92}
$$

$$
\begin{aligned}
\frac{\partial F_{oi,j}}{\partial S_{oi,j}} = {} & \left(\Delta p_o|_{i+\frac{1}{2}} - \gamma_o|_{i+\frac{1}{2}}\,\Delta Z|_{i+\frac{1}{2}}\right)\frac{\partial T_{oi+\frac{1}{2}}}{\partial S_{oi,j}} - \left(\Delta p_o|_{i-\frac{1}{2}} - \gamma_o|_{i-\frac{1}{2}}\,\Delta Z|_{i-\frac{1}{2}}\right)\frac{\partial T_{oi-\frac{1}{2}}}{\partial S_{oi,j}} \\
& + \left(\Delta p_o|_{j+\frac{1}{2}} - \gamma_o|_{j+\frac{1}{2}}\,\Delta Z|_{j+\frac{1}{2}}\right)\frac{\partial T_{o\,j+\frac{1}{2}}}{\partial S_{oi,j}} - \left(\Delta p_o|_{j-\frac{1}{2}} - \gamma_o|_{j-\frac{1}{2}}\,\Delta Z|_{j-\frac{1}{2}}\right)\frac{\partial T_{o\,j-\frac{1}{2}}}{\partial S_{oi,j}} \\
& - r_s|_{i+\frac{1}{2}}\,T_g|_{i+\frac{1}{2}}\frac{\partial Pc_{go\,i,j}}{\partial S_{oi,j}} - r_s|_{i+\frac{1}{2}}\left(T_g|_{i+\frac{1}{2}}\,\Delta Z|_{i+\frac{1}{2}}\right)\frac{\partial \gamma_{g\,i+\frac{1}{2}}}{\partial S_{oi,j}} \\
& + r_s|_{i+\frac{1}{2}}\left(\Delta p_o|_{i+\frac{1}{2}} + \Delta Pc_{go}|_{i+\frac{1}{2}} - \gamma_g|_{i+\frac{1}{2}}\,\Delta Z|_{i+\frac{1}{2}}\right)\frac{\partial T_{g\,i+\frac{1}{2}}}{\partial S_{oi,j}} \\
& + T_g|_{i+\frac{1}{2}}\left(\Delta p_o|_{i+\frac{1}{2}} + \Delta Pc_{go}|_{i+\frac{1}{2}} - \gamma_g|_{i+\frac{1}{2}}\,\Delta Z|_{i+\frac{1}{2}}\right)\frac{\partial r_{s\,i+\frac{1}{2}}}{\partial p_{g\,i,j}}\frac{\partial Pc_{go\,i,j}}{\partial S_{oi,j}} \\
& - r_s|_{i-\frac{1}{2}}\,T_g|_{i-\frac{1}{2}}\frac{\partial Pc_{go\,i,j}}{\partial S_{oi,j}} + r_s|_{i-\frac{1}{2}}\left(T_g|_{i-\frac{1}{2}}\,\Delta Z|_{i-\frac{1}{2}}\right)\frac{\partial \gamma_{g\,i-\frac{1}{2}}}{\partial S_{oi,j}} \\
& - r_s|_{i-\frac{1}{2}}\left(\Delta p_o|_{i-\frac{1}{2}} + \Delta Pc_{go}|_{i-\frac{1}{2}} - \gamma_g|_{i-\frac{1}{2}}\,\Delta Z|_{i-\frac{1}{2}}\right)\frac{\partial T_{g\,i-\frac{1}{2}}}{\partial S_{oi,j}} \\
& - T_g|_{i-\frac{1}{2}}\left(\Delta p_o|_{i-\frac{1}{2}} + \Delta Pc_{go}|_{i-\frac{1}{2}} - \gamma_g|_{i-\frac{1}{2}}\,\Delta Z|_{i-\frac{1}{2}}\right)\frac{\partial r_{s\,i-\frac{1}{2}}}{\partial p_{g\,i,j}}\frac{\partial Pc_{go\,i,j}}{\partial S_{oi,j}}
\end{aligned}
$$

$$
- r_s|_{j+\frac{1}{2}} \, T_g|_{j+\frac{1}{2}} \, \frac{\partial Pc_{go\,i,j}}{\partial S_{o\,i,j}} - r_s|_{j+\frac{1}{2}} \left(T_g|_{j+\frac{1}{2}} \, \Delta Z|_{j+\frac{1}{2}} \right) \frac{\partial \gamma_{g\,j+\frac{1}{2}}}{\partial S_{o\,i,j}}
$$

$$
+ r_s|_{j+\frac{1}{2}} \left(\Delta p_o|_{j+\frac{1}{2}} + \Delta Pc_{go}|_{j+\frac{1}{2}} - \gamma_g|_{j+\frac{1}{2}} \, \Delta Z|_{j+\frac{1}{2}} \right) \frac{\partial T_{g\,j+\frac{1}{2}}}{\partial S_{o\,i,j}}
$$

$$
+ T_g|_{j+\frac{1}{2}} \left(\Delta p_o|_{j+\frac{1}{2}} + \Delta Pc_{go}|_{j+\frac{1}{2}} - \gamma_g|_{j+\frac{1}{2}} \, \Delta Z|_{j+\frac{1}{2}} \right) \frac{\partial r_{s\,j+\frac{1}{2}}}{\partial p_{g\,i,j}} \frac{\partial Pc_{go\,i,j}}{\partial S_{o\,i,j}}
$$

$$
- r_s|_{j-\frac{1}{2}} \, T_g|_{j-\frac{1}{2}} \, \frac{\partial Pc_{go\,i,j}}{\partial S_{o\,i,j}} + r_s|_{j-\frac{1}{2}} \left(T_g|_{j-\frac{1}{2}} \, \Delta Z|_{j-\frac{1}{2}} \right) \frac{\partial \gamma_{g\,j-\frac{1}{2}}}{\partial S_{o\,i,j}}
$$

$$
- r_s|_{j-\frac{1}{2}} \left(\Delta p_o|_{j-\frac{1}{2}} + \Delta Pc_{go}|_{j-\frac{1}{2}} - \gamma_g|_{j-\frac{1}{2}} \, \Delta Z|_{j-\frac{1}{2}} \right) \frac{\partial T_{g\,j-\frac{1}{2}}}{\partial S_{o\,i,j}}
$$

$$
- T_g|_{j-\frac{1}{2}} \left(\Delta p_o|_{j-\frac{1}{2}} + \Delta Pc_{go}|_{j-\frac{1}{2}} - \gamma_g|_{j-\frac{1}{2}} \, \Delta Z|_{j-\frac{1}{2}} \right) \frac{\partial r_{s\,j-\frac{1}{2}}}{\partial p_{g\,i,j}} \frac{\partial Pc_{go\,i,j}}{\partial S_{o\,i,j}}
$$

$$
+ \frac{\partial q_{o\,i,j}}{\partial S_{o\,i,j}} + r_s|_{i,j} \, \frac{\partial q_{g\,i,j}}{\partial S_{o\,i,j}}
$$

$$
- \frac{Vr|_{i,j}}{\Delta t} \left\{ \phi \left(b_o - b_g r_s + r_s(1 - S_o)\frac{\partial b_g}{\partial S_o} + b_g(1 - S_o)\frac{\partial r_s}{\partial S_o} \right) \right\} \tag{8.93}
$$

$$
\frac{\partial F_{o\,i,j}}{\partial S_{o\,i+1,j}} = \left(\Delta p_o|_{i+\frac{1}{2}} - \gamma_o|_{i+\frac{1}{2}} \, \Delta Z|_{i+\frac{1}{2}} \right) \frac{\partial T_{o\,i+\frac{1}{2}}}{\partial S_{o\,i+1,j}}
$$

$$
+ r_s|_{i+\frac{1}{2}} \, T_g|_{i+\frac{1}{2}} \, \frac{\partial Pc_{go\,i+1,j}}{\partial S_{o\,i+1,j}} - r_s|_{i+\frac{1}{2}} \left(T_g|_{i+\frac{1}{2}} \, \Delta Z|_{i+\frac{1}{2}} \right) \frac{\partial \gamma_{g\,i+\frac{1}{2}}}{\partial S_{o\,i+1,j}}
$$

$$
+ r_s|_{i+\frac{1}{2}} \left(\Delta p_o|_{i+\frac{1}{2}} + \Delta Pc_{go}|_{i+\frac{1}{2}} - \gamma_g|_{i+\frac{1}{2}} \, \Delta Z|_{i+\frac{1}{2}} \right) \frac{\partial T_{g\,i+\frac{1}{2}}}{\partial S_{o\,i+1,j}} \tag{8.94}
$$

$$
+ T_g|_{i+\frac{1}{2}} \left(\Delta p_o|_{i+\frac{1}{2}} + \Delta Pc_{go}|_{i+\frac{1}{2}} - \gamma_g|_{i+\frac{1}{2}} \, \Delta Z|_{i+\frac{1}{2}} \right) \frac{\partial r_{s\,i+\frac{1}{2}}}{\partial p_{g\,i+1,j}} \frac{\partial Pc_{go\,i+1,j}}{\partial S_{o\,i+1,j}}
$$

$$
\frac{\partial F_{o\,i,j}}{\partial S_{o\,i,j+1}} = \left(\Delta p_o|_{j+\frac{1}{2}} - \gamma_o|_{j+\frac{1}{2}} \, \Delta Z|_{j+\frac{1}{2}} \right) \frac{\partial T_{o\,j+\frac{1}{2}}}{\partial S_{o\,i,j+1}}
$$

$$
+ r_s|_{j+\frac{1}{2}} \, T_g|_{j+\frac{1}{2}} \, \frac{\partial Pc_{go\,i,j+1}}{\partial S_{o\,i,j+1}} - r_s|_{j+\frac{1}{2}} \left(T_g|_{j+\frac{1}{2}} \, \Delta Z|_{j+\frac{1}{2}} \right) \frac{\partial \gamma_{g\,j+\frac{1}{2}}}{\partial S_{o\,i,j+1}}
$$

$$
+ r_s|_{j+\frac{1}{2}} \left(\Delta p_o|_{j+\frac{1}{2}} + \Delta Pc_{go}|_{j+\frac{1}{2}} - \gamma_g|_{j+\frac{1}{2}} \, \Delta Z|_{j+\frac{1}{2}} \right) \frac{\partial T_{g\,j+\frac{1}{2}}}{\partial S_{o\,i,j+1}} \tag{8.95}
$$

$$
+ T_g|_{j+\frac{1}{2}} \left(\Delta p_o|_{j+\frac{1}{2}} + \Delta Pc_{go}|_{j+\frac{1}{2}} - \gamma_g|_{j+\frac{1}{2}} \, \Delta Z|_{j+\frac{1}{2}} \right) \frac{\partial r_{s\,j+\frac{1}{2}}}{\partial p_{g\,i,j+1}} \frac{\partial Pc_{go\,i,j+1}}{\partial S_{o\,i,j+1}}
$$

Due to space reasons, the expressions pertaining to gas component are not presented, but similar expressions are developed.

Definition of the partial derivatives at the borders: $i \pm 1/2$ and $j \pm 1/2$

A. Terms of specific weight

Because of the dependence on oil saturation, and oil pressure, the partial derivatives for the gas phase are determined by applying the chain rule

$$\frac{\partial \gamma_g}{\partial p_o} = \frac{\partial \gamma_g}{\partial p_g} \frac{\partial p_g}{\partial p_o} = \frac{\partial \gamma_g}{\partial p_g} \tag{8.96}$$

$$\frac{\partial \gamma_g}{\partial S_o} = \frac{\partial \gamma_g}{\partial p_g} \frac{\partial p_g}{\partial Pc_{go}} \frac{\partial Pc_{go}}{\partial S_o} = \frac{\partial \gamma_g}{\partial p_g} \frac{\partial Pc_{go}}{\partial S_o} \tag{8.97}$$

For both phases $p = o, g$, the terms $i + 1/2$ for specific weight are given by

$$\gamma_p|_{i+\frac{1}{2}} = \frac{g}{g_c} \left(\frac{1}{1 + \left(\Delta d_x|_{i+1} / \Delta d_x|_i \right)} \right) \rho_p|_{i+1}$$
$$+ \frac{g}{g_c} \left(1 - \frac{1}{1 + \left(\Delta d_x|_{i+1} / \Delta d_x|_i \right)} \right) \rho_p|_i \tag{8.98}$$

And we have for the oil component

$$\frac{\partial \gamma_{oi+\frac{1}{2}}}{\partial p_{oi,j}} = \left(1 - \frac{1}{1 + \left(\Delta d_x|_{i+1} / \Delta d_x|_i \right)} \right) \left(\frac{g}{g_c} \right) \left(\frac{\partial \rho_o}{\partial p_o} \right)_{i,j} \quad \frac{\partial \gamma_{oi+\frac{1}{2}}}{\partial S_{oi,j}} = 0 \tag{8.99}$$

$$\frac{\partial \gamma_{oi+\frac{1}{2}}}{\partial p_{oi+1}} = \left(\frac{1}{1 + \left(\Delta d_x|_{i+1} / \Delta d_x|_i \right)} \right) \left(\frac{g}{g_c} \right) \left(\frac{\partial \rho_o}{\partial p_o} \right)_{i+1} \quad \frac{\partial \gamma_{oi+\frac{1}{2}}}{\partial S_{oi+1}} = 0 \tag{8.100}$$

for the gas component

$$\frac{\partial \gamma_{gi+\frac{1}{2}}}{\partial p_{oi,j}} = \left(1 - \frac{1}{1 + \left(\Delta d_x|_{i+1} / \Delta d_x|_i \right)} \right) \left(\frac{g}{g_c} \right) \left(\frac{\partial \rho_g}{\partial p_g} \right)_{i,j} \tag{8.101}$$

$$\frac{\partial \gamma_{gi+\frac{1}{2}}}{\partial S_{oi,j}} = \left(1 - \frac{1}{1 + \left(\Delta d_x|_{i+1} / \Delta d_x|_i \right)} \right) \left(\frac{g}{g_c} \right) \left(\frac{\partial \rho_g}{\partial p_g} \frac{\partial Pc_{go}}{\partial S_o} \right)_{i,j} \tag{8.102}$$

$$\frac{\partial \gamma_{gi+\frac{1}{2}}}{\partial p_{oi+1}} = \left(\frac{1}{1 + \left(\Delta d_x|_{i+1} / \Delta d_x|_i \right)} \right) \left(\frac{g}{g_c} \right) \left(\frac{\partial \rho_g}{\partial p_g} \right)_{i+1} \tag{8.103}$$

$$\frac{\partial \gamma_{gi+\frac{1}{2}}}{\partial S_{oi+1}} = \left(\frac{1}{1 + \left(\Delta d_x|_{i+1} / \Delta d_x|_i \right)} \right) \left(\frac{g}{g_c} \right) \left(\frac{\partial \rho_g}{\partial p_g} \frac{\partial Pc_{go}}{\partial S_o} \right)_{i+1} \tag{8.104}$$

for the terms in $i - 1/2$ and for $p = o, g$

$$\gamma_p|_{i-\frac{1}{2}} = \frac{g}{g_c} \left(\frac{1}{1 + \left(\Delta d_x|_{i+1} / \Delta d_x|_i \right)} \right) \rho_p|_i + \frac{g}{g_c} \left(1 - \frac{1}{1 + \left(\Delta d_x|_{i+1} / \Delta d_x|_i \right)} \right) \rho_p|_{i-1} \tag{8.105}$$

we have for the oil

$$\frac{\partial \gamma_{oi-\frac{1}{2}}}{\partial p_{oi,j}} = \left(\frac{1}{1 + \left(\Delta d_x|_i / \Delta d_x|_{i-1} \right)} \right) \left(\frac{g}{g_c} \right) \left(\frac{\partial \rho_o}{\partial p_o} \right)_{i,j} \quad \frac{\partial \gamma_{pi-\frac{1}{2}}}{\partial S_{oi,j}} = 0 \tag{8.106}$$

$$\frac{\partial \gamma_{o\,i-\frac{1}{2}}}{\partial p_{o\,i-1}} = \left(1 - \frac{1}{1 + \left(\Delta d_x|_i / \Delta d_x|_{i-1}\right)}\right)\left(\frac{g}{g_c}\right)\left(\frac{\partial \rho_o}{\partial p_o}\right)_{i-1} \quad \frac{\partial \gamma_{o\,i-\frac{1}{2}}}{\partial S_{o\,i-1}} = 0 \qquad (8.107)$$

For the gas

$$\frac{\partial \gamma_{g\,i-\frac{1}{2}}}{\partial p_{o\,i,j}} = \left(\frac{1}{1 + \left(\Delta d_x|_i / \Delta d_x|_{i-1}\right)}\right)\left(\frac{g}{g_c}\right)\left(\frac{\partial \rho_g}{\partial p_g}\right)_{i,j}$$

$$\frac{\partial \gamma_{g\,i-\frac{1}{2}}}{\partial S_{o\,i,j}} = \left(\frac{1}{1 + \left(\Delta d_x|_i / \Delta d_x|_{i-1}\right)}\right)\left(\frac{g}{g_c}\right)\left(\frac{\partial \rho_g}{\partial p_g}\frac{\partial Pc_{go}}{\partial S_o}\right)_{i,j} \qquad (8.108)$$

$$\frac{\partial \gamma_{g\,i-\frac{1}{2}}}{\partial p_{o\,i-1}} = \left(1 - \frac{1}{1 + \left(\Delta d_x|_i / \Delta d_x|_{i-1}\right)}\right)\left(\frac{g}{g_c}\right)\left(\frac{\partial \rho_g}{\partial p_g}\right)_{i-1}$$

$$\frac{\partial \gamma_{g\,i-\frac{1}{2}}}{\partial S_{o\,i-1}} = \left(1 - \frac{1}{1 + \left(\Delta d_x|_i / \Delta d_x|_{i-1}\right)}\right)\left(\frac{g}{g_c}\right)\left(\frac{\partial \rho_g}{\partial p_g}\frac{\partial Pc_{go}}{\partial S_o}\right)_{i-1} \qquad (8.109)$$

For direction and partials on $j \pm 1$, will have similar expressions replacing the indices i by j.

B.   Terms of solubility

For the terms in $i + 1/2$ gas solubility ratio in oil, R_s, is given by

$$R_s|_{i+\frac{1}{2}} = \left(\frac{1}{1 + \left(\Delta d_x|_{i+1} / \Delta d_x|_i\right)}\right) R_s|_{i+1} + \left(1 - \frac{1}{1 + \left(\Delta d_x|_{i+1} / \Delta d_x|_i\right)}\right) R_s|_i \qquad (8.110)$$

so

$$\frac{\partial R_{s\,i+\frac{1}{2}}}{\partial p_{o\,i,j}} = \left(1 - \frac{1}{1 + \left(\Delta d_x|_{i+1} / \Delta d_x|_i\right)}\right)\left(\frac{\partial R_s}{\partial p_o}\right)_{i,j} \quad \frac{\partial R_{s\,i+\frac{1}{2}}}{\partial S_{o\,i,j}} = 0 \qquad (8.111)$$

$$\frac{\partial R_{s\,i+\frac{1}{2}}}{\partial p_{o\,i+1}} = \left(\frac{1}{1 + \left(\Delta d_x|_{i+1} / \Delta d_x|_i\right)}\right)\left(\frac{\partial R_s}{\partial p_o}\right)_{i+1} \quad \frac{\partial R_{s\,i+\frac{1}{2}}}{\partial S_{o\,i+1}} = 0 \qquad (8.112)$$

For the terms $i - 1/2$ it is given by

$$R_s|_{i-\frac{1}{2}} = \left(\frac{1}{1 + \left(\Delta d_x|_i / \Delta d_x|_{i-1}\right)}\right) R_s|_i + \left(1 - \frac{1}{1 + \left(\Delta d_x|_i / \Delta d_x|_{i-1}\right)}\right) R_s|_{i-1} \qquad (8.113)$$

so

$$\frac{\partial R_{s\,i-\frac{1}{2}}}{\partial p_{o\,i,j}} = \left(\frac{1}{1 + \left(\Delta d_x|_i / \Delta d_x|_{i-1}\right)}\right)\left(\frac{\partial R_s}{\partial p_o}\right)_{i,j} \quad \frac{\partial R_{s\,i-\frac{1}{2}}}{\partial S_{o\,i,j}} = 0 \qquad (8.114)$$

$$\frac{\partial R_{s\,i-\frac{1}{2}}}{\partial p_{o\,i-1}} = \left(1 - \frac{1}{1 + \left(\Delta d_x|_i / \Delta d_x|_{i-1}\right)}\right)\left(\frac{\partial R_s}{\partial p_o}\right)_{i-1} \quad \frac{\partial R_{s\,i-\frac{1}{2}}}{\partial S_{o\,i-1}} = 0 \qquad (8.115)$$

For direction and partials on $j \pm 1$, will have similar expressions replacing the indices i by j.

C.   Terms of vaporization

For terms of vaporization similar expressions as for R_s are obtained, by replacing r_s.

For direction and partials on $j \pm 1$, will have similar expressions replacing the indices i by j.

D. Advective transmissivity terms

Because of the dependence on oil saturation and oil pressure, the partial derivatives for the gas phase are determined by applying the chain rule

$$\frac{\partial T_g}{\partial p_o} = \frac{\partial T_g}{\partial p_g}\frac{\partial p_g}{\partial P_o} = \frac{\partial T_g}{\partial p_g} \tag{8.116}$$

$$\frac{\partial T_g}{\partial S_o} = \frac{\partial T_g}{\partial p_g}\frac{\partial p_g}{\partial Pc_{go}}\frac{\partial Pc_{go}}{\partial S_o} = \frac{\partial T_g}{\partial p_g}\frac{\partial Pc_{go}}{\partial S_o} \tag{8.117}$$

By the other side applying the "upstream" criterium the partial derivative of transmissivity terms in $i + 1/2$ for the oil phase are given by

$$\frac{\partial T_{o\,i+\frac{1}{2}}}{\partial p_{o\,i,j}} = (1-\delta_o)\left(\left(\frac{A}{\Delta d_x}\right)_{i+\frac{1}{2}}(k)_{i+\frac{1}{2}}\left(\frac{k_{ro}}{\mu_o^2}\left(\mu_o\frac{\partial b_o}{\partial p_o} - b_o\frac{\partial \mu_o}{\partial p_o}\right)\right)\Big|_{i,j}\right) \tag{8.118}$$

$$\frac{\partial T_{o\,i+\frac{1}{2}}}{\partial S_{o\,i,j}} = (1-\delta_o)\left(\left(\frac{A}{\Delta d_x}\right)_{i+\frac{1}{2}}(k)_{i+\frac{1}{2}}\left(\frac{b_o}{\mu_o}\left(\frac{\partial k_{ro}}{\partial S_o}\right)\right)\Big|_{i,j}\right) \tag{8.119}$$

if $(\Delta p_o - \gamma_o \Delta Z)|_{i+\frac{1}{2}} < 0 \quad \delta_o = 0$; otherwise $\delta_o = 1$

$$\frac{\partial T_{o\,i+\frac{1}{2}}}{\partial p_{o\,i+1}} = (1-\delta_o)\left(\left(\frac{A}{\Delta d_x}\right)_{i+\frac{1}{2}}(k)_{i+\frac{1}{2}}\left(\frac{k_{ro}}{\mu_o^2}\left(\mu_o\frac{\partial b_o}{\partial p_o} - b_o\frac{\partial \mu_o}{\partial p_o}\right)\right)\Big|_{i+1}\right) \tag{8.120}$$

$$\frac{\partial T_{o\,i+\frac{1}{2}}}{\partial S_{o\,i+1}} = (1-\delta_o)\left(\left(\frac{A}{\Delta d_x}\right)_{i+\frac{1}{2}}(k)_{i+\frac{1}{2}}\left(\frac{b_o}{\mu_o}\left(\frac{\partial k_{ro}}{\partial S_o}\right)\right)\Big|_{i+1}\right) \tag{8.121}$$

if $(\Delta p_o - \gamma_o \Delta Z)|_{i+\frac{1}{2}} \geq 0 \quad \delta_o = 0$; if not $\delta_o = 1$.

For the terms in $i - 1/2$

$$\frac{\partial T_{o\,i-\frac{1}{2}}}{\partial p_{o\,i,j}} = (1-\delta_o)\left(\left(\frac{A}{\Delta d_x}\right)_{i-\frac{1}{2}}(k)_{i-\frac{1}{2}}\left(\frac{k_{ro}}{\mu_o^2}\left(\mu_o\frac{\partial b_o}{\partial p_o} - b_o\frac{\partial \mu_o}{\partial p_o}\right)\right)\Big|_{i,j}\right) \tag{8.122}$$

$$\frac{\partial T_{o\,i-\frac{1}{2}}}{\partial S_{o\,i,j}} = (1-\delta_o)\left(\left(\frac{A}{\Delta d_x}\right)_{i-\frac{1}{2}}(k)_{i-\frac{1}{2}}\left(\frac{b_o}{\mu_o}\left(\frac{\partial k_{ro}}{\partial S_o}\right)\right)\Big|_{i,j}\right) \tag{8.123}$$

if $(\Delta p_o - \gamma_o \Delta Z)|_{i-\frac{1}{2}} > 0 \quad \delta_o = 0$; if not $\delta_o = 1$

$$\frac{\partial T_{o\,i-\frac{1}{2}}}{\partial p_{o\,i-1}} = (1-\delta_o)\left(\left(\frac{A}{\Delta d_x}\right)_{i-\frac{1}{2}}(k)_{i-\frac{1}{2}}\left(\frac{k_{ro}}{\mu_o^2}\left(\mu_o\frac{\partial b_o}{\partial p_o} - b_o\frac{\partial \mu_o}{\partial p_o}\right)\right)\Big|_{i-1}\right) \tag{8.124}$$

$$\frac{\partial T_{o\,i-\frac{1}{2}}}{\partial S_{o\,i-1}} = (1-\delta_o)\left(\left(\frac{A}{\Delta d_x}\right)_{i-\frac{1}{2}}(k)_{i-\frac{1}{2}}\left(\frac{b_o}{\mu_o}\left(\frac{\partial k_{ro}}{\partial S_o}\right)\right)\Big|_{i-1}\right) \tag{8.125}$$

if $(\Delta p_o - \gamma_o \Delta Z)|_{i-\frac{1}{2}} \leq 0 \quad \delta_o = 0$; if not $\delta_o = 1$

Similar expressions have been developed in the vertical direction, considering the replace of the index i by j, and also to consider their respective areas and distances.

For the terms in $i + 1/2$ gas phase

$$\frac{\partial T_{g\,i+\frac{1}{2}}}{\partial p_{o\,i,j}} = \left(1 - \delta_g\right) \left(\left(\frac{A}{\Delta d_x}\right)_{i+\frac{1}{2}} (k)_{i+\frac{1}{2}} \left(\frac{k_{rg}}{\mu_g^2} \left(\mu_g \frac{\partial b_g}{\partial p_g} - b_g \frac{\partial \mu_g}{\partial p_g} \right) \right) \Bigg|_{i,j} \right)$$

$$\frac{\partial T_{g\,i+\frac{1}{2}}}{\partial S_{o\,i,j}} = \left(1 - \delta_g\right) \left(\frac{\partial T_{g\,i+\frac{1}{2}}}{\partial p_{o\,i,j}} \right) \left(\frac{\partial Pc_{og}}{\partial S_o} \right)_{i,j} \tag{8.126}$$

if $\left(\Delta p_o + \Delta Pc_{og} - \gamma_g \Delta Z\right)\big|_{i+\frac{1}{2}} < 0$ $\delta_g = 0$, otherwise $\delta_g = 1$.

$$\frac{\partial T_{g\,i+\frac{1}{2}}}{\partial p_{o\,i+1}} = \left(1 - \delta_g\right) \left(\left(\frac{A}{\Delta d_x}\right)_{i+\frac{1}{2}} (k)_{i+\frac{1}{2}} \left(\frac{k_{rg}}{\mu_g^2} \left(\mu_g \frac{\partial b_g}{\partial p_g} - b_g \frac{\partial \mu_g}{\partial p_g} \right) \right) \Bigg|_{i+1} \right)$$

$$\frac{\partial T_{g\,i+\frac{1}{2}}}{\partial S_{o\,i+1}} = \left(1 - \delta_g\right) \left(\frac{\partial T_{g\,i+\frac{1}{2}}}{\partial p_{o\,i+1}} \right) \left(\frac{\partial Pc_{og}}{\partial S_o} \right)_{i+1} \tag{8.127}$$

if $\left(\Delta p_o + \Delta Pc_{og} - \gamma_g \Delta Z\right)\big|_{i+\frac{1}{2}} \geq 0$ $\delta_g = 0$, otherwise $\delta_g = 1$.

For the terms in $i - 1/2$

$$\frac{\partial T_{p\,i-\frac{1}{2}}}{\partial p_{o\,i,j}} = \left(1 - \delta_p\right) \left(\left(\frac{A}{\Delta d_x}\right)_{i-\frac{1}{2}} (k)_{i-\frac{1}{2}} \left(\frac{k_{rp}}{\mu_p^2} \left(\mu_p \frac{\partial b_p}{\partial p_p} - b_p \frac{\partial \mu_p}{\partial p_p} \right) \right) \Bigg|_{i,j} \right) \tag{8.128}$$

$$\frac{\partial T_{p\,i-\frac{1}{2}}}{\partial S_{o\,i,j}} = \left(1 - \delta_p\right) \left(\left(\frac{A}{\Delta d_x}\right)_{i-\frac{1}{2}} (k)_{i-\frac{1}{2}} \left(\frac{b_p}{\mu_p} \left(\frac{\partial k_{rp}}{\partial S_o} \right) \right) \Bigg|_{i,j} \right) \tag{8.129}$$

if $\left(\Delta p_p - \gamma_p \Delta Z\right)\big|_{i-\frac{1}{2}} > 0$ $\delta_p = 0$, otherwise $\delta_p = 1$, for $p = o, g$

$$\frac{\partial T_{p\,i-\frac{1}{2}}}{\partial p_{o\,i-1}} = \left(1 - \delta_p\right) \left(\left(\frac{A}{\Delta d_x}\right)_{i-\frac{1}{2}} (k)_{i-\frac{1}{2}} \left(\frac{k_{rp}}{\mu_p^2} \left(\mu_p \frac{\partial b_p}{\partial p_p} - b_p \frac{\partial \mu_p}{\partial p_p} \right) \right) \Bigg|_{i-1} \right) \tag{8.130}$$

$$\frac{\partial T_{p\,i-\frac{1}{2}}}{\partial S_{o\,i-1}} = \left(1 - \delta_p\right) \left(\left(\frac{A}{\Delta d_x}\right)_{i-\frac{1}{2}} (k)_{i-\frac{1}{2}} \left(\frac{b_p}{\mu_p} \left(\frac{\partial k_{rp}}{\partial S_o} \right) \right) \Bigg|_{i-1} \right) \tag{8.131}$$

if $\left(\Delta p_p - \gamma_p \Delta Z\right)\big|_{i-\frac{1}{2}} \leq 0$ $\delta_p = 0$, otherwise $\delta_p = 1$, for $p = o, g$.

Similar expressions have been developed in the vertical direction, considering the replace of the index i by j, and also to consider their respective areas and distances.

E.   Source terms

The management of the expressions for the source terms depends on the boundary conditions selected. The calculations for the derivatives of the source terms will be presented in the section of the handling of boundary conditions, Section 8.2.7.

F.   Accumulation term

Management of the porosity expressions. The variation of porosity can be expressed in terms of the compressibility of the rock, Cr, by

$$\phi\big|_{i,j}^{n+1} = \left(\phi^n(1 + Cr(p^{n+1} - p^n))\right)_{i,j} \tag{8.132}$$

thus

$$\frac{\partial \phi_{i,j}}{\partial p_{i,j}}\Bigg|^{n+1} = \left(\phi^n Cr\right)_{i,j} \tag{8.133}$$

if we use the pore volume at the beginning of time step (n) we have $Vp|_{i,j}^{n} = Vr|_{i,j}\,\phi^{n}$ so the terms of accumulation for the oil component relative to the oil pressure will be

$$\frac{Vr|_{i,j}}{\Delta t}\left(S_o b_o + (1-S_o)\,b_g r_s\right)\frac{\partial\phi}{\partial p_{o i,j}}$$

$$+\frac{Vr|_{i,j}}{\Delta t}\left\{\phi\left(S_o\frac{\partial b_o}{\partial p_{o i,j}} + (1-S_o)\left(b_g\frac{\partial r_s}{\partial p_{o i,j}} + r_s\frac{\partial b_g}{\partial p_{o i,j}}\right)\right)\right\}$$

$$=\frac{Vp|_{i,j}^{n}}{\Delta t}\left\{\left(S_o b_o + (1-S_o)\,b_g r_s\right)Cr\right\}_{i,j}$$

$$+\frac{Vp|_{i,j}^{n}}{\Delta t}\left\{\left(1+Cr\left(p^{n+1}-p^{n}\right)\right)\left(S_o\frac{\partial b_o}{\partial p_{o i,j}} + (1-S_o)\left(b_g\frac{\partial r_s}{\partial p_{o i,j}} + r_s\frac{\partial b_g}{\partial p_{o i,j}}\right)\right)\right\}_{i,j}$$

(8.134)

for the oil component respect to saturation

$$\frac{Vr|_{i,j}}{\Delta t}\left\{\phi\left(b_o - b_g r_s + r_s(1-S_o)\frac{\partial b_g}{\partial S_o} + b_g(1-S_o)\frac{\partial r_s}{\partial S_o}\right)\right\}$$

(8.135)

$$=\frac{Vp|_{i,j}^{n}}{\Delta t}\left(1+Cr\left(p^{n+1}-p^{n}\right)\right)\left(b_o - b_g r_s + r_s(1-S_o)\frac{\partial b_g}{\partial S_o} + b_g(1-S_o)\frac{\partial r_s}{\partial S_o}\right)_{i,j}$$

for the gas component respect to the pressure

$$\frac{Vr|_{i,j}}{\Delta t}\left\{\left(S_o b_o R_s + (1-S_o)\,b_g\right)\frac{\partial\phi}{\partial p_{o i,j}}\right\}$$

$$+\frac{Vr|_{i,j}}{\Delta t}\left\{\phi\left(S_o\left(b_o\frac{\partial R_s}{\partial p_{o i,j}} + R_s\frac{\partial b_o}{\partial p_{o i,j}}\right) + (1-S_o)\frac{\partial b_g}{\partial p_{o i,j}}\right)\right\}$$

$$=\frac{Vp|_{i,j}^{n}}{\Delta t}\left\{\left(S_o b_o R_s + (1-S_o)\,b_g\right)Cr\right\}_{i,j}$$

$$+\frac{Vp|_{i,j}^{n}}{\Delta t}\left\{\left(1+Cr\left(p^{n+1}-p^{n}\right)\right)\left(S_o\left(b_o\frac{\partial R_s}{\partial p_{o i,j}} + R_s\frac{\partial b_o}{\partial p_{o i,j}}\right) + (1-S_o)\frac{\partial b_g}{\partial p_{o i,j}}\right)\right\}_{i,j}$$

(8.136)

for the gas component respect to saturation

$$\frac{Vr|_{i,j}}{\Delta t}\left\{\phi\left(R_s b_o - b_g + (1-S_o)\frac{\partial b_g}{\partial S_o}\right)\right\}$$

$$=\frac{Vp|_{i,j}^{n}}{\Delta t}\left(1+Cr\left(p^{n+1}-p^{n}\right)\right)\left(R_s b_o - b_g + (1-S_o)\frac{\partial b_g}{\partial S_o}\right)_{i,j}$$

(8.137)

8.2.5.2 *Matrix for the tracer:* $\mathbf{M}_T$

By analyzing the system for tracer component, we can see that the nonlinear flux terms are eliminated because the two-phase matrix was already solved for the unknowns p_o and S_o, that is

$T(p,S)^{n+1}\left[\Delta p + \Delta Pc(S) - \gamma(p,S)\,\Delta Z\right]^{n+1}$ are known for the advective term,
$q(p,S)^{n+1}$ are known for the source terms, and
$(\phi\,b\,S)^{n+1}$ are known in the terms of accumulation.

The nonlinearities in the system are only caused by the tracer concentration

$\Upsilon(p)^{n+1}\left[\Delta x_T\right]^{n+1}$ the diffusive term.

Thus, the nonlinearities that occur in the functions of tracer component residues are resolved by a fully implicit scheme (TI) for the mole fraction, which are evaluated at time level $n + 1$, to solve the tracer concentration

$$F_T|_{i,j}^{n+1} \rightarrow \left\{ [\Delta x_T]^{n+1} \right\} \tag{8.138}$$

In this way the elements of the Jacobian IT for the tracer component will be configured for the following derivatives respect to concentration,

$$
\begin{aligned}
\frac{\partial F_{T\,i,j}}{\partial x_{T\,i,j-1}} = &- T_o|_{j-\frac{1}{2}} \left(\Delta p_o|_{j-\frac{1}{2}} - \gamma_o|_{j-\frac{1}{2}} \Delta Z|_{j-\frac{1}{2}} \right) \frac{\partial x_{T\,j-\frac{1}{2}}}{\partial x_{T\,i,j-1}} + [\Upsilon_{To} b_o]_{j-\frac{1}{2}} \\
&- R_{ro} \left[K_T T_g \right]_{j-\frac{1}{2}} \left(\Delta p_o|_{j-\frac{1}{2}} + \Delta P c_{go}|_{j-\frac{1}{2}} - \gamma_g|_{j-\frac{1}{2}} \Delta Z|_{j-\frac{1}{2}} \right) \frac{\partial x_{T\,j-\frac{1}{2}}}{\partial x_{T\,i,j-1}} \\
&+ R_{ro} \left[\Upsilon_{Tg} b_g \right]_{j-\frac{1}{2}} K_T|_{j-1}
\end{aligned}
\tag{8.139}
$$

$$
\begin{aligned}
\frac{\partial F_{T\,i,j}}{\partial x_{T\,i-1,j}} = &- T_o|_{i-\frac{1}{2}} \left(\Delta p_o|_{i-\frac{1}{2}} - \gamma_o|_{i-\frac{1}{2}} \Delta Z|_{i-\frac{1}{2}} \right) \frac{\partial x_{T\,i-\frac{1}{2}}}{\partial x_{T\,i-1,j}} + [\Upsilon_{To} b_o]_{i-\frac{1}{2}} \\
&- R_{ro} \left[K_T T_g \right]_{i-\frac{1}{2}} \left(\Delta p_o|_{i-\frac{1}{2}} + \Delta P c_{go}|_{i-\frac{1}{2}} - \gamma_g|_{i-\frac{1}{2}} \Delta Z|_{i-\frac{1}{2}} \right) \frac{\partial x_{T\,i-\frac{1}{2}}}{\partial x_{T\,i-1,j}} \\
&+ R_{ro} \left[\Upsilon_{Tg} b_g \right]_{i-\frac{1}{2}} K_T|_{i-1}
\end{aligned}
\tag{8.140}
$$

$$
\begin{aligned}
\frac{\partial F_{T\,i,j}}{\partial x_{T\,i,j}} = & \; T_o|_{i+\frac{1}{2}} \left(\Delta p_o|_{i+\frac{1}{2}} - \gamma_o|_{i+\frac{1}{2}} \Delta Z|_{i+\frac{1}{2}} \right) \frac{\partial x_{T\,i+\frac{1}{2}}}{\partial x_{T\,i,j}} - [\Upsilon_{To} b_o]_{i+\frac{1}{2}} \\
&- T_o|_{i-\frac{1}{2}} \left(\Delta p_o|_{i-\frac{1}{2}} - \gamma_o|_{i-\frac{1}{2}} \Delta Z|_{i-\frac{1}{2}} \right) \frac{\partial x_{T\,i-\frac{1}{2}}}{\partial x_{T\,i,j}} - [\Upsilon_{To} b_o]_{i-\frac{1}{2}} \\
&+ T_o|_{j+\frac{1}{2}} \left(\Delta p_o|_{j+\frac{1}{2}} - \gamma_o|_{j+\frac{1}{2}} \Delta Z|_{j+\frac{1}{2}} \right) \frac{\partial x_{T\,j+\frac{1}{2}}}{\partial x_{T\,i,j}} - [\Upsilon_{To} b_o]_{j+\frac{1}{2}} \\
&- T_o|_{j-\frac{1}{2}} \left(\Delta p_o|_{j-\frac{1}{2}} - \gamma_o|_{j-\frac{1}{2}} \Delta Z|_{j-\frac{1}{2}} \right) \frac{\partial x_{T\,j-\frac{1}{2}}}{\partial x_{T\,i,j}} - [\Upsilon_{To} b_o]_{j-\frac{1}{2}} - R_{ro} \left[\Upsilon_{Tg} b_g \right] \\
&+ R_{ro}[K_T T_g]_{i+\frac{1}{2}} \left(\Delta p_o|_{i+\frac{1}{2}} + \Delta P c_{go}|_{i+\frac{1}{2}} - \gamma_g|_{i+\frac{1}{2}} \Delta Z|_{i+\frac{1}{2}} \right) \frac{\partial x_{T\,i+\frac{1}{2}}}{\partial x_{T\,i,j}}\bigg|_{i+\frac{1}{2}} K_T|_{i,j} \\
&- R_{ro}[K_T T_g]_{i-\frac{1}{2}} \left(\Delta p_o|_{i-\frac{1}{2}} + \Delta P c_{go}|_{i-\frac{1}{2}} - \gamma_g|_{i-\frac{1}{2}} \Delta Z|_{i-\frac{1}{2}} \right) \frac{\partial x_{T\,i-\frac{1}{2}}}{\partial x_{T\,i,j}} \\
&- R_{ro}[\Upsilon_{Tg} b_g]_{i-\frac{1}{2}} K_T|_{i,j} \\
&+ R_{ro}[K_T T_g]_{j+\frac{1}{2}} \left(\Delta p_o|_{j+\frac{1}{2}} + \Delta P c_{go}|_{j+\frac{1}{2}} - \gamma_g|_{j+\frac{1}{2}} \Delta Z|_{j+\frac{1}{2}} \right) \frac{\partial x_{T\,j+\frac{1}{2}}}{\partial x_{T\,i,j}} \\
&- R_{ro}[\Upsilon_{Tg} b_g]_{j+\frac{1}{2}} K_T|_{i,j} \\
&- R_{ro}[K_T T_g]_{j-\frac{1}{2}} \left(\Delta p_o|_{j-\frac{1}{2}} + \Delta P c_{go}|_{j-\frac{1}{2}} - \gamma_g|_{j-\frac{1}{2}} \Delta Z|_{j-\frac{1}{2}} \right) \frac{\partial x_{T\,j-\frac{1}{2}}}{\partial x_{T\,i,j}} \\
&- R_{ro}[\Upsilon_{Tg} b_g]_{j-\frac{1}{2}} K_T|_{i,j} \\
&+ \left(q_o|_{i,j} + x_T|_{i,j} \frac{\partial q_{o\,i,j}}{\partial x_{T\,i,j}} \right) + \left((R_{ro} x_T K_T)_{i,j} \frac{\partial q_{g\,i,j}}{\partial x_{T\,i,j}} + (R_{ro} K_T q_g)_{i,j} \right) \\
&- \frac{Vr|_{i,j}}{\Delta t} \left\{ \phi \left(S_o b_o + (1 - S_o) K_T R_{ro} b_g \right) \right\}
\end{aligned}
\tag{8.141}
$$

$$\frac{\partial F_{Ti,j}}{\partial x_{T\,i+1,j}} = T_o|_{i+\frac{1}{2}} \left(\Delta p_o|_{i+\frac{1}{2}} - \gamma_o|_{i+\frac{1}{2}} \Delta Z|_{i+\frac{1}{2}} \right) \frac{\partial x_{T\,i+\frac{1}{2}}}{\partial x_{T\,i+1,j}} + [\Upsilon_{To} b_o]_{i+\frac{1}{2}}$$

$$+ R_{ro} \left[K_T T_g \right]_{i+\frac{1}{2}} \left(\Delta p_o|_{i+\frac{1}{2}} + \Delta Pc_{go}|_{i+\frac{1}{2}} - \gamma_g|_{i+\frac{1}{2}} \Delta Z|_{i+\frac{1}{2}} \right) \frac{\partial x_{T\,i+\frac{1}{2}}}{\partial x_{T\,i+1,j}} \qquad (8.142)$$

$$+ R_{ro} \left[\Upsilon_{Tg} b_g \right]_{i+\frac{1}{2}} K_T|_{i+1}$$

$$\frac{\partial F_{Ti,j}}{\partial x_{T\,i,j+1}} = T_o|_{j+\frac{1}{2}} \left(\Delta p_o|_{j+\frac{1}{2}} - \gamma_o|_{j+\frac{1}{2}} \Delta Z|_{j+\frac{1}{2}} \right) \frac{\partial x_{T\,j+\frac{1}{2}}}{\partial x_{T\,i,j+1}} + [\Upsilon_{To} b_o]_{j+\frac{1}{2}}$$

$$+ R_{ro} \left[K_T T_g \right]_{j+\frac{1}{2}} \left(\Delta p_o|_{j+\frac{1}{2}} + \Delta Pc_{go}|_{j+\frac{1}{2}} - \gamma_g|_{j+\frac{1}{2}} \Delta Z|_{j+\frac{1}{2}} \right) \frac{\partial x_{T\,j+\frac{1}{2}}}{\partial x_{T\,i,j+1}} \qquad (8.143)$$

$$+ R_{ro} \left[\Upsilon_{Tg} b_g \right]_{j+\frac{1}{2}} K_T|_{j+1}$$

Definition of the partial derivatives at the borders: $i \pm 1/2$ and $j \pm 1/2$

A.   Terms of mole fractions

To determine the concentration partial derivative of the tracer, we need to apply the "upstream" criterium and the terms obtained are:

For the terms $i + 1/2$

$$\frac{\partial x_{T\,i+\frac{1}{2}}}{\partial x_{T\,i,j}} = (1 - \delta_o) \qquad \frac{\partial x_{T\,i+\frac{1}{2}}}{\partial x_{T\,i+1,j}} = (\delta_o) \qquad (8.144)$$

if $\left(x_T|_{i+1} - x_T|_i \right) < 0$ $\delta_o = 0$, otherwise $\delta_o = 1$.

For the terms in $i - 1/2$

$$\frac{\partial x_{T\,i-\frac{1}{2}}}{\partial x_{T\,i,j}} = (1 - \delta_o) \qquad \frac{\partial x_{T\,i-\frac{1}{2}}}{\partial x_{T\,i-1,j}} = (\delta_o) \qquad (8.145)$$

if $\left(x_T|_i - x_T|_{i-1} \right) > 0$ $\delta_o = 0$, otherwise $\delta_o = 1$

Similar expressions are developed in the vertical direction for the partial in $j \pm 1$, only the index i is replaced by j.

B.   Source terms

The management of the expressions for the source terms depends on the boundary conditions that are selected. The calculations for the derivatives of the source terms will be presented in the section treating the boundary conditions, Section 2.8.

C.   Accumulation term

Because the pressure and saturation fields are known, the variation of porosity is know as well, thus the accumulation term then is

$$\frac{Vr|_{i,j}}{\Delta t} \{\phi(S_o b_o + (1 - S_o) K_T R_{ro} b_g)\} = \left[\frac{Vp|_{i,j}^{n+1}}{\Delta t} \{(S_o^{n+1} b_o^{n+1} + (1 - S_o^{n+1}) K_T^{n+1} R_{ro} b_g^{n+1})\} \right]_{i,j} \qquad (8.146)$$

8.2.6   Solution scheme

As discussed in Section 8.2.4, the methodology to solve the unknowns is (i) first solve the two-phase equation system for pressure and saturation, and then (ii) solve the tracer concentration. The solution of each system is described hereafter.

8.2.6.1 Solution for the $\mathbf{M}_{bf}^{(v)} \mathbf{d}_{bf}^{(v+1)} = -\mathbf{F}_{bf}^{(v)}$ system

We present the solution of the system using the Newton-Raphson method considering their general solution, known as "Total Implicit solution", as well as two other particular solutions or variants: the Semi-Implicit and the IMPES.

General method of Fully Implicit Newton-Raphson (Total Implicit - TI).

As presented in Section 8.2.5, the dependence on two-phase matrix can define the "nature" of the terms that constitute the Jacobian of the partial derivatives of residual functions in its most general form, the fully implicit Jacobian (TI). Consequently, the Jacobian can be expressed as the sum of the following submatrices,

$$[J]_{TI} = [T] + [T']_{p,S} + [Pc']_S + [\gamma']_{p,S} + [Q']_{p,S} + [(\phi bS)']_{p,S} \tag{8.147}$$

where

$[T]$ is a submatrix of terms of transmissivity
$[T']_{p,S}$ is a submatrix of partial derivatives from transmissivity
$[Pc']_S$ is a submatrix of partial derivatives from capillary pressure
$[\gamma']_{p,S}$ is a submatrix of partial derivatives from specific weight
$[Q']_{p,S}$ is a submatrix of partial derivatives from the source terms
$[(\phi bS)']_{p,S}$ is the matrix of partial derivatives from the accumulation term

As discussed in Section 8.2.6 the implicit terms for the Total Implicit Method (TI) are given by

$$F|_{i,j}^{n+1} \rightarrow \left\{ T(p,S)^{n+1} [\Delta p + \Delta Pc(S) - \gamma(p,S) \Delta Z]^{n+1}, q(p,S)^{n+1}, (\phi bS)^{n+1} \right\} \tag{8.148}$$

By treating the matrix representation of the system we get for the i, j-node

$$
\left(\begin{bmatrix} fb_{op} & fb_{os} \\ fb_{gp} & fb_{gs} \end{bmatrix} \begin{bmatrix} cb_{op} & cb_{os} \\ cb_{gp} & cb_{gs} \end{bmatrix} \begin{bmatrix} ab_{op} & ab_{os} \\ ab_{gp} & ab_{gs} \end{bmatrix} \begin{bmatrix} bb_{op} & bb_{os} \\ bb_{gp} & bb_{gs} \end{bmatrix} \begin{bmatrix} eb_{op} & eb_{os} \\ eb_{gp} & eb_{gs} \end{bmatrix} \right)_{i,j}^{(v)}
\begin{pmatrix}
\begin{bmatrix} \delta p_o \\ \delta S_o \end{bmatrix}_{j-1}^{(v+1)} \\
\begin{bmatrix} \delta p_o \\ \delta S_o \end{bmatrix}_{i-1}^{(v+1)} \\
\begin{bmatrix} \delta p_o \\ \delta S_o \end{bmatrix}_{i}^{(v+1)} \\
\begin{bmatrix} \delta p_o \\ \delta S_o \end{bmatrix}_{i+1}^{(v+1)} \\
\begin{bmatrix} \delta p_o \\ \delta S_o \end{bmatrix}_{j+1}^{(v+1)}
\end{pmatrix}
$$

$$
= -\begin{bmatrix} F_o \\ F_g \end{bmatrix}_{i,j}^{(v)} \tag{8.149}
$$

By solving the derivatives, considering that all terms are implicit, we obtain a fully implicit system, as an iterative process

$$[J]_{TI}^v \, \delta u^{v+1} = -F^v \tag{8.150}$$

Which is an $IJ \times IJ$ system that can be solved iteratively to obtain $\delta p_{oi,j}^{n+1}$ y $\delta S_{oi,j}^{n+1}$ (subject to a preset tolerance for the values calculated in subsequent iterations). In this way the pressure and saturation values of each cell at time $n + 1$ are obtained with the following expressions, in the last iteration

$$p_o|_{i,j}^{n+1} = p_o|_{i,j}^n + \delta p_o|_{i,j}^{n+1} \quad \text{and} \quad S_o|_{i,j}^{n+1} = S_o|_{i,j}^n + \delta S_o|_{i,j}^{n+1} \qquad (8.151)$$

8.2.6.2   *Semi-implicit method (SI)*

In this method it is assumed that the nonlinearities produced by the transmissivity, density and source terms are small compared to those caused by the saturation terms. It is consistent with the here considered fully implicit solution scheme, providing that the solution is obtained in the first iteration (first iteration of fully implicit scheme).

The dependence on pressure-saturation is then given exactly as in TI

$$[J]_{TI} = [T] + [T']_{p,S} + [Pc']_S + [\gamma']_{p,S} + [\Upsilon] + [\Upsilon']_{p,S} + [b']_p + [Q']_{p,S} + [(\phi bS)']_{p,S} \quad (8.152)$$

however, the iterative process is only performed once, and the solution is as follows

$$[J]_{TI}^n \, \delta u^{n+1} = -F^n, \qquad (8.153)$$

which is an $IJ \times IJ$ system solved in the first TI iteration for $\delta p_o|_{i,j}^{n+1}$ and $\delta S_o|_{i,j}^{n+1}$.

8.2.6.3   *IMPES method (Implicit pressure-explicit saturation)*

This method considers that all parameters that depend on the saturation are explicit (at time level n), and only the pressures are considered as implicit (this is a less computer demanding solution option). With this assumption the dependency expression for the residual functions is given as follows

$$F|_{i,j\,IMPES}^{n+1} \rightarrow \{T(p,S)^{n+1}[\Delta p^{n+1} + \Delta Pc(S)^n - \gamma(p,S)^n \Delta Z], q(p,S)^n, (\phi bS)^{n+1}\} \quad (8.154)$$

Therefore, the Jacobian in the IMPES only consider the transmissivity and accumulation terms,

$$[J]_{IMPES} = [T] + [T']_{p,S} + [(\phi bS)']_{p,S} \qquad (8.155)$$

By solving the system, in the first iteration, we have

$$[J]_{IMPES}^n \, \delta u^{n+1} = -F^n \qquad (8.156)$$

Using a matrix representation we have for the i,j-node

$$\left(\begin{bmatrix} fb_{op} & 0 \\ fb_{gp} & 0 \end{bmatrix} \begin{bmatrix} cb_{op} & 0 \\ cb_{gp} & 0 \end{bmatrix} \begin{bmatrix} ab_{op} & ab_{os} \\ ab_{gp} & ab_{gs} \end{bmatrix} \begin{bmatrix} bb_{op} & 0 \\ bb_{gp} & 0 \end{bmatrix} \begin{bmatrix} eb_{op} & 0 \\ eb_{gp} & 0 \end{bmatrix} \right)_{i,j}^n \begin{pmatrix} \begin{bmatrix} \delta p_o \\ \delta S_o \end{bmatrix}_{j-1}^{n+1} \\ \begin{bmatrix} \delta p_o \\ \delta S_o \end{bmatrix}_{i-1}^{n+1} \\ \begin{bmatrix} \delta p_o \\ \delta S_o \end{bmatrix}_{i}^{n+1} \\ \begin{bmatrix} \delta p_o \\ \delta S_o \end{bmatrix}_{i+1}^{n+1} \\ \begin{bmatrix} \delta p_o \\ \delta S_o \end{bmatrix}_{j+1}^{n+1} \end{pmatrix}$$

$$= -\begin{bmatrix} F_o \\ F_g \end{bmatrix}_{i,j\,IMPES}^n \qquad (8.157)$$

Due to the structure of the previous system, it decouples the calculation of δS_o, i.e.,

$$[\delta S_o]_{i,j}^{n+1} = -\left[F_{g\,IMPES}^*\right]_{i,j}^n - \left[\, fb_{gp}^* \;\; cb_{gp}^* \;\; ab_{gp}^* \;\; bb_{gp}^* \;\; eb_{gp}^* \,\right]_{i,j}^n \begin{bmatrix} \delta p_{o\,j-1} \\ \delta p_{o\,i-1} \\ \delta p_{o\,i,j} \\ \delta p_{o\,i+1} \\ \delta p_{o\,j+1} \end{bmatrix}^{n+1} \qquad (8.158)$$

where

$$\left[ab_{gp}^*\right]_{i,j} = \left[\frac{ab_{gp}}{ab_{gs}}\right]_{i,j} \;;\quad \left[bb_{gp}^*\right]_{i,j} = \left[\frac{bb_{gp}}{ab_{gs}}\right]_{i,j} \qquad (8.159)$$

$$\left[cb_{gp}^*\right]_{i,j} = \left[\frac{cb_{gp}}{ab_{gs}}\right]_{i,j} \;;\quad \left[eb_{gp}^*\right]_{i,j} = \left[\frac{eb_{gp}}{ab_{gs}}\right]_{i,j} \qquad (8.160)$$

$$\left[fb_{gp}^*\right]_{i,j} = \left[\frac{fb_{gp}}{ab_{gs}}\right]_{i,j} \;;\quad \left[F_{g\,IMPES}^*\right]_{i,j} = \left[\frac{ab_{gp}}{ab_{gs}}\right]_{i,j} \qquad (8.161)$$

δS_o^{n+1} can be expressed as

$$[\delta S_o]_{i,j}^{n+1} = -\left[F_g^*\right]_{i,j}^n - fb_{gp}^*\big|_{i,j}^n \,\delta p_o\big|_{i,j-1}^{n+1} - cb_{gp}^*\big|_{i,j}^n \,\delta p_o\big|_{i-1,j}^{n+1}$$
$$- ab_{gp}^*\big|_{i,j}^n \,\delta p_o\big|_{i,j}^{n+1} - bb_{gp}^*\big|_{i,j}^n \,\delta p_o\big|_{i+1,j}^{n+1} - eb_{gp}^*\big|_{i,j}^n \,\delta p_o\big|_{i,j+1}^{n+1} \qquad (8.162)$$

And the solution for the oil component, δp_o^{n+1}, can be expressed as

$$fb_{op}\big|_{i,j}^n \,|\delta p_o|_{i,j-1}^{n+1} + cb_{op}\big|_{i,j}^n \,\delta p_o\big|_{i-1,j}^{n+1} + ab_{op}\big|_{i,j}^n \,\delta p_o\big|_{i,j}^{n+1}$$
$$+ ab_{os}\big|_{i,j}^n \,\delta S_o\big|_{i,j}^{n+1} + bb_{op}\big|_{i,j}^n \,\delta p_o\big|_{i+1,j}^{n+1} + eb_{op}\big|_{i,j}^n \,\delta p_o\big|_{i,j+1}^{n+1} = -[F_o]_{i,j}^n \qquad (8.163)$$

By substituting the oil saturation expression in the oil component equation, an expression involving only terms of pressure is finally obtained:

$$fb_{op}^*\big|_{i,j}^n \,\delta p_o\big|_{i,j-1}^{n+1} + cb_{op}^*\big|_{i,j}^n \,\delta p_o\big|_{i-1,j}^{n+1} + ab_{op}^*\big|_{i,j}^n \,\delta p_o\big|_{i,j}^{n+1}$$
$$+ bb_{op}^*\big|_{i,j}^n \,\delta p_o\big|_{i+1,j}^{n+1} + eb_{op}^*\big|_{i,j}^n \,\delta p_o\big|_{i,j+1}^{n+1} = -\left[F_o^*\right]_{i,j}^n \qquad (8.164)$$

where

$$\left[ab_{op}^*\right]_{i,j} = \left[ab_{op} - ab_{os}ab_{gp}^*\right]_{i,j} \;;\quad \left[bb_{op}^*\right]_{i,j} = \left[bb_{op} - ab_{os}bb_{gp}^*\right]_{i,j} \qquad (8.165)$$

$$\left[cb_{op}^*\right]_{i,j} = \left[cb_{op} - ab_{os}cb_{gp}^*\right]_{i,j} \;;\quad \left[eb_{op}^*\right]_{i,j} = \left[eb_{op} - ab_{os}eb_{gp}^*\right]_{i,j} \qquad (8.166)$$

$$\left[fb_{op}^*\right]_{i,j} = \left[fb_{op} - ab_{os}fb_{gp}^*\right]_{i,j} \;;\quad \left[F_o^*\right]_{i,j} = \left[F_o - ab_{os}F_g^*\right]_{i,j} \qquad (8.167)$$

Which is a system of equations with $IJ \times IJ$ unknowns, that can solve the values obtained $\delta p_o\big|_{i,j}^{n+1}$, and later replaced in the system for the obtained $\delta S_o\big|_{i,j}^{n+1}$. The pressure and saturation values are obtained using the following expressions

$$p_o\big|_{i,j}^{n+1} = p_o\big|_{i,j}^n + \delta p_o\big|_{i,j}^{n+1} \;;\quad S_o\big|_{i,j}^{n+1} = S_o\big|_{i,j}^n + \delta S_o\big|_{i,j}^{n+1} \qquad (8.168)$$

8.2.6.4   *Solution for the* $\mathbf{M}_T^{(n+1)}\mathbf{d}_T^{(n+1)} = -\mathbf{F}_T^{(n+1)}$ *system*

By analyzing the parameter dependence with respect to the tracer, we can define the nature of the factors that constitute the $\mathbf{M}_T$ matrix. Because we already solved the two-phase flow system, the pressure and oil saturation values are known at time level $n + 1$. These values are to be integrated into the tracer system to define the matrix coefficients $\mathbf{M}_T$.

$$[M_T]_{TI} = T + p + Pc + \gamma + \Upsilon + b + Q + (\phi b S) \tag{8.169}$$

treating the matrix representation of the system we have for i, j-node

$$[fT_{ox}]_{i,j}^{n+1} \ [cT_{ox}]_{i,j}^{n+1} \ [aT_{ox}]_{i,j}^{n+1} \ [bT_{ox}]_{i,j}^{n+1} \ [eT_{ox}]_{i,j}^{n+1} \ \begin{matrix} [\delta x_T]_{j-1}^{n+1} \\ [\delta x_T]_{i-1}^{n+1} \\ [\delta x_T]_i^{n+1} \\ [\delta x_T]_{i+1}^{n+1} \\ [\delta x_T]_{j+1}^{n+1} \end{matrix} = -[F_T]_{i,j}^n \tag{8.170}$$

The tracer component system is then a set of linear equations, which can be solved to find the tracer mole fraction in the liquid phase:

$$x_T|_{i,j}^{n+1} = x_T|_{i,j}^n + \delta x_T|_{i,j}^{n+1} \tag{8.171}$$

8.2.7 Treating the boundary conditions

8.2.7.1 Wells

The total volume rate (at surface conditions) associated to a production well is given by the sum of the volume of oil and gas, and this is linked to the flowing bottom hole pressure.

$$q_{TOT}(t) = q_{oTOT}(t) + q_{gTOT}(t) \Rightarrow p_{wf}(t) \tag{8.172}$$

Therefore, the boundary conditions for any injector or producer well can be handled through the superficial flow rate, or through the bottom-hole pressure. The general relationship that links rate and pressure in a cell is given by

$$q_p = Tp^* \left(\Delta p_p - \gamma_p \Delta Z\right) \quad \text{for } p = o, g \tag{8.173}$$

where T^* represents an equivalent transmissivity depending on the case being treated.

With the two types of conditions, the following production/injection cases must be considered

1. Oil producer well is maintained at constant flow rate.
2. Gas producer well set at a constant flow rate.
3. Liquid injector well is kept at constant flow rate.
4. Gas injector well maintained at constant flow rate.
5. Well producer/injector is set at constant bottom-hole pressure.

To handle these conditions we also have to define two different cell types (see Figure 8.5):

- "wall of well" to simulate the flow in an area of influence of well (radial flow, r, z) or flow in a vertical section between two wells (linear flow, x, z),
- "well cell" to simulate a specific location for a well (areal flow, x, y) in a cell.

The definitions of terms and the derivatives for different cases and conditions will be developed in subsequent paragraphs.

Required surface rate, q_{pTOT} (options 1,2,3 and 4).

The sign defines if the rate corresponds to a producer (minus sign) or to an injector well (plus sign). Expressing the flow rates in terms of total rate at standard conditions (@SC), and considering the solubility and vaporization relationship for the phases we have

$$q_{oTOT} = q_o + r_s q_g \quad \text{for the oil component}$$
$$q_{gTOT} = q_g + R_s q_o \quad \text{for the gas component}$$

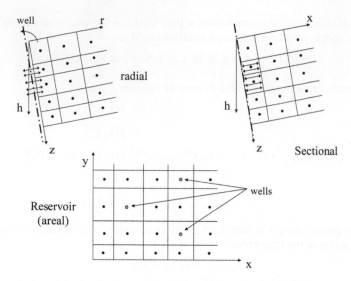

Figure 8.5 Description of the three different boundary condition considered.

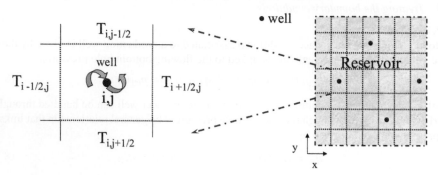

Figure 8.6 Transmissivity consideration scheme for each cell.

A. "Well-Cell" type (areal reservoir simulation)

The capacity fractions in terms of transmissivity of each component can be expressed in this case as:

$$\frac{q_o}{q_{oTOT}} = \frac{\sum T_o}{\sum T_o + r_s \sum T_g} \quad \frac{q_g}{q_{gTOT}} = \frac{\sum T_g}{\sum T_g + R_s \sum T_o}; \quad \frac{q_g}{q_o} = \frac{\sum T_g}{\sum T_o} \tag{8.174}$$

where the $\sum$ corresponds to the transmissivity terms at the boundary of a well cell, as shown in Figure 8.6.

$$\sum T_p = T_p\big|_{i,j-\frac{1}{2}} + T_p\big|_{i-\frac{1}{2},j} + T_p\big|_{i+\frac{1}{2},j} + T_p\big|_{i,j+\frac{1}{2}} \tag{8.175}$$

Given the superficial volumetric total oil rate (options 1 or 5), the source term in the well i, j-cell is expressed as

$$q_o = \frac{q_{oTOT} \sum T_o}{\sum T_o + r_s \sum T_g} q_g = \frac{\sum T_g}{\sum T_o} \left(\frac{q_{oTOT} \sum T_o}{\sum T_o + r_s \sum T_g} \right). \tag{8.176}$$

If the superficial volumetric total gas rate is given (option 2 or 6), it follows

$$q_g = \frac{q_{gTOT} \sum T_g}{\sum T_g + R_s \sum T_o} q_o = \frac{\sum T_o}{\sum T_g} \left(\frac{q_{gTOT} \sum T_g}{\sum T_g + R_s \sum T_o} \right) \tag{8.177}$$

To evaluate the partial derivatives respect to oil pressure and oil saturation we use, for the oil phase:

$$\frac{\partial q_o}{\partial p_o} = \frac{q_{oTOT}}{\left(\sum T_o + r_s \sum T_g\right)^2} \left(\left(\sum T_o + r_s \sum T_g\right) \frac{\partial \sum T_o}{\partial p_o}\right)$$
$$- \frac{q_{oTOT}}{\left(\sum T_o + r_s \sum T_g\right)^2} \left(\sum T_o \left(\frac{\partial \sum T_o}{\partial p_o} + r_s \frac{\partial \sum T_g}{\partial p_o}\right)\right)$$

(8.178)

$$\frac{\partial q_o}{\partial S_o} = \frac{q_{oTOT}}{\left(\sum T_o + r_s \sum T_g\right)^2} \left(\left(\sum T_o + r_s \sum T_g\right) \frac{\partial \sum T_o}{\partial S_o}\right)$$
$$- \frac{q_{oTOT}}{\left(\sum T_o + r_s \sum T_g\right)^2} \left(\sum T_o \left(\frac{\partial \sum T_o}{\partial S_o} + r_s \frac{\partial \sum T_g}{\partial S_o}\right)\right)$$

(8.179)

and for gas phase:

$$\frac{\partial q_g}{\partial p_o} = \left(\frac{\sum T_g}{\sum T_o}\right) \frac{\partial q_o}{\partial p_o} + \frac{q_o}{\left(\sum T_o\right)^2} \left(\sum T_o \frac{\partial \sum T_g}{\partial p_o} - \sum T_g \frac{\partial \sum T_o}{\partial p_o}\right)$$

(8.180)

$$\frac{\partial q_g}{\partial S_o} = \left(\frac{\sum T_g}{\sum T_o}\right) \frac{\partial q_o}{\partial S_o} + \frac{q_o}{\left(\sum T_o\right)^2} \left(\sum T_o \frac{\partial \sum T_g}{\partial S_o} - \sum T_g \frac{\partial \sum T_o}{\partial S_o}\right)$$

(8.181)

In the case the condition is set for a gas producing well (a given gas rate), we have for the oil phase

$$\frac{\partial q_o}{\partial p_o} = \left(\frac{\sum T_o}{\sum T_g}\right) \frac{\partial q_g}{\partial p_o} + \frac{q_g}{\left(\sum T_g\right)^2} \left(\sum T_g \frac{\partial \sum T_o}{\partial p_o} - \sum T_o \frac{\partial \sum T_g}{\partial p_o}\right)$$

(8.182)

$$\frac{\partial q_o}{\partial S_o} = \left(\frac{\sum T_o}{\sum T_g}\right) \frac{\partial q_g}{\partial S_o} + \frac{q_g}{\left(\sum T_g\right)^2} \left(\sum T_g \frac{\partial \sum T_o}{\partial S_o} - \sum T_o \frac{\partial \sum T_g}{\partial S_o}\right)$$

(8.183)

and for the gas phase,

$$\frac{\partial q_g}{\partial p_o} = \frac{q_{gTOT}}{\left(\sum T_g + R_s \sum T_o\right)^2} \left(\left(\sum T_g + R_s \sum T_o\right) \frac{\partial \sum T_g}{\partial p_o}\right)$$
$$- \frac{q_{gTOT}}{\left(\sum T_g + R_s \sum T_o\right)^2} \left(\sum T_g \left(\frac{\partial \sum T_g}{\partial p_o} + R_s \frac{\partial \sum T_g}{\partial p_o}\right)\right)$$

(8.184)

$$\frac{\partial q_g}{\partial S_o} = \frac{q_{gTOT}}{\left(\sum T_g + R_s \sum T_o\right)^2} \left(\left(\sum T_g + R_s \sum T_o\right) \frac{\partial \sum T_g}{\partial S_o}\right)$$
$$- \frac{q_{gTOT}}{\left(\sum T_g + R_s \sum T_o\right)^2} \left(\sum T_g \left(\frac{\partial \sum T_g}{\partial S_o} + R_s \frac{\partial \sum T_o}{\partial S_o}\right)\right)$$

(8.185)

where

$$\frac{\partial \sum T_p}{\partial p_o} = \frac{\partial T_{pi,j-\frac{1}{2}}}{\partial p_{oi,j}} + \frac{\partial T_{pi-\frac{1}{2},j}}{\partial p_{oi,j}} + \frac{\partial T_{pi+\frac{1}{2},j}}{\partial p_{oi,j}} + \frac{\partial T_{pi,j+\frac{1}{2}}}{\partial p_{oi,j}}$$

(8.186)

$$\frac{\partial \sum T_p}{\partial S_o} = \frac{\partial T_{pi,j-\frac{1}{2}}}{\partial S_{oi,j}} + \frac{\partial T_{pi-\frac{1}{2},j}}{\partial S_{oi,j}} + \frac{\partial T_{pi+\frac{1}{2},j}}{\partial S_{oi,j}} + \frac{\partial T_{pi,j+\frac{1}{2}}}{\partial S_{oi,j}}$$

(8.187)

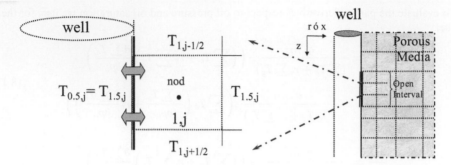

Figure 8.7　Transmissivity considerations for a "well wall" cell.

A flowing relationship for the bottom hole pressure in the well can be defined. This is based on the fact that oil/gas rates in a producer/injection cell are associated to a fall/rise in bottom hole pressure, and that the pressure depends on the characteristics of the production zone, the permeability and the size of the cell. These relationships read

$$P_{wf}\big|_{i,j} = \left[p_o - \frac{q_o B_o \mu_o}{2\pi \langle \mathbf{k}\rangle h \lambda_o} \ln\left(\frac{r_0}{r_w}\right) \right]_{i,j} \quad \text{with } r_0\big|_{i,j} = \left[0.28 \frac{\sqrt{\sqrt{\frac{k_y}{k_x}}\Delta x^2 + \sqrt{\frac{k_x}{k_y}}\Delta y^2}}{\sqrt[4]{\frac{k_x}{k_y}} + \sqrt[4]{\frac{k_y}{k_x}}} \right]_{i,j}$$

(8.188)

where $\langle k\rangle$ is the average permeability of the cell, h the average thickness of the producing zone, and r_w the radius of the well.

B.　"Well-wall" cell type (radial or linear simulation of vertical section)
When this condition type is employed, the "well wall" cell is defined by the boundaries corresponding to the well production interval. This means, that the index j of the "well wall" cell is associated to an index w, which refers to the location of the open interval.

$$j \Rightarrow w \quad w = 1, 2, 3, \dots, nid,$$

here nid corresponds to the total number of cells that are open to the flow in the well wall. We also know that the total rate in each phase is distributed among the cells that define the location of the open interval (as illustrated in Figure 8.7), i.e.

$$q_{pTOT} = \sum_{w=1}^{nid} q_{pTOT\,j\Rightarrow w} \quad \text{for } p = o, g$$

(8.189)

Thus the total cell rate expressed in terms of a fractions of the production capacity (transmissivity) of each component is given by

$$\frac{q_{oTOT}\big|_{j\Rightarrow w}}{q_{oTOT}} = \frac{T_o\big|_{0.5,\,j\Rightarrow w}}{\sum T_o\big|_{0.5}}, \frac{q_{gTOT}\big|_{j\Rightarrow w}}{q_{oTOT}\big|_{j\Rightarrow w}} = \frac{\sum T_g\big|_{0.5}}{\sum T_o\big|_{0.5}}$$

(8.190)

Here the sums comprehends the transmissivity boundary $i = 0.5$ for all the cells that comprise the open interval,

$$\sum T_p\big|_{0.5} = \sum_{w=1}^{nid} T_p\big|_{0.5,\,j\Rightarrow w} \quad \text{for } p = o, g$$

(8.191)

This conceptualization also applies to the case of vertical section simulation between two wells, the boundary condition is identically applied to the second well (for the right side well of the section) considering for the boundary transmissivity at $I + 1/2$.

After determining the rate that corresponds to each cell in the open interval (the fractions of the total well rate), the same expressions that were developed for "well cell" type can be applied. This can be done by considering

$$q_{o\,TOT} = q_{o\,TOTO}|_{j\Rightarrow w} \sum T_p = \sum T_p|_{0.5} \tag{8.192}$$

To calculate the bottom-hole pressure in the well we must make some considerations. Since the well rate is distributed among the cells that correspond to the open interval, the flowing bottom hole pressure in the well is defined by the average value weighted in terms of cell size in z direction. The pressure calculated at the boundary is determined by

$$p_o|_{i=0.5,j\Rightarrow w} = \left(\frac{q_o}{T_{0.5}} + [p_{o1} - \gamma_o \, (Z_1 - Z_{0.5})] \right)_{i,j\Rightarrow w} \tag{8.193}$$

Knowing pressures of each cell in the open interval, the average value of the flowing bottom hole pressure is given by

$$p_{wf} = \frac{\sum p_o|_{i=0.5} \, \Delta z|_{1,j\Rightarrow w}}{\sum \Delta z|_{1,j\Rightarrow w}} \tag{8.194}$$

The pressures on the boundary $I + 1/2$ (for the right well in the simulation of vertical section) has similar expression considering the boundary at $I + 1/2$.

The bottom-hole pressure required, P_{wf} (option 5).

By using conditions on the bottom hole pressure, the production/injection rate is that obtained directly by solving the equation system. As discussed in previous paragraphs, in the case of "well cell" type, the calculation of the rate involves the well characteristics and dimensions of the cell. If the cell is "well wall" type, the P_{wf} defines the pressure on the border 0.5 (or $I + 1/2$).

C. "Well-Cell" type (areal reservoir simulation)
As presented in the last section, cell pressure and bottom hole pressure depend on the characteristics of the production zone, the permeability and the size of the cell. Given a bottom hole pressure and cell pressure, the calculation of the flow rate is obtained by

$$q_o\,(t)|_{i,j} = \frac{2\pi \langle k \rangle h \lambda_o}{B_o \mu_o \ln \left(\frac{r_0}{r_w} \right)} \, (p_o(t) - P_{wf})|_{i,j} \tag{8.195}$$

D. "Well-wall" cell type (radial or linear simulation of vertical section)
Given a bottom hole pressure, and knowing the relationship between rate and pressure

$$q_p = T_p|_{0.5} \, (\Delta p_p - \gamma_p \Delta Z) \tag{8.196}$$

For each "well wall" cell, the rate will be determined for the oil component, by

$$q_o|_{1,j\Rightarrow w} = T_o|_{0.5,j\Rightarrow w} \, (p_o|_{1,j\Rightarrow w} - p_{wf} - \gamma_o|_{0.5,j\Rightarrow w} \, \Delta Z|_{0.5,j\Rightarrow w}), \tag{8.197}$$

and for the gas component, by

$$q_g|_{1,j\Rightarrow w} = T_g|_{0.5,j\Rightarrow w} \, \left(p_g|_{1,j\Rightarrow w} - p_{wf} + \Delta Pc_{go}|_{0.5,j\Rightarrow w}\right) - T_g|_{0.5,j\Rightarrow w} \, \left(\gamma_g|_{0.5,j\Rightarrow w} \, \Delta Z|_{0.5,j\Rightarrow w}\right) \tag{8.198}$$

This rate will be considered constant during each simulated time interval, and has to be treated as the equations developed for the case of constant flow.

The total rate in the well will be determined by the sum of rates from cells in the open interval

$$q_{pTOT} = \sum_{w=1}^{nid} q_p \big|_{0.5, j \Rightarrow w} \tag{8.199}$$

8.2.7.2   *Tracer*

As previously defined, the total volume flow (at surface conditions) associated to a well is given by the sum of the volume flows of oil and gas. In the tracer case, it is incorporated into the equation system through the gas phase. In a producer well the tracer can appear in both phases.

- Injector Well (s)

  The management of tracer concentration in the injector will be as follows:

  If the rate is for gas injection (q_g is given), the term $x_T K_T R_{ro} q_g$ in the equation of the tracer component is known, and it holds $x_T K_T = 1$, i.e. the tracer is fully integrated in the gas injected ($y_T = 1$), additionally we have to consider that $q_o = 0$.

- Production Well (s)

  After solving the flow equation system, the producing well is defined by the bottom hole pressure, so the source term in the producing cell will be determined by for gas phase

$$V_{Tg} = \frac{x_T K_T R_{ro}(q_g + R_s q_o)}{\Delta t} \tag{8.200}$$

for the oil phase

$$V_{To} = \frac{x_T(q_o + r_s q_g)}{\Delta t}. \tag{8.201}$$

8.2.8   *Initial conditions for the fluid flow and the tracer systems*

8.2.8.1   *Initial conditions for the two-phase flow system, $p_o(x, y, t = 0)$, $S_o(x, y, t = 0)$*

To determine the pressure and saturation in the porous media, fluids are considered to be in equilibrium, including capillarity and gravity. In this case the initial pressure and saturation values in the cells will be based on the capillarity properties of the rock and the densities of the phases, as well as the depths of the contacts.

A reference pressure and its corresponding depth are required to calculate the initial values. For the x and y we have

$$\frac{\partial z}{\partial x} = \frac{\partial p_p}{\partial x} = 0; \quad \frac{\partial z}{\partial y} = \frac{\partial p_p}{\partial y} = 0 \tag{8.202}$$

And for z direction

$$p_p(z) = p_{\text{pref}} + \frac{g}{g_c} \bar{\rho}_p (z - z_{\text{ref}}) \quad \text{for } p = o, g \tag{8.203}$$

Thus, for a given reference pressure, associated with the oil phase, at a depth, you can set the function of pressure respect to the depth. In the contact we have this particular expression,

$$p_{g,zcgo} = p_{o,zcgo} + Pc_{ego}(S_g = 1). \tag{8.204}$$

Once the phase pressure as a function of depth is defined, the phase saturation can be determined by assuming capillary pressure equilibrium

$$Pc_{go}(S_g(z)) = p_g(z) - p_o(z) \Rightarrow S_g(z) \tag{8.205}$$

Thus we will have the initial conditions for pressure and for saturation

$$S_p(x, y, z, t = 0); \quad p_p(x, y, z, t = 0) \quad \text{for } p = o, g \tag{8.206}$$

8.2.8.2 *Initial condition for tracer, $x_T(t = 0)$*

As seen in the development of the model, we consider that the tracer is injected as a molar fraction, x_T, in one of the phases introduced in the well; so the initial condition is that there is no tracer in the system:

$$x_T(x, y, t = 0) = 0 \tag{8.207}$$

where x is r or x, and y is y or z, depending on which case is being simulated.

8.3   VALIDATION OF BIPHASIC FLOW SYSTEM

At the present moment the code for the biphasic model has been already validated, and the tracer model is under development. An analytical solution from literature (Mathews 1967) was used to compare an academic study case that considers a single well pressure test.

Analytical solutions for transient flow presented by Dake (1978) were used in the oil well pressure analysis. It is given by

$$p_{wf} = p_i - \frac{q\mu}{12.56kh} \left(\ln \frac{2.246kt}{\phi\mu cr_w^2} \right) \tag{8.208}$$

The information used for the validation is presented in the Table 8.1.

The biphasic model results were compared with the pressure well test results. Figure 8.8 shows this comparison. It can be seen that numerical results agree with those obtained from a field case (Matthews 1967), as they appropriately match with data from a single well pressure test during a drawdown test in an oil well.

Table 8.1   Information used to compared simulation and reference data.

Property	Description	Value	Unit
p_i	Initial well pressure	1895	psi
p_w	flooding well pressure	t function	psi
t	time		hrs
q	Rate	800	STB/day
m	Viscosity	1	cp
B	Volume factor	1.25	frac
k	Permeability	96	mD
h	Net width	8	ft
r_w	Well radius	0.33	ft
f	Porosity	0.1	frac
Ct	Compressibility	1.77E−05	1/psi
re	Drainage radius	1067	ft
V_r	Rock volume	28613416.3	ft^3
V_p	Pore volume	2861341.6	ft^3
So	Oil Saturation	0.65	frac
Vo	Oil volume at reservoir conditions	1859872.1	ft^3@rc
Vo@sc	Surface Oil volume	1487897.6	ft^3

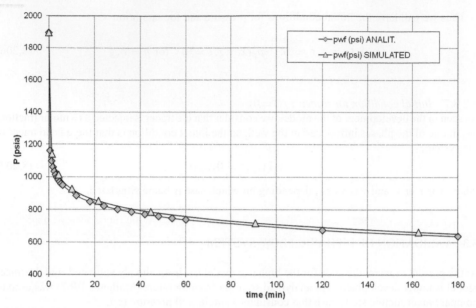

Figure 8.8   Comparison between analytical solution for a drawdown well test, and the numerical results from the biphasic code developed here in the well simulator case.

8.4   CONCLUSIONS

As seen in the numerical results, the mass transport is well simulated, concluding that the advection term in the model is already validated. We are in the trend of making the diffusion term validation, in order to valid the tracer option. The results that can also validate the transport by diffusion at the time of this publication, is not available yet.

REFERENCES

Dake, L.P. (1978) *Fundamentals of Reservoir Engineering*. Amsterdam, Elsevier.
Deuflhard, P. (2004) *Newton Methods for Nonlinear Problems. Affine Invariance and Adaptive Algorithms*. Springer Series in Computational Mathematics, Vol. 35. Berlin, Springer.
Hildebrand, F.B. (1968) *Finite-Difference Equations and Simulations*. New Jersey, Prentice-Hall.
Levy, H. & Lessman, F. (1992) *Finite Difference Equations*. New York, Dover.
Matthews, C.S. & Russell, D.G. (1967) *Pressure Buildup and Flow Tests in Wells*. SPE Monograph Series Vol. 1, Society of Petroleum Engineers, Texas, USA.
Zemel, B. (1995) *Tracers in the oil field*. Developments in Petroleum sciences, 43. Amsterdam, Elsevier.

Section 3:
Statistical and stochastic characterization

CHAPTER 9

A 3D geostatistical model of Upper Jurassic Kimmeridgian facies distribution in Cantarell oil field, Mexico

R. Casar-González, M.A. Díaz-Viera, G. Murillo-Muñetón, L. Velasquillo-Martínez,
J. García-Hernández & E. Aguirre-Cerda

9.1 INTRODUCTION

The oil reservoirs are the result of a complex succession of geological, physical, chemical, and biological process acting during the geological past. Those processes have created a unique distribution of rocks with petrophysical properties that make up the current petroleum reservoirs. Although these processes may be modeled, we do not understand completely all the details and their interactions. Moreover we do not have the complete data set to provide a unique actual distribution of facies (rock types) and their petrophysical properties within the reservoir. Nevertheless, we can try to create numerical models that represent the result of all these processes and the characteristics of the reservoir rock. Therefore, we attempt to make numerical models consistent with the conceptual idea of the geological setting and also honor all the available information. Thus, reservoir modeling is a multidisciplinary task which involves diverse disciplines such as geology, geophysics, geostatistics and stochastic simulation.

To address the problems of reservoir modeling, the oil industry is applying today an integrated geological and petrophysical modeling strategy, where geostatistical techniques play a crucial role. This is because, those stochastic techniques allow us, in straight forward manner, to combine information from diverse sources and analyze the influence of different parameters involved in the model. The pursued objective in applying these techniques is to arrive to realistic and reliable models and to assess the degree of uncertainty associated with them.

In the process of characterizing a reservoir, it is assumed that the known facts are never comprehensive enough to describe a field in a complete and accurate manner. Thus, under this approach, geostatistical methods represent a practical tool, now commonly used in oil industry. Through this it is possible to obtain maps of reservoir characteristics and heterogeneities. These representations are consistent with the known data and geological knowledge of the processes that formed the deposits, plus it provides the elements for an analysis of the uncertainty associated with the models.

One of the reasons that sometimes one does not understand the real potential of geostatistical modeling is that there is a widespread misconception that applying the stochastic models, chance is being played with, the same way as if one launched a coin to decide whether at a given location there is one or another facies. Actually, the stochastic modeling which allows it to properly handle the uncertainty that is about the complex geological and petrophysical reservoir from the systematic integration of all the hard data (outcrops, cores, well logs, seismic, etc.) and the soft data (conceptual geological model). In this work we applied a systematic methodology for modeling the geology of the reservoir and the geostatistical simulation methods was used to distribute the sedimentary facies, however is advisable to apply the same methodology for the petrophysivcal properties, as they are closely linked to the sedimentary facies.

The main goal of this study was to create a 3D geostatistical model for the Upper Jurassic (Kimmeridgian) (UJK) sedimentary facies distribution of the Akal block across the Cantarell oil field. The study area is located in the northern portion of Cantarell oil field. This field lies on the continental shelf of the Gulf of Mexico (Campeche Bay area), around 75 kilometers northwest

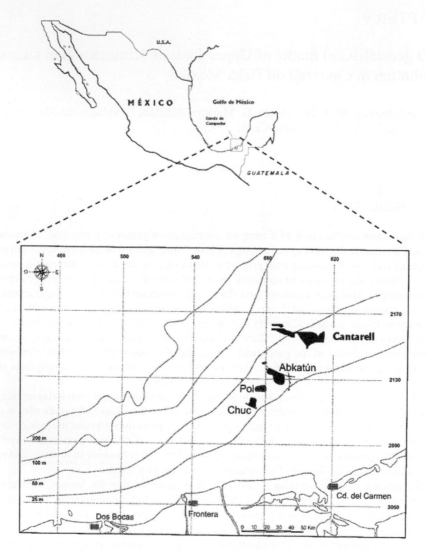

Figure 9.1   Cantarell oil field is located on the continental shelf of the Gulf of Mexico in the Campeche Bay area, 75 km northwest direction from Ciudad del Carmen, Campeche (PEMEX Exploración y Producción 1999).

from Ciudad del Carmen, Campeche, in water depths ranging from 35 to 50 meters (Fig. 9.1). The length of the field is approximately 10 km and its width is 7 km in average. The oil field consists of Akal, Nohoch, Chac and Kutz blocks, covering an area of approximately 162 square kilometers. Cantarell is still one of the most important oil-producing fields in Mexico, and is considered the eighth largest field in the world (PEMEX Exploración y Producción 1999).

The aim of the developed model was to shed light on the sites to explore in the field with better chances of finding the producing sedimentary facies. The northern part of the field account with a enough information, particular of wells drilled, while the southern part of the field is lacking of information. So, one of the objectives was to determine the potential production of the southern part of the field.

In this paper it was combined two geostatistical simulation methods, Gaussian Truncated and Sequential Indicator, resulting in a good strategy from the standpoint of geostatistics methodologies and numerical and computational effort. This strategy allowed for taking advantage of each method, reproducing very accurately the spatial variability observed in the UJK facies, described at well logs and consistent with sedimentological and stratigraphic interpretation. At the same time managed to achieve high computational efficiency, which is associated with the strategy commonly used to divide the complex problem in simpler subproblems, which facilitated its resolution.

There are some paper related to this work, an example is a geostatistical 3D model of oolite shoals in the southwest of Kansas (Qi et al. 2007). The objective of this work was to quantify the geometry and spatial distribution of an oolitic reservoirs and the continuity of flow units using a geostatistical approach. The authors report that the uncertainty and variability of the external and internal geometry of the limestone oolitic reservoirs were explored using stochastic simulation methods, and producing multiple equally probabilistic realizations. Also they note that the results models are useful for understanding the geometry of oolitic shoals complexes and the methodology can be implemented in complex depositional systems where there are not a good seismic data. The geostatistical methodology used in this paper is similar than the used in this work.

9.2   METHODOLOGICAL ASPECTS OF GEOLOGICAL AND PETROPHYSICAL MODELING

The geological and petrophysical model is a fundamental part in reservoir modeling. Since, it allows us to understand which are the main mechanisms that control the fluid flow and transport in the reservoir. This basically consists of obtaining a representative petrophysical property distribution of the reservoir starting from a conceptual geological model. This stage of the oil field modeling, also known as static reservoir characterization, includes collaborative and integrated work of many earth sciences disciplines such as geology, geophysics, petrophysics, and so on (Yarus & Chambers 1994).

A geological and petrophysical model typically involves two stages. Firstly, it is established a quantitative geological model and subsequently a petrophysical model is built based on the former. Below, there is a brief description of the most important methodological aspects that are considered in geological and petrophysical modeling.

9.2.1   *The geological model*

The definition of a quantitative geological model is one of the most important task in a reservoir study due to the amount of work and the impact it has in the reservoir behavior prediction. It comprises the following steps (Cosentino 2001):

1. Structural model: Structural model is the identification of the basic geometric framework of the hydrocarbon trap; it includes the definition of the boundaries. It is about the construction of the structural map of the top and bottom of the hydrocarbon accumulation and the identification and interpretation of the faults affecting the reservoir.
2. Stratigraphic model: The stratigraphic model is responsible for the definition of the internal structure of the reservoir; it is about the construction of a stratigraphic grid and also the definition of the main reservoir flow units. From the stratigraphic point of view the main issue is the definition of the internal geometry and architecture of the different stratigraphic units and layers. Layers may have different geometrical forms, such as parallel, proportional, truncated, etc.
3. Lithological model: The lithological model represents a powerful tool to guide the distribution of petrophysical properties according to lithology definition, since in most cases these two

issues are closely related. The lithological model consists in the classification or grouping of facies types in lithotypes or petrophysical classes.

4. Reservoir heterogeneities: Reservoir heterogeneities are geological features that range from small to large scale and may not be significant from a strictly static reservoir characterization but have a significant impact on the flow of fluids. The relationship between the heterogeneity of the reservoir and the dynamic parameters of the field is one of the key points of a comprehensive study as this determines the level of detail and accuracy that can be achieved in the geological description.

The development of a quantitative geological model should be guided by a conceptual sedimentological model. This conceptual model provides an assessment of the stratigraphic settings, and semi-quantitative geometric parameters, which will be used for the process of stochastic modeling (shape and dimension of the sedimentary bodies, thickness of layers, dimension of stratigraphic units, etc.).

In a lithological model the concept of facies plays a fundamental role and is particularly suitable for detailed studies of reservoir. Facies can be considered as a practical elementary volume of the reservoir and represent the basic building block of geological models in 3D. The facies are a tool for transferring geological information through the different stages of a study. Description and classification of lithofacies, commonly is made from rock samples in core and its goal is the classification of reservoir rock from the standpoint of lithology and depositional environment. A simpler study is reduced to the definition of two types of facies: reservoir and not reservoir facies. When you have good information, that is, when it is possible to identify a larger number of facies, we can try a more sophisticated approach based on multivariate statistical treatment of data, like cluster analysis or main components. When the reservoir petrophysical properties are not directly controlled by the facies (lithotypes), an alternative approach in terms of petrophysical classes (grouping petrophysical properties) can be used.

It is important to remark here that our interest is not in how the types of facies are defined, but how to build 3D realistic facies distributions so that they can later be used in making decisions during the modeling field. The facies must have significant control over the functions of the porosity, permeability, and/or saturation; otherwise, the modeling of 3D distribution of facies is of little benefit because the uncertainty is not reduced and the resulting models will not have a predictive power.

The modern way to modeling the spatial distribution of facies is based on the stochastic approach. The stochastic modeling approach is particularly suitable for the description of the reservoir, allowing the integration of a variety of quantitative (hard) and qualitative (soft) data, and also gives us a measure of the uncertainty of it (Kelkar & Perez 2002).

Currently, there are a variety of geostatistical stochastic simulation models for reservoir characterization (Deutsch 2002). The most commonly used methods are of two types:

1. Cell-based models: A variable is considered to be simulated as a realization of a continuous random function whose distribution (usually Gaussian) is characterized with different thresholds (cutoffs), which identify different facies or different petrophysical classes. These methods work best in the presence of facies associations that vary smoothly, as is often the case of coastal or marine deltaic deposits. It makes no assumption about the shape of sedimentary bodies.

2. Object-based (or boolean) models: Generate spatial distributions of sedimentary bodies which are obtained by superimposing the simplified geometries such as plates, discs or sinusoids, typically used in simulations of river environment, deltaic and turbiditic. Object parameters (orientation, sinuosity, length, width, etc.), can be estimated based on the sedimentological model, the seismic data, a similar surface and the interpretations of well tests. In some depositional environments, especially the meandering river, where sand channels are the main destination site, these models can produce very realistic images of the facies architecture of the site.

9.2.2   *The petrophysical model*

Petrophysical properties such as porosity and permeability are modeled within each facies and each layer of the reservoir. The petrophysical property values are assigned within a facies so as to reproduce the representative statistical characteristics (histogram, variogram, correlation with another variable) of the property for this facies. This can be done using geostatistical simulations. Gaussian simulation techniques are usually employed for this purpose.

Because of the objective of this study, it was restricted to obtaining only the geological model, which will be described in some detail below.

9.3   CONCEPTUAL GEOLOGICAL MODEL

9.3.1   *Geological setting*

The stratigraphic column in the Cantarell oil field spans from the Upper Jurassic (Oxfordian) to the Pliocene-Holocene. The representative stratigraphic column consists of thick layers of sedimentary rocks that include the following rock types: for the Upper Jurassic evaporites, carbonate rocks, bituminous and shaly limestone. The Lower and Middle Cretaceous are represented by dolostones in part shaly, and the Upper Cretaceous-Lower Paleocene (included the Cretaceous-Tertiary boundary) is made up of shaly limestone and dolomitized calcareous breccias. The Tertiary is made up of thick calcareous siliciclastic succession containing mudstone, siltstone, sandstone and limestone. The Pliocene-Holocene is a sedimentary succession of poorly consolidated clays and sands (Angeles-Aquino 1988 and 1996).

Structurally, the Cantarell oil field is a anticline with a thrust in the northeastern flank and affected also by strike-slip and normal faulting. The allochthonous thrusting block is the Akal block and the autochthonous block is the Sihil block The Akal block represents a structural-stratigraphic oil trap. The Upper Jurassic Tithonian (UJT) is the main source rock of hydrocarbons and is characterized by a high content of organic matter (Guzmán-Vega 2001). The main oil-producing stratigraphic horizons include: the Upper Jurassic (Kimmeridgian), Lower and Middle Cretaceous and the Cretaceous/Paleocene boundary facies. The thickness of pay zone ranges from 141 to 908 meters. The main reservoir corresponds to the Cretaceous-Tertiary boundary breccia which is a naturally fractured dolomitized calcareous breccia succession where diagenetic processes have increased significantly the porosity and permeability. This unit is composed of dolomitized angular lithoclasts derived mainly from shallow water facies, with porosities values from 8 to 12 percent and permeabilitys values ranges from 3 to 5 darcies (PEMEX Exploración y Producción 1999). Another important petroleum target in this field is the UJK ooid facies which are the main subject of this paper.

9.3.2   *Sedimentary model and stratigraphic framework*

Recently, a sedimentologic-stratigraphic study of the Upper Jurassic (Kimmeridgian) sedimentary facies in Akal block of the Cantarell field, conducted by Murillo-Muñetón et al. (2007, 2008) concluded that this unit represents a mixed carbonate/siliciclastic system. The reservoir facies correspond to ooid grainstones and packstones, with oncolithic-skeletal packstone. These facies are intercalated with fine-grained facies with carbonate (mudstone and skeletal wackestones commonly with clay and/or silt) and siliciclastic (mainly shale and siltstone and minor amounts of fine to very fine grained arkose). The sedimentological model established indicates that the ooid grainstone and packstone/grainstone facies correspond to high-energy shallow subtidal deposits (shoreface) that were developed as carbonate sand shoals accumulated at the margin of a homoclinal carbonate ramp system. The fine-grained carbonate facies (mudstone and wackestone) are interpreted as deposits of open sea (below the normal wave-base level) and probably shallow subtidal settings (lagoon or embayment). Whereas oncolithic-bioclastic packstones are also considered lagoonal deposits. Moreover, the fine terrigenous facies (shales) were deposited in a relatively deep quiet marine environment (below the normal wave-base level in the outer platform), some were deposited in an environment of intertidal to subtidal. A few thin horizons of reddish brown terrigenous (shales, siltstones and sandstones) represent alluvial/supratide deposits, while very thin layers of arkoses are considered shallow marine deposits (possibly protected, like a lagoon with hypersaline conditions) and some are rather coastal bars (beach to shoreface).

According to Murillo-Muñetón et al. (2008), the stratigraphic architecture of the UJK facies resulted in the compartmentalization of the reservoir horizons due to the interaction of several factors. These factors include tectonic subsidence (associated with the opening events of the Gulf of Mexico and possibly halokinetic activity), sedimentation rate and eustacy. The Kimmeridgian facies stack forming cycles or parasequences of small hierarchy whose thickness varies from a few meters to tens meters and these in turn are stacked forming at least 3 depositional sequences (probably of 3rd order). The thickness varies from 70 to 120 m and are composed of shale, mudstone and wackestone (transgressive systems tract), with subordinate amounts of packstones and grainstones. At the top, there were dominated by ooid packstone and grainstone facies (highstand systems tract), and to a lesser extent by fine clay and carbonate facies. An additional process that contributed notably to that the horizons of ooids grainstones and packstones were a reservoir rock was the diagenesis, mainly the dolomitization. During this process it was generated intercrystalline secondary porosity in the packstones and grainstones facies that were originally compact without primary porosity due to precipitation of calcareous cements during earlier diagenetic stages. Apparently, the fractures were not a major control in the petrophysical quality of these facies.

The following stratigraphic units and subunits based on the stratigraphic-sedimentologic model were used for constructing the 3D geostatistical model. From bottom to top, there are four units: B, C, D and E; the latter is subdivided into three subunits E1, E2 and E3 (Fig. 9.2). The oldest horizon is Unit B, and its age is Oxfordian, consists of shales that were deposited in a relatively deep marine setting, probably below the normal wave-base level under transgressive conditions. The unit C is composed of carbonate sand shoals (ooid grainstones and packstones/grainstones). There are also skeletal-oncolithic packstones that correspond to a lagoon environment. Towards the south of the area shales dominate in Unit C and the volumetric ratio grainstones/packstones decreases significantly. Unit C represents the establishment of a ramp-type carbonate platform dominated by shallow-water sedimentation. This system is the result of slowdown in the subsidence rate and/or due to a eustatic highstand time. The next deposit is Unit D that represents transgressive marine event which terminated with the shallow-water carbonate sedimentation event of Unit C. Unit D consists mainly of fine grained terrigenous and carbonate sediments accumulated in relatively deep water setting, most probably below the normal wave action. This unit also includes a carbonate horizon (subunit D), and represents the returns to the shallow-water carbonate sedimentation probably due to a decrease in the tectonic subsidence. However, the ooids shoals were restricted exclusively to the northwest part of the field. The top of Unit D represents sedimentation conditions similar to those of its basal facies. That is, relative sea level rise and consequently occurred the deposition of fine-grained terrigenous and carbonate sediments.

It is likely that there was a repetition of a slowdown in the rate of tectonic subsidence and possibly a eustatic highstand time and again the conditions were adequate for the establishment of a ramp-type carbonate platform represented by the subunit E1. Similar to carbonate subunit D, this new carbonate platform developed mainly to the northwest part of the field. During the deposition of the subunit E2, a marine transgression ocurred again, likely due to an increase in subsidence rate and/or a eustatic rise. Subunit E2 is represented by shales, lime mudstone facies of deep water (outer plataform) and shaly wackestones in a protected shallow-water environment (lagoon). During this time also some reddish fine-grained terrigenous including very thin layers of arkoses were deposited. For the deposition of subunit E3, the sedimentation conditions return to shallow-water conditions (carbonated ramp), probability due to a decrease in the subsidence rate and possibly a eustatic highstand time. It is important to note that this time was dominated by facies of packstones to grainstones. The Kimmeridgian time dominated by shallow-water sedimentation

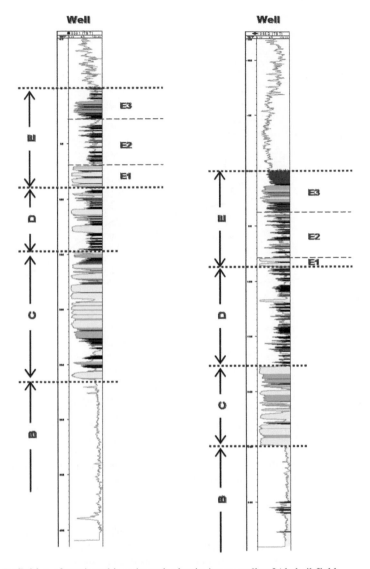

Figure 9.2   Definition of stratigraphic units and subunits in two wells of Akal oil field.

was terminated by the beginning of the great marine transgression during Tithonian time in the Campeche Bay area.

9.3.3 *The conceptual geological model definition*

The conceptual geological model definition starts with the compilation of the geological information available, that include, conventional cores, well logs and well log images. The cores showing the facies of the Kimmeridgian were described and studied in detail, defining the various facies and making an interpretation of its depositional environment. Also it was established the diagenetic history of the different facies.

The interpretation of the well logs was carried on and also was calibrated with the information of the core. The study includes the analysis of the patterns of fractures in cores and well log images.

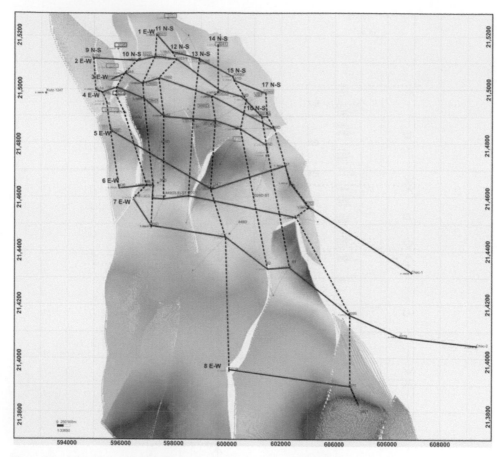

Figure 9.3   Map of the oil field showing well distribution and stratigraphic correlation sections: 8 sections are oriented east-west and 9 are oriented north-south direction, respectively.

Stratigraphic correlation sections were built in order to establish the stratigraphic correlation and have better idea about facies distribution. With all this information, it was developed a stratigraphic-sedimentological conceptual model that was used in this study. For the stratigraphic correlation sections, 65 wells were considered, particularly those to reach or cut off part of the Kimmeridgian rocks (Fig. 9.3). The main idea is that having stratigraphic correlation sections, it provides additional information about the geometry, thickness and lateral extent of the different facies. Of a total of 17 stratigraphic correlation sections, 8 are oriented approximately east-west direction and 9 are oriented north-south direction, respectively (Fig. 9.3). The definition of the sections was trying to cover the largest area that had wells with data.

9.3.4   *Analysis of the structural sections*

Four structural sections were constructed, these structural sections served as control sections to be compared with the results of the 3D geostatistical model. The construction of the structural sections was based in the interpretation of the stratigraphic correlation sections. The structural section shows the paths of the wells, the stratigraphic units, the fault model, and exhibit the structural complexity presented in the Akal block (Fig. 9.4). In the western portion of the structural section it defines two structural blocks bounded by normal faults, the presence of these faults

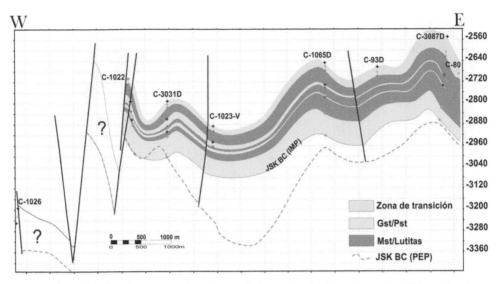

Figure 9.4   Structural section for stratigraphic correlation section 2 oriented west-east. This section shows the structural top of the carbonate base provided by the technical staff of Cantarell field (UJK BC PEP dashed line) and the base of carbonates UJK obtained by wells (UJK BC IMP). Continuous red lines indicate the faults that affected the Akal block. Depths in meters (true vertical depth).

and the absence of a well to confirm this, not allowing the interpretation of the distribution of stratigraphic units, proposed in the correlation stratigraphic section Murillo-Muñetón et al. (2008).

9.3.5   *Description of the stratigraphic correlation sections*

An analysis of the stratigraphic correlation sections suggests the following: Unit C has the largest thickness, varying from 50 to more than 100 m and exhibits a slight decrease in thickness with a corresponding increase of the fine clays and carbonates facies to the south and east of the field. It is difficult to define a clear pattern in the thickness of this unit; we can deduce that the best development was in the north-center of the field and decreases to south (where there is little information of wells). Carbonate subunit D reaches a maximum thickness of 22 m. However, it shows a clear decrease in thickness towards the east and south of the field until it disappears and changes laterally to fine clay and carbonate facies. The subunit E1 have the same behavior as the carbonate subunit D, which reaches a maximum thickness of 37 m and decreases its thickness toward the east and south of the field where it disappears and changes laterally to the fine clay and carbonate facies. With respect to the subunit E3, it corresponds to a carbonate horizon mainly consisting of clay and packstone facies and lower proportion of grainstone and mudstone. This horizon covers practically the entire field and in general terms becomes shally (decrease grainstones and increase mudstones) to the south and to the east of the field (Fig. 9.5).

9.3.6   *Lithofacies definition*

The information of sixty five wells which drilled totally or partially the Upper Jurassic Kimmeridgian (UJK) in the Akal block was initially considered to build the geostatistical model. However, some wells have problems related to missing information, structural problems and others. After a preliminary analysis, it was decided to exclude some wells; therefore, the model was built with the information of forty one wells. Seventeen cores from eight wells were described and studied in order to define sedimentary facies. From core analysis a total of seven facies

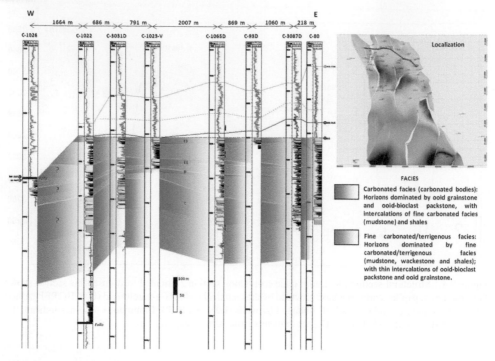

Figure 9.5　Stratigraphic correlation section 2 oriented E-W.

Table 9.1　Facies of the UJK, definition, abbreviation and code.

Facies	Abbreviation	Code
Shale and limonitic shale greenish gray	Lut/lut-limo	0
Mudstone silty-shaly dolomitic and mudstone dolomitic; cycles of shales and dolomites silty/silty calcareous	Mst arc dol.	1
Wackestone dolomitic with mudstone/wackestone shaly dolomited	Mst/wst dol.	2
Packstone totally dolomitic	Pst ooid dol.	3
Packstone/grainstone	Gst/pst ooid dol.	4
Grainstone of ooids totally dolomitic	Gst ooid dol.	5
Arkosa	arcosas	6

were identified. For facies definition at well log scale was used a standard well log suite, which include: gamma-ray (GR, CGR), resistivity (LLD, LLS, and MSFL), density (RHOB), neutron porosity (NPHI) and sonic (DT) logs. A value range was established for each utilized well log based on its response, that can be associated with specific identified facies at core scale. The core information was used to calibrate the facies definition process at well log scale. This calibration procedure was useful to validate the log curve ranges which define each facies. This procedure was programmed in order to automate the facies identification process at well log scale. Once the automatic process finished, it was necessary a visual check to corroborate the interpretation, and where it was necessary further corrections were made. As result, a facies well log was defined in each one of the 41 wells. The facies definition and their corresponding code are shown in Table 9.1.

9.4   GEOSTATISTICAL MODELING

All the reservoir modeling was done in the software Petrel™ (Schlumberger). An important task before the geostatistical modeling is to create the structural model of the reservoir, meaning the geometry (architecture) of the reservoir. The structural model includes the position and characteristics of main stratigraphic horizons and faults. It is common that this information come from a structural interpretation of seismic information.

9.4.1   *Zone partition*

The stratigraphic interval of the UJK was divided in three zones. The task of dividing in zones has the objective to have better control during the modeling process, because it is performed separately. The criteria to divide the stratigraphic intervals were based in the cluster of horizons as a coherent way, meaning that the zones must have a homogeneous thickness respect each other. The practical aspect to create zones is related to avoid the mixing of bodies or beds from different stratigraphic horizons during the application of the simulation methodologies. The zones were established as follows (see Figure 9.6).

- Zone 1: The UJK/UJT transition and subunits E3 and E2.
- Zone 2: Subunit E1 and unit D.
- Zone 3: Unit C.

9.4.2   *Stratigraphic grid definition*

The next step was the construction of the stratigraphic grid. The stratigraphic grid consists in dividing the zones in a given number of layers of the same thickness. In this case the layering process was done from the top (cell layering is parallel to the top of the UJK surface) and was buit proportionally. The zone 1 was divided in 9 layers, the zone 2 in 8 and the zone 3 in 10 layers, so we have 27 layers for the UJK stratigraphic interval. The mean value of the cells height is 10 meters. A statistics of cell heights and a histogram is presented in Table 9.2.

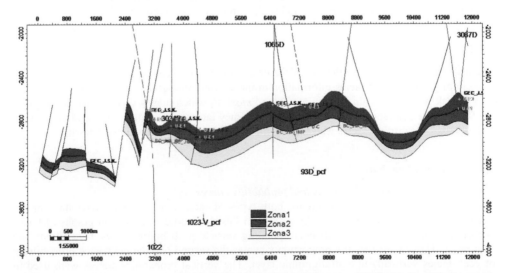

Figure 9.6   The stratigraphic interval of the UJK was divided in three zones. The figure shows the tree zones in the second stratigraphic section oriented east-west.

Table 9.2   Statistics of the cells height per zone.

Statistics	Zone 1 (9 layers)	Zone 2 (8 layers)	Zone 3 (10 layers)	Global
Number of cells	693,315	616,280	770,350	2,079,945
Minimum	1.19	1.08	0.38	0.38
Maximum	32.69	33.24	26.04	33.24
Mean	10.06	10.01	10.28	10.13
Standard deviation	3.87	3.8	3.76	3.81
Variance	14.99	14.47	14.13	14.53

9.4.3   *CA facies classification*

From the original seven facies described above and identified in the wells, the facies number seven, arkosa, was eliminated because of his negligible percentage in the reservoir and the rest was grouped in two main classes named as:

- Carbonate facies (carbonate bodies): Horizons dominated by ooid grainstone and ooid-bioclast packstone, with intercalations of fine carbonate facies (mudstone) and shales. This category, basically comprises facies codes 3, 4 and 5 in Table 9.2.
- Fine carbonate/terrigenous facies: Horizons dominated by fine carbonate/terrigenous facies (mudstone, wackestone and shales); with thin intercalations of ooid-bioclast packstone and ooid grainstone. This category, basically comprises facies codes 0, 1 and 2 in Table 9.2.

Henceforth, we will use the term "CA facies" when we refer to the latter classification consisting of two facies, while the term "original facies" will be used in the case of the seven facies previously defined in Table 9.1. The idea behind to create the last classification of facies in two categories corresponds to a simulation strategy which will consist in modeling the facies in two steps. The first step will be modeling the carbonate facies as carbonate bodies and the fine carbonate/terrigenous facies using the stochastic simulation algorithm Sequential Indicator Simulation (SIS). The second step will be modeling the six original facies constrains to the two last categories with the Truncated Gaussian Simulation (TGS) algorithm. The idea to use this approach is to ensure the obtention of a 3D model consistent with the conceptual sedimentology-stratigraphic model.

9.4.4   *Facies upscaling process*

The next step was to scale up the log of the original facies in wells from centimeters to meters scale in order to have the facies information in the stratigraphic grid. The methodology used to do this task was an upscaling technique named *most of*, which means that every cell will have the facies value corresponding to the highest frequency facies value within each grid model cell. In Figure 9.7 the original facies well log and the scale up results for a well are shown.

9.4.5   *Statistical analysis*

9.4.5.1   *Vertical proportions graphs and probability curves*

A statistical analysis was performed for both types of facies, the original facies and for the CA facies for each zone separately. As a result of this step vertical proportion graphs of facies occurrence were estimated and its corresponding vertical probability curves were fitted. The first one tells us how varies the proportion whereas the second one the ocurrence probabilility of each facies vertically in a stratigraphic layer. The vertical probability curves were used in the geostatistical simulation methods as constraints that must be reproduced in the 3D facies distribution by the stochastic model.

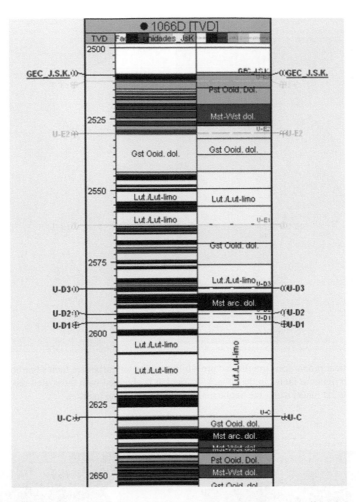

Figure 9.7   Original facies well log (left), and the scale up results for the same well (right).

Figure 9.8 shows the vertical proportions graphs and probability curves for scaled and simulated CA facies in the zone 3, respectively. The last ones, are obtained from applying the Sequential Indicator method to CA facies. Similarly, in Figures 9.9 and 9.10 the same curves for scaled and simulated original facies in the zone 3 are shown. Figure 9.9 is for fine carbonate/terrigenous facies whereas Figure 9.10 is for carbonate facies (carbonate bodies). The simulated facies are resulted of applying the Gaussian Truncated method to the original facies per each class of CA facies. In all cases it is observed that the 3D model reproduces very well both the vertical facies proportions and the probability curves.

9.4.5.2   *Variography analysis*
One of the most important parts of the geostatistical modeling is known as spatial correlation or variography analysis. The variography analysis is a geostatistical procedure for variogram estimation and modeling. The variogram is a function that describes the spatial dependence present in the data in a specific direction. In the context of geological modeling it is considered as a key tool that allows us to quantify the geological continuity (Caers 2005).

A variography analysis for both CA and original facies was performed in order to obtain a variogram model in horizontal and in vertical directions. After the variography analysis in

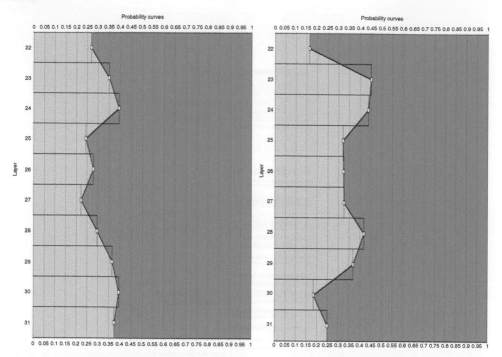

Figure 9.8   Vertical proportions graphs and probability curves for carbonate facies (carbonate bodies) and fine carbonate/terrigenous facies in the zone 3. On the left is obtained with the scaled records, on the right with the resulting 3D model using the Sequential Indicator simulation method.

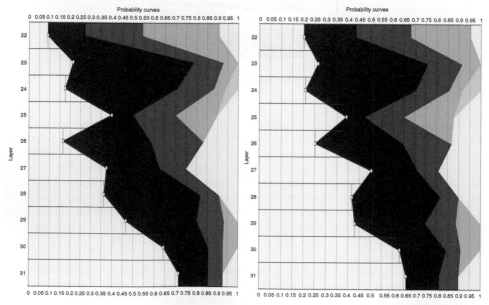

Figure 9.9   Vertical proportions graphs and probability curves of original facies for fine carbonate/ terrigenous facies in the zone 3. On the left is obtained with the scaled records, on the right with the resulting 3D model using the Truncated Gaussian simulation method.

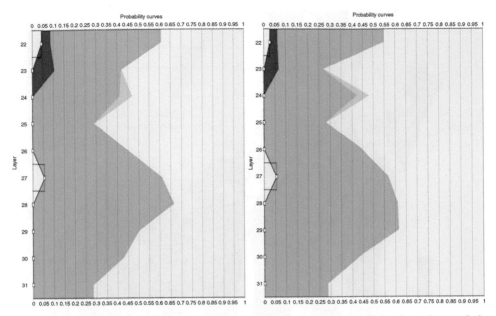

Figure 9.10   Vertical proportions graphs and probability curves of original facies for carbonate facies (carbonate bodies) in the zone 3. On the left is obtained with the scaled records, on the right with the resulting 3D model using the Truncated Gaussian method.

horizontal direction, we concluded that the resulting estimated variograms are unreliable, since there are not enough data for a statistically representative estimation. The solution to this problem was to propose or design variogram models to reflect the expected behavior according to the sedimentary-stratigraphic interpretation previously obtained.

As it is well known the main variogram parameter is the range, which controls the spatial continuity extension of correlation blocks for a given random function (Díaz-Viera & Casar-González 2006). Here facies are modeled as random discrete functions by indicator variables, so that a correlation block may represent the average extension of geological bodies.

For CA facies, anisotropic spherical variogram models in the horizontal direction with large ranges in comparison with the field dimensions were proposed, because we are expecting large spatial continuity of carbonate bodies. In particular, the selected variogram parameters were 10,000 meters as a major range in the north-south direction, and 6500 meters as a minor range in the east-west direction. As in vertical direction we have more data, the calculated variograms reflect more accurate the spatial correlation of thickness for the carbonate bodies. The resulting vertical ranges obtained by the variography analysis vary from 15 to 60 meters. They are in correspondence with the mean value of carbonate body thickness within each zone.

In the case of original facies, the variogram model associated with the most abundant or predominant facies within each CA facies class was chosen, so that, it was representative for the rest of the facies. The resulting variogram models considered in the simulations of the original facies were anisotropic Gaussian models with the same range of 1050 meters in the horizontal direction for all zones with the exception of the second one, which has a range of 630 meters; whereas the vertical ranges varying from 13 to 20 meters, in correspondence with the mean thickness value of predominant facies within each zone.

9.4.5.3   *CA facies thickness maps*
As it was mentioned before the unit UJK is composed of carbonate facies and fine carbonate/terrigenous facies. Now, it is intended to quantify the relationship between the thickness of the carbonate facies (packstone and ooids grainstone) against the fine carbonate/terrigenous facies.

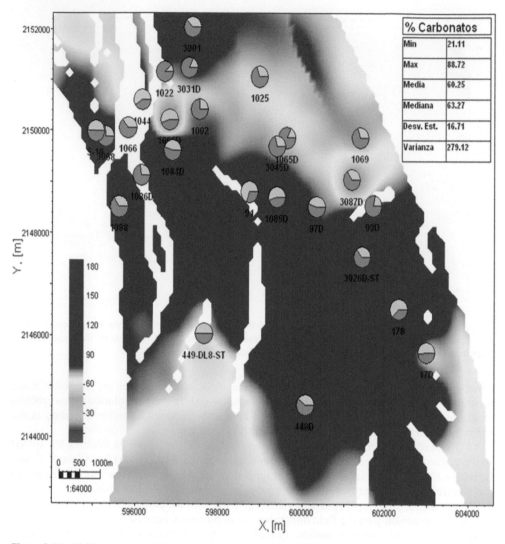

Figure 9.11 Thickness map of the unit C, the location of the wells has a circular chart (pie) that shows the percentage of carbonate facies against the percentage of fine carbonate/terrigenous facies with respect to the thickness of the body C.

There were constructed thickness maps in order to establish the trends that show such relationships. Those maps highlight the areas dominated by carbonate facies, where it is expected to find the highest values of porosity and contrast the areas with high percentage of fine carbonate/terrigenous facies with lower values of porosity. The information interpreted through the maps establishes the most favorable areas for oil exploration, as the proposed relationships are correlated with the petrophysical quality of the rock. This kind of maps can be used for geological interpretation of the prevailing spatial trends in the facies behavior, and also may serve as a fundamental tool to guide the facies spatial distribution for the stochastic modeling in the study region.

For the purpose described above a thickness map was constructed for each one of the carbonate units of interest (C, D, E1 and E3). As an example, here in Figure 9.11 the thickness map for the unit C is shown. It can be observed, that in the well locations have been inserted circular charts that show the relation between the percentage of carbonates versus the percentage of fine carbonate/terrigenous facies with respect to the thickness of the body C. The unit C has an average thickness of 87 m and a maximum content of 89% of carbonate facies, with an average of 60% of carbonate facies. In this case it can be argued that this unit has a clear predominance of carbonate facies. The map shows two areas dominated by a high carbonate facies content. One of the areas is located at the north-west boundary of the field, in which wells have carbonate facies percentages ranging from 62 to 89%. The other location with high values is at the north-east of the study area. The central portion of the map shows the lowest values, with percentages from 37 to 48%. Finally, the southern portion of the field shows values of less than 60% of carbonate facies. An important feature observed is when the thickness is lower the percent of carbonate facies is higher; and when the unit become more argillaceous (fine carbonate/terrigenous facies), the thickness is higher. This means that there is an inverse dependence between the unit thickness and its carbonate facies content.

9.4.6 *Geostatistical simulations*

In summary, the 3D facies distribution was performed following the strategy mention above:

- Firstly, CA facies (carbonate bodies and the fine carbonate/terrigenous facies) are modeled using the Sequential Indicator Simulation (SIS) method.
- Secondly, the original facies are modeled constrained to the two previous CA facies categories using Truncated Gaussian Simulation (TGS) method.

The two simulation methods used here are stochastic simulations, subsequent all realizations are different, but all have the same probability, in others words, all the realizations are equiprobable, which allow us to evaluate different scenarios.

The Sequential Indicator Simulation method is a cell based model suitable for simulating categorical variables such as facies. Sequential indicator simulation is based on a sequential simulation approach, which uses indicator kriging estimator to model the prior cumulative probability density function (cdf) at each unsampled location (Deutsch & Journel 1998). In particular, its performance is optimal when is applied to the case of two categories or types of facies (0 and 1) and when it is not clearly known a particular geometric shape. Sequential Indicator Simulation is more appropriated for use where either the shape of particular facies bodies is uncertain or where the user has a number of trends which will control the facies type, e.g. when using a seismic attribute to control the probability of occurrence of a certain facies.

Additionally, the Sequential Indicator Simulation procedure is conditioned with probability maps in the horizontal direction, which were built from thickness maps presented in the previous subsection. The construction of probability maps was carried out by normalizing thickness values with respect to its maximum, thus obtaining a map with values ranging between 0 and 1. These probability maps allow us to control trends, size and shape of the fine carbonate/terrigenous facies in the horizontal direction according with the well data and the conceptual sedimentological model.

While the Truncated Gaussian simulation method is a generalization of sequential indicator that performs optimally for the case of a few categories or types of facies whose spatial relationships are not fully established. Truncated Gaussian Simulation method has been first designed to provide stochastic images of sedimentary geology, mostly in fluvio-deltaic environments. This method is recommended to apply in systems where there is a natural transition through a sequence of facies. Typical examples include carbonate environments and fluvial sequences.

The basic principle of the method consists on simulating a stationary Gaussian random function with a given variogram model. This variogram model is selected in a way that when the simulated Gaussian random function is truncated at given thresholds the indicator variogram models of each resulting categories are reproduced (Chilés & Delfiner 1999). The thresholds are choosen by the proportions of each rocktype. The variogram of the Gaussian random function controls the contact shape between categories (Lantuejoul 2002). If the variogram is parabolic at the origin (e.g. the Gaussian variogram), the contact perimeter between classes will have finite length and smooth contour, but if the variogram is linear at the origin (e.g. exponential variogram)

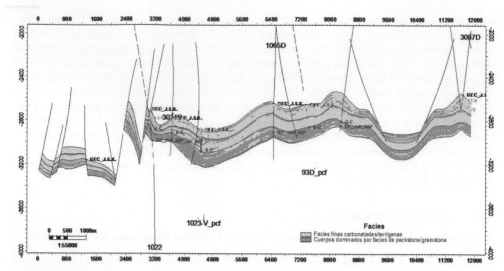

Figure 9.12   Distribution of carbonate facies (carbonate bodies) in the second stratigraphic section oriented east-west using Sequential Indicator simulation method.

the contact perimeter between classes has infinite length and fractal structure. In our case the variograms observed have not linear behavior at the origin, so it was modeled with Gaussian and spherical variograms models, therefore the contact between facies present finite length and smooth contours.

The application of the method involves, first, choosing which facies codes are to be included in the sequence and in what order. Next, specify the global proportions for each of the facies and must be given one variogram for all the different facies. It was chosen the variogram model that correspond with the most representative facies for each case.

An example of a realization using SIS for the CA facies distribution in the second stratigraphic section oriented east-west can be seen in Figure 9.12 whereas a realization using TGS for the original facies distribution using TGS in the same stratigraphic section is presented in Figure 9.13.

Figure 9.12, shows the distribution of carbonate bodies dominated by packstone/grainstone facies (carbonate facies) and the fine carbonate/terrigenous facies. It can be seen how the realization shows reproduces the observed facies presented at the correlation sections. It can been see for zone 1 of the model it reproduces the subunits E3 and E1, for zone 2 shows the carbonate body D, and the zone 3 contains the body C with a great continuity. It was found that the simulation method selected reproduces consistently the distribution of facies shows at the stratigraphic correlation sections (see Figure 9.5).

In Figure 9.13, it shows the distribution of the original facies, it was found that in each zone the simulation process reproduced the facies that were observed in the wells, according and consistent with the interpretation stratigraphic correlation sections, even more the simulation method reproduces the proportion of each facies.

In Figures 9.14 and 9.15 is shown a layer sequence of CA and original facies distribution maps obtained using Sequential Indicator and Truncated Gaussian simulation methods in the zone 3, respectively. The maps are top views from 22 to 31 layers of the simulated 3D model.

Figure 9.14 shows in yellow the carbonate bodies dominated by packstone/grainstone facies (carbonate facies) and in gray the fine carbonate/terrigenous facies. This map shows the predominance of the carbonate facies in the northern part of the area and it is possible to see how it diminishes with depth.

Figure 9.15 shows a sequence of top views of the distribution of the original facies. In red, orange and yellow colors the best oil producer facies corresponding to packstone/grainstone

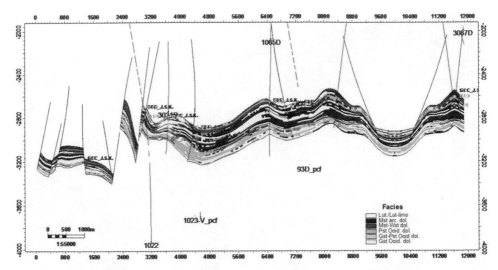

Figure 9.13   Distribution of facies in the second stratigraphic section oriented east-west using the Truncated Gaussian simulation method.

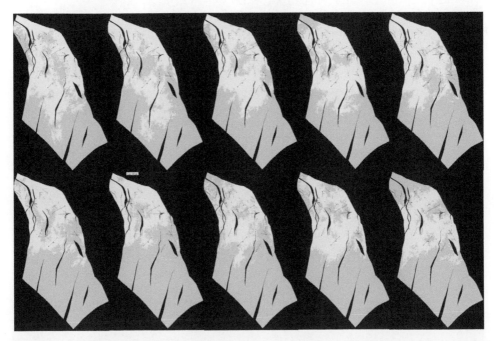

Figure 9.14   Sequence of CA facies distribution maps obtained using Sequential Indicator simulation method in the zone 3. The maps are top views from 22 to 31 layers of the simulated 3D model.

carbonate facies. That facies are concentrated in the north and center region of the area, while the poor producer facies with a predominance of the fine carbonate/terrigenous facies (blue colors) are present mostly in the south of the oil field.

Figure 9.16 shows a 3-dimensional perspective which shows the distribution of the original facies, it can be seen some sections interlinked, seven sections in E-W direction and four sections in N-S direction.

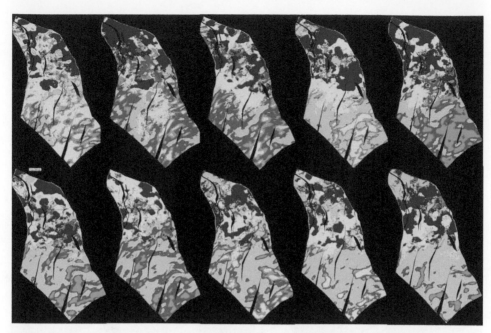

Figure 9.15   Sequence of original facies distribution maps obtained using Truncated Gaussian simulation method in the zone 3. The maps are top views from 22 to 31 layers of the simulated 3D model.

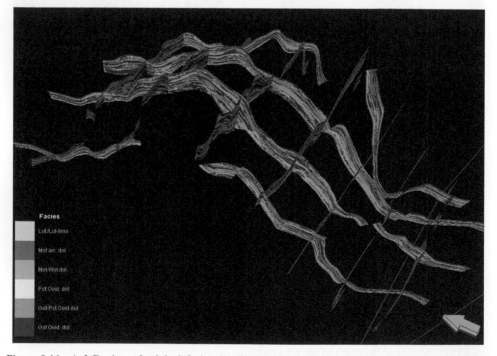

Figure 9.16   A 3-D view of original facies distribution obtained using Truncated Gaussian simulation method, seven sections in E-W direction and four sections in N-S direction.

9.5 CONCLUSIONS

The Upper Jurassic Kimmeridgian (UJK) of Akal field block of Cantarell contains four carbonate rocks bodies with the characteristics to be a hydrocarbon tramp, they are called: unit C, subunits D carbonate, E1 and E3. These bodies are composing mainly by packstone facies of ooids and grainstone facies with thin intercalations of carbonate and terrigenous fine facies horizons.

Unit C have the largest thickness (TVD, true vertical thickness), varies from 50 to 155 meters and decreases discretely to the South and East of the field with a corresponding increase in fine carbonate/terrigenous facies. The unit C shows a better development in the North-Center of the field. The absent of information to the South portion of the area, does not allow an interpretation about its possible extension to this region.

This paper introduces the Upper Jurassic lithostratigraphic nomenclature for Akal block of Cantarell Field, the subunit D carbonate. This body is dominated by clean carbonates and is part of the unit D. Subunit D is underlain and overlain by facies packages of fine carbonate/terrigenous facies from the same unit D. This subunit D reaches a carbonate maximum thickness (TVD) of 22 meters.

The combined application of two simulation methods: Sequential Indicator Simulation and Truncated Gaussian Simulation is a creative approach, not common in geostatistics applied to the description of reservoirs. This approach gave us an excellent results both from the standpoint of distribution of facies, the main objective and for the numerical and computational efficient. This strategy allowed exploit the relative advantages of each method, which means to accurately reproduction of the spatial variability of facies, observed at well logs and consistent with sedimentology and stratigraphic interpretation. The facies distribution model should be considered the first step input for a model distribution of petrophysical properties.

The recommendation to achieve a model distribution of petrophysical properties is the creation of petrophysical classes, defined by statistical analysis of the petrophysical information and the consistency relationship with the facies distribution proposed in this paper.

REFERENCES

Angeles-Aquino, F.J. (1988) Estudio estratigráfico sedimentológico del Jurásico Superior en la Sonda de Campeche. *Revista Ingeniería Petrolera*, 28, 45–54.

Angeles-Aquino, F.J. (1996) *Estratigrafia de las rocas del Jurásico Superior del Subsuelo en la Sonda de Campeche, Golfo de México*. Escuela Superior de Ingeniería y Arquitectura, Instituto Politécnico Nacional, Tesis de Maestría, México.

Caers, J. (2005) *Petroleum Geostatistics*. Society of Petroleum Engineers.

Chilès, J.P. & Delfiner, P. (1999) *Geostatistics: Modeling Spatial Uncertainty*. John Wiley and Sons., Inc. Wiley Series in Probability and Statistics, New York, U. S. A.

Consentino, L. (2001) *Integrated Reservoir Studies*. Editions TECHNIP, Paris, France.

Deutsch, C.V. & Journel, A.G. (1998) *GSLIB: Geostatistics Software Library and User's Guide*. Oxford University Press. New York, U.S.A.

Deutsch, C.V. (2002) *Geostatistical Reservoir Modeling*. New York, Oxford University Press.

Díaz-Viera, M.A. & Casar-González, R. (2006) *Manual del Instructor del Curso: Modelación Geológica Petrofísica de Yacimientos*. Instituto Mexicano del Petróleo, México, D.F., México.

Guzmán-Vega, M.A. Castro-Ruiz, L. Roman-Ramos, J.R. Medrano-Morales, L. Clara-Valdes, L. Vazquez-Covarrubias, E. & Ziga-Rodriguez, G. (2001) El origen del petróleo en las subprovincias mexicanas del golfo de México. *Boletín de la Asociación Mexicana de Geólogos Petroleros, A.C.* XLIX (1–2). México, D.F. México.

Kelkar, M. & Pérez, G. (2002) *Applied Geostatistics for Reservoir Characterization*. Society of Petroleum Engineers.

Lantuejoul, Ch. (2002) *Geostatistical Simulation: Models and Algorithms*. Berlin Springer Verlag. New York, U.S.A.

Murillo-Muñetón, G., Grajales-Nishimura, J.M., Velasquillo-Martínez, L., Casar-González, R., Xu, S.,Díaz-Viera, M.A., Maldonado-Susano, M. del C., Díaz-Rodríguez, F., García-Hernández, J. & Aguirre Cerda,

E. (2007) *Interpretación geológica en horizontes productores del Cretácico y Jurásico en el Campo Cantarell.* Instituto Mexicano del Petróleo, Proyecto F.30686. Internal report. México, D.F., México.

Murillo-Muñetón, G., Velasquillo-Martínez, L., Casar-González, R., Díaz-Viera, M.A., Maldonado-Susano, M. del C., Díaz-Rodríguez, F., García-Hernández, J. & Aguirre-Cerda, E. (2008) *Actualización geológica y modelado geoestadístico de distribución de facies del Jurásico Superior Kimmeridgiano en el campo Cantarell.* Instituto Mexicano del Petróleo, Proyecto F.30798. Internal report. México, D.F., México.

PEMEX Exploración y Producción. (1999) *Las Reservas de hidrocarburos de México. Volumen I.* Primera Edición. PEMEX, Exploración y Producción, México, D.F., México.

Qi, L., Carr, T.R. & Goldstein, R.H. (2007) Geostatistical three-dimensional modeling of oolite shoals, St. Louis Limestone, southwest Kansas. *AAPG Bulletin*, 91 (1), 69–96.

Schlumberger (2007) *PETREL™: Reservoir Modeling Software*. Schlumberger. France.

Yarus, J.M. & Chambers, R.L. (ed.) (1994) *Stochastic Modeling and Geostatistics: Principles, Methods, and Case Studies*. American Association of Petroleum Geologists. U.S.A.

CHAPTER 10

Trivariate nonparametric dependence modeling of petrophysical properties

A. Erdely, M.A. Díaz-Viera & V. Hernández-Maldonado

10.1 INTRODUCTION

Assessment of formation permeability is a complex and challenging problem that plays a key role in oil reservoir forecasts and optimized reservoir management. Generally, permeability evaluation is performed using porosity-permeability relationships obtained by integrated analysis of various petrophysical parameters from cores and well logs. In carbonate double-porosity formations with complex microstructure of pore space this problem becomes more difficult because the permeability usually does not depend on the total porosity, but on classes of porosity, such as vuggular and fracture porosity (secondary porosity). Even more, in such cases permeability is directly related to the connectivity structure of the pore system. This fact makes permeability prediction a challenging task.

Dependence relationships among petrophysical random variables, such as permeability and porosity, are usually nonlinear and complex, and therefore those statistical tools that rely on assumptions of linearity and/or normality are not suitable in this case. The use of copulas for modeling petrophysical dependencies is not new, see Díaz-Viera & Casar-González (2005) where t-copulas have been used for this purpose. But expecting a single copula family to be able to model any kind of bivariate dependency seems to be still too restrictive, at least for the petrophysical variables under consideration in this work. Therefore, it has been adopted a nonparametric approach by the use of the Bernstein copula, see Sancetta & Satchell (2004) and Sancetta (2007).

10.1.1 *The problem of modeling the complex dependence pattern between porosity and permeability in carbonate formations*

According to Balan (1995) by far the most used permeability predictor is the porosity-permeability relationship. It has long been assumed that most reservoir rocks show a reasonably linear relationship between these parameters in a semi-log scale, which allows for the estimation of permeability when a porosity profile is available. This normally requires a calibration data set that is represented by one or more key wells where comprehensive information is available in terms of core and log data. This calibration data set is used to build the predictor and to test the reliability of the results.

The regression approach, using statistical instead of deterministic formalism, tries to predict a conditional average, or expectation of permeability, corresponding to a given set of parameters. A different predictive equation must be established for each new area or new field. The main drawback of traditional regression methods is that the complex variability of data may not be effectively captured just in terms of variance or standard deviation (which may not even exist), and therefore the predicted permeability profile will be ineffective in reproducing extreme values.

What is important to note in this case is that the predicted permeability profile will be effective in estimating the average characteristics of the true profile, but will be ineffective in estimating the extreme values. These extreme values, from a fluid flow point of view, are the most important parts of the distribution, since they may represent either high permeability streaks or impermeable barriers.

Reservoir rocks show a wide spectrum of porosity-permeability relationships. In some formations, like for example homogeneous clastic rocks, these relationships show very low dispersion and can be reasonably used for prediction purposes. In other cases, as it is frequently for carbonates, this relationship is very loose and does not allow any safe regression, under traditional statistical tools.

On the other hand, model-free function estimators like artificial neural networks are very flexible tools for recognizing and reproducing the pattern of permeability distribution, but require a time consuming "learning" process which strongly depends on the amount and quality of available data.

A competitive and more systematic method for predicting permeability may be achieved by applying stochastic joint simulations, in which the correct specification of dependence pattern in the bivariate porosity-permeability distribution is crucial. According to Deutsch (1994) this approach basically consisted on an annealing geostatistical porosity-permeability cosimulation using their empirical joint distribution. A modification of the previous methodology was proposed by Díaz-Viera & Casar-González (2005) where the basic idea was to apply a t-copula bivariate distribution instead of the empirical one, in which the permeability-porosity observed dependence pattern is specified through a rank correlation measure such as Kendall's tau or Spearman's rho.

10.1.2   *Trivariate copula and random variables dependence*

According to Sklar's Theorem, see Sklar (1959), the underlying *trivariate copula* associated to a trivariate random vector (X, Y, Z) represents a functional link between the joint probability distribution D and the univariate marginal distributions F, G and H, respectively:

$$D(x, y, z) = C(F(x), G(y), H(z)) \qquad (10.1)$$

for all x, y, z in the extended real numbers system, where $C : [0, 1]^3 \rightarrow [0, 1]$ is unique whenever X, Y and Z are continuous random variables. Therefore, all the information about the dependence between random variables is contained in their corresponding copula. Several properties may be derived for copulas, see Schweizer & Sklar (1983) and Nelsen (2006), and among there is an immediate corollary from Sklar's Theorem: X, Y and Z are independent continuous random variables if and only if their underlying copula is $\Pi(u, v, w) = uvw$. Another interesting property is the fact that copulas are invariant under strictly increasing transformations of the random variables: the copula for (X, Y, Z) is the same than the one for $(g_1(X), g_2(Y), g_3(Z))$, where the g_i functions are strictly increasing and well defined in the range of the corresponding random variables.

Let $\mathcal{S} := \{(x_1, y_1, z_1), \dots, (x_n, y_n, z_n)\}$ be observations of a random vector (X, Y, Z). Empirical estimates may be obtained for the marginal distributions of X, Y and Z by means of

$$F_n(x) = \frac{1}{n} \sum_{m=1}^{n} \mathbb{I}\{x_m \leq x\}, \;\; G_n(y) = \frac{1}{n} \sum_{m=1}^{n} \mathbb{I}\{y_m \leq y\}, \;\; H_n(z) = \frac{1}{n} \sum_{m=1}^{n} \mathbb{I}\{z_m \leq z\}, \qquad (10.2)$$

where $\mathbb{I}$ stands for an indicator function which takes a value equal to 1 whenever its argument is true, and 0 otherwise. It is well-known, see for example Billingsley (1995), that the empirical distribution F_n is a consistent estimator of F, that is, $F_n(t)$ converges almost surely to $F(t)$ as $n \rightarrow \infty$, for all t.

Similarly, from Deheuvels (1979) we have the *empirical copula*, a function C_n with domain $\{i/n : i = 0, 1, \dots, n\}^3$ defined as

$$C_n \left(\frac{i}{n}, \frac{j}{n}, \frac{k}{n} \right) = \frac{1}{n} \sum_{m=1}^{n} \mathbb{I}\{rank(x_m) \leq i, \; rank(y_m) \leq j, \; rank(z_m) \leq k\} \qquad (10.3)$$

and its convergence to the true copula C has also been proved, see Fermanian et al. (2004). The empirical copula is not a copula, since it is only defined on a finite grid, not in the whole unit cube $[0, 1]^3$, but by Sklar's Theorem C_n may be extended to a copula.

10.2   TRIVARIATE DATA MODELING

A copula-based nonparametric approach is proposed to model the relationship between permeability, porosity and shear wave velocity (S-waves) of the double porosity carbonate formations of a South Florida Aquifer in the western Hillsboro Basin of Palm Beach County, Florida.

The characterization of this aquifer for the borehole and field scales is given in Parra et al. (2001) and Parra & Hackert (2002), and a hydrogeological situation is described by Bennett et al. (2002). The interpretation of the borehole data and determination of the matrix and secondary porosity and secondary-pore types (shapes of spheroids approximating secondary pores) were presented by Kazatchenko et al. (2006a), where to determine the pore microstructure of aquifer carbonate formations the authors applied the petrophysical inversion technique that consists in minimizing a cost function that includes the sum of weighted square differences between the experimentally measured and theoretically calculated logs as in Kazatchenko et al. (2004).

In this case the following well logs were used for joint simultaneous inversion as input data: resistivity log, transit times of the P- and S-waves (acoustic log), total porosity (neutron log), and formation density (density log). To calculate the theoretical acoustic and resistivity logs the double-porosity model for describing carbonate formations was applied: Kazatchenko et al. (2006b).

This model treats carbonate rocks as a composite material that consists of a homogeneous isotropic matrix (solid skeleton and matrix pore system) where the secondary pores of different shapes are embedded. The secondary pores were approximated by spheroids with variable aspect ratios to represent different secondary porosity types: vugs (close-to-sphere shapes), quasi-vugs (oblate spheroids), channels (prolate spheroids), and microfractures (flattened spheroids). For computing the effective properties the symmetrical self-consistent method of the effective medium approximation was used.

In this paper it has been used the results of inversion obtained by Kazatchenko et al. (2006a) for carbonate formations of South Florida Aquifer that includes the following petrophysical characteristics: matrix porosity, secondary vugular and crack porosities. It should be noted that the secondary-porosity system of this formation has complex microstructure and corresponds to a model with two types of pore shapes: cracks (flattened ellipsoids) with the overall porosity of 2% and vugs (close to sphere) with the porosity variations in the range of 10–30%. Such a secondary-porosity model can be interpreted as the interconnection of microfractures and channels vugular formation.

The relative vugular porosities (PHIV), that is vugular porosity divided by matrix porosity, is modeled as an absolutely continuous random variable X with unknown marginal distribution function F, shear wave velocity (VS meas = velocity of S-waves measured) as an absolutely continuous random variable Y with unknown marginal distribution function G, and permeability (K) as an absolutely continuous random variable Z with unknown marginal distribution function H. Trivariate observations from the random vector (X, Y, Z) are obtained from Kazatchenko et al. (2006a). For continuous random variables, the use of the empirical distribution function estimates Eq. (10.2) is not appropriate since F_n is a step function, and therefore discontinuous, so a smoothing technique is needed: a smooth estimation of the marginal quantile function $Q(u) = F^{-1}(u) = \inf\{x : F(x) \geq u\}, 0 \leq u \leq 1$, which is possible by means of Bernstein polynomials as in Muñoz-Pérez & Fernández-Palacín (1987):

$$\tilde{Q}_n(u) = \sum_{m=0}^{n} \frac{1}{2}(x_m + x_{m+1})\binom{n}{m}u^m(1 - u)^{n-m}, \tag{10.4}$$

and the analogous case for marginals G and H in terms of values y_m and z_m obtaining $\widetilde{R}_n(v)$ and $\widetilde{S}_n(w)$, respectively. For a smooth estimation of the underlying copula it has been used the Bernstein copula as in Sancetta & Satchell (2004) and Sancetta (2007):

$$\widetilde{C}(u,v,w) = \sum_{i=0}^{n}\sum_{j=0}^{n}\sum_{k=0}^{n} C_n\left(\frac{i}{n},\frac{j}{n},\frac{k}{n}\right)\binom{n}{i}u^i(1-u)^{n-i}\binom{n}{j}v^j(1-v)^{n-j}\binom{n}{k}w^j(1-w)^{n-k}$$

(10.5)

for every (u,v,w) in the unit cube $[0,1]^3$, and where C_n is as defined in Eq. (10.3).

10.3 NONPARAMETRIC REGRESSION

The main objetive in regression models is to explain/predict a random variable of interest (permeability) in terms of some other explanatory variables (relative vugular porosity and shear wave velocity, for example). A nonparametric approach is recommended when the data does not exhibit a "nice" behavior that might suggest some parametric models to be fitted, and/or when the assumptions of parametric candidates are too strong to be considered realistic in a particular case. Since the copula approach – parametric or not – leads to the estimation of the joint probability distribution of the involved variables, from this last one it is possible to obtain/estimate the conditional distribution of the variable of interest given certain values of the explanatory variables, and therefore point and interval estimates may be derived by means of a regression curve or regression surface, instead of imposing a functional form to such curve or surface as it happens, for example, in classical multiple linear regression, and which may happen to be unrealistic. But a no-free-lunch principle appplies: under the copula approach, intensive computational issues arise if many variables are involved.

Given multivariate data, it is common to start choosing as explanatory variables those who exhibit higher dependence with the variable that is to be explained. Pearson's linear correlation coefficient has widely been used for this purpose, but unfortunately it has many flaws that may induce to misleading conclusions, see Embrechts et al. (1999, 2003) for a discussion why it should not be considered as a dependence measure for general purposes, specially when normality and/or moment existence and/or linear dependence are unrealistic assumptions, which happens to be the case of the data under consideration. Therefore, dependence has been measured in terms of the dependence index Φ proposed by Hoeffding (1940), which satisfies all desirable properties for a dependence measure for continuous random variables, see for example Nelsen (2006).

From all the possible explanatory variables for permeability (K) in Kazatchenko et al. (2006a), for the first explanatory random variable it was chosen relative vugular porosity (PHIV) since it exhibited the highest dependence $\Phi(\text{PHIV}, \text{K}) = 0.71$ (on a $[0,1]$ scale). In choosing a second explanatory random variable it is preferred, in addition to have a high dependence with permeability, the lowest possible dependence with the first explanatory variable (PHIV), otherwise it would mean that it is quite similar to it and it will add no significant information to what the first one already can provide. Under this criteria, the second best choice was the share wave velocity (VS meas), with $\Phi(\text{VS meas}, \text{K}) = 0.60$ and $\Phi(\text{PHIV}, \text{VS meas}) = 0.55$. At this point, no more explanatory variables are considered since intensive computational issues arise when dealing with a 4-dimensional Bernstein copula. Hence, we model a trivariate random vector $(X,Y,Z) = (\text{PHIV}, \text{VS meas}, \text{K})$ under a nonparametric copula approach, and by conditioning we may obtain two regression curves (K given PHIV and K given VS meas) and one regression surface (K given PHIV and VS meas jointly).

For a value x in the range of the random variable X and $0 < \alpha < 1$ let $z = \beta_\alpha(x)$ denote a solution to the equation $\mathbb{P}(Z \leq z | X = x) = \alpha$. Then the graph of $z = \beta_\alpha(x)$ is the *α-quantile regression curve* of Z conditional on $X = x$. The particular case $\alpha = 0.5$ is the *median regression curve*. It has been proved in Nelsen (2006) that

$$\mathbb{P}(Z \leq z | X = x) = c_u(w)\big|_{u=F(x), w=H(z)},$$

(10.6)

where

$$c_u(w) = \frac{\partial \widetilde{C}(u, w)}{\partial u},$$ (10.7)

and $\widetilde{C}(u, w) = \widetilde{C}(u, 1, w)$, the bivariate case of Eq. (10.5). The above result (10.6) leads to the following algorithm in Nelsen (2006) to obtain the α-quantile regression curve of Z conditional on $X = x$.

ALGORITHM 1

1. Set $c_u(w) = \alpha$.
2. Solve for the regression curve $w = \gamma_\alpha(u)$.
3. Replace u by $\widetilde{Q}_n^{-1}(x)$ and w by $\widetilde{S}_n^{-1}(z)$, see Eq. (10.4).
4. Solve for the regression curve $z = \beta_\alpha(x)$.

In Figure 10.1 it is shown a scatter plot of relative vugular porosity (PHIV) versus permeability (K), along with three regression curves: $\alpha = 0.5$ (median) which represents a point estimate of K given PHIV values, and with $\alpha = 0.1, 0.9$ which represent 80% probability bands for the point estimates. In Figure 10.2 it is shown the log-scale values of K in spatial form (in terms of depth) and point estimates of log K given spatial values of PHIV. We have the analogous cases for shear wave velocity (VS meas) in Figures 10.3 and 10.4.

For a value x in the range of the random variable X, a value y in the range of the random variable Y and $0 < \alpha < 1$ let $z = \beta_\alpha(x, y)$ denote a solution to the equation $\mathbb{P}(Z \le z \mid X = x, Y = y) = \alpha$. Then the graph of $z = \beta_\alpha(x, y)$ is the α-quantile regression surface of Z conditional on $X = x$ and $Y = y$. Let

$$c_{uv}(w) = \frac{\partial \widetilde{C}(u, v, w)}{\partial u \, \partial v} \bigg/ \frac{\partial \widetilde{C}(u, v, 1)}{\partial u \, \partial v}$$ (10.8)

where $\widetilde{C}$ is obtained by Eq. (10.5). Then:

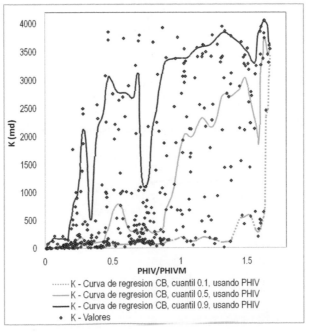

Figure 10.1   Scatterplot of relative vugular porosity (PHIV) and permeability (K) data, with $\alpha = 0.1, 0.5$, 0.9 quantile regression curves.

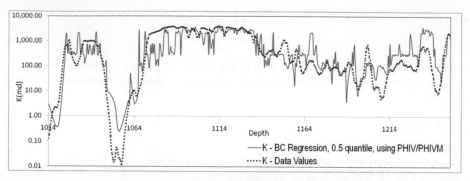

Figure 10.2   Log-scale permeability (K) spatial data (in terms of depth) and spatial median regression curve given relative vugular porosity (PHIV).

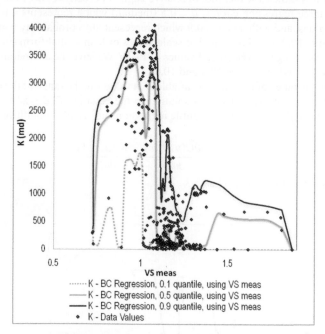

Figure 10.3   Scatterplot of shear wave velocity (VS meas) and permeability (K) data, with $\alpha = 0.1, 0.5, 0.9$ quantile regression curves.

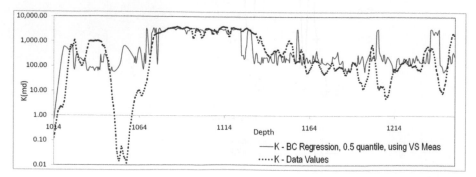

Figure 10.4   Log-scale permeability (K) spatial data (in terms of depth) and spatial median regression curve given shear wave velocity (VS meas).

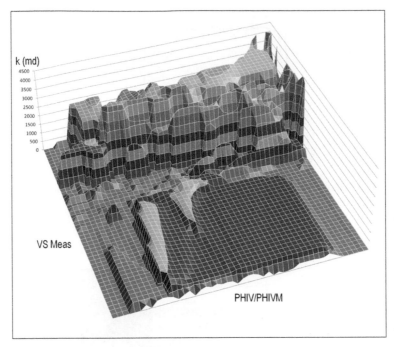

Figure 10.5   Median regression surface for permeability (K) given relative vugular porosity (PHIV) and shear wave velocity (VS meas).

ALGORITHM 2

1. Set $c_{uv}(w) = \alpha$.
2. Solve for the regression surface $w = \gamma_\alpha(u, v)$.
3. Replace u by $\widetilde{Q}_n^{-1}(x)$, v by $\widetilde{R}_n^{-1}(y)$, and w by $\widetilde{S}_n^{-1}(z)$, see Eq. (10.4).
4. Solve for the regression surface $z = \beta_\alpha(x, y)$.

In Figure 10.5 it is shown the median regression surface for permeability (K) given relative vugular porosity (PHIV) and shear wave velocity (VS meas), and in Figure 10.6 the log-scale values of K in spatial form (in terms of depth) and point estimates of log K given spatial values of PHIV and VS meas.

As a descriptive measure of the goodness of fit of predicted values of log-permeability (log K) given values of the explanatory variables PHIV and VS meas, the mean squared error (MSE) has been calculated in each case:

$$\text{MSE}(\log K \mid \text{PHIV}) = \frac{1}{n} \sum_{m=1}^{n} [z_m - \beta_{0.5}(x_m)]^2$$

$$\text{MSE}(\log K \mid \text{VS meas}) = \frac{1}{n} \sum_{m=1}^{n} [z_m - \beta_{0.5}(y_m)]^2$$

$$\text{MSE}(\log K \mid \text{PHIV, VS meas}) = \frac{1}{n} \sum_{m=1}^{n} [z_m - \beta_{0.5}(x_m, y_m)]^2$$

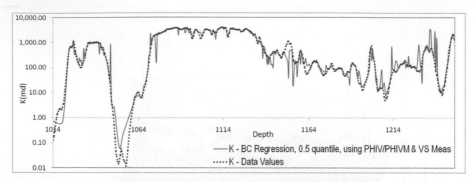

Figure 10.6   Log-scale permeability (K) spatial data (in terms of depth) and spatial median regression curve given relative vugular porosity (PHIV) and shear wave velocity (VS meas).

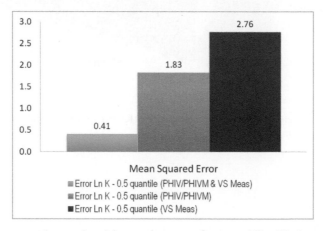

Figure 10.7   Mean squared error of spatial regression curves for permeability (K) given: a) relative vugular porosity (PHIV) and shear wave velocity (VS meas) jointly, b) relative vugular porosity alone, c) shear wave velocity alone.

where the lower the MSE value, the better the goodness of fit. As expected, in Figure 10.7 we have that the spatially predicted log K values have the lowest MSE when using both explanatory variables than each one alone. Also, the MSE of PHIV alone is better than VS meas alone, which was also expected since PHIV has a higher dependence value with K ($\Phi = 0.71$) that VS meas with K ($\Phi = 0.60$).

10.4   CONCLUSIONS

From a methodological point of view, this approach provides a very flexible statistical research tool to investigate the existing complex dependence relationships of petrophysical properties such as porosity and permeability, without imposing strong assumptions of linearity or log-linearity, and/or normality when modeling them as random variables, not even the existence of first or second moments of the variables involved. The only assumption has been the random variables to be jointly absolutely continuous, and thereafter the data is allowed to speak by itself about the dependence structure.

The methodology used in this work has the main advantage of being a straightforward way to perform nonparametric quantile regression, which is useful in obtaining conditional point and interval estimates for Z given $X = x$ and $Y = y$ without imposing or assuming functional relationships among the variables involved.

All the information about the dependence structure is contained in the underlying copula, and its estimation is being used, instead of the extreme information reduction that is done by the use of numerical measures such as the linear correlation coefficient, which under the presence of nonlinear dependence may become useless and/or quite misleading, see Embrechts et al. (1999, 2003). The nonparametric regression obtained is useful to confirm or to question prior ideas about relationships among variables, or even in proposing an apropriate model to explain such relations.

In relation with the geostatistical applications this method opens a promising line of research to model in a nonparametric fashion the intrinsic spatial dependence of random functions overcoming the restriction imposed by a linear co-regionalization models. This methology may be extended to more than three variables, but some intensive computing issues need to be solved efficiently.

REFERENCES

Balan, B., Mohaghegh, S. & Ameri, S. (1995) State-of-the-art in permeability determination from well log data: Part 1. A comparative study, model development. *SPE 30978.*

Bennett, W.M., Linton, P.F. & Rectenwald, E.E. (2002) Hydrologic investigation of the Floridian aquifer system, western Hillsboro Basin, Palm Beach County, Florida. *Technical Publication WS-8.* South Florida Management District.

Billingsley, P. (1995) *Probability and Measure.* 3rd edition. New York, Wiley.

Deheuvels, P. La fonction de dépendance empirique et ses propriétés. Un test nonparamétrique d'independance, *académie royale de belgique. bulletin de la classe des sciences (5),* 65 (6), 274–292.

Deutsch, C.V. & Cockerham, P.W. (1994) Geostatistical modeling of permeability with annealing cosimulation (ACS). *SPE 28413.*

Díaz-Viera, M. & Casar-González, R. (2005) Stochastic simulation of complex dependency patterns of petrophysical properties using t-copulas. *Proceedings of IAMG'05: GIS and Spatial Analysis,* 2, 749–755.

Embrechts, P., McNeil, A. & Straumann, D. (1999) Correlation: Pitfalls and alternatives. *Risk Magazine,* 5, 69–71.

Embrechts, P., Lindskog, F. & McNeil, A.J. (2003) Modeling dependence with copulas and applications to risk management. In: Rachev, S. (ed.) *Handbook of Heavy-Tailed Distributions in Finance.* New York, Elsevier. pp. 329–384.

Fermanian, J-D., Radulović, D. & Wegcamp, M. (2004) Weak convergence of empirical copula processes. *Bernoulli,* 10, 547–560.

Hoeffding, W. (1941) Scale-invariant correlation theory. In: Fisher, N.I. & Sen, P.K. (eds.). *The Collected Works of Wassily Hoeffding.* New York, Springer. pp. 57–107.

Kazatchenko, E., Markov, M. & Mousatov, A. (2004) Joint inversion of acoustic and resistivity data for carbonate microstructure evaluation. *Petrophysics,* 45, 130–140.

Kazatchenko, E., Markov, M., Mousatov, A. & Parra, J. (2006a) Carbonate microstructure determination by inversion of acoustic and electrical data: application to a South Florida Aquifer, *Journal of Applied Geophysics,* 59, 1–15.

Kazatchenko, E., Markov, M. & Mousatov, A. (2006b) Simulation of the acoustical velocities, electrical and thermal conductivities using unified pore structure model of double-porosity carbonate rocks. *Journal of Applied Geophysics,* 59, 16–35.

Muñoz-Pérez, J. & Fernández-Palacín, A. (1987) Estimating the quantile function by Bernstein polynomials. *Computational Statistics and Data Analysis,* 5, 391–397.

Nelsen, R.B. (2006) *An Introduction to Copulas.* 2nd edition. New York, Springer.

Parra, J.O., Hackert, C.L., Collier, H.A. & Bennett, M. (2001) A methodology to integrate magnetic resonance and acoustic measurements for reservoir characterization. *Report DOE/BC/ 15203-3.* Tulsa, Oklahoma, National Petroleum Technology Office, Department of Energy.

Parra, J.O. & Hackert, C.L. (2002) Permeability and porosity images based on crosswell reflection seismic measurements of a vugular carbonate aquifer at the Hillsboro site, South Florida. *72nd Annual Meeting of the Society of Exploration Geophysicists.* Paper VCD P1.2.

Sancetta, A. (2007) Nonparametric estimation of distributions with given marginals via Bernstein–Kantorovic polynomials: L_1 and pointwise convergence theory. *Journal of Multivariate Analysis*, 98, 1376–1390.

Sancetta, A. & Satchell, S. (2004) The Bernstein copula and its applications to modeling and approximations of multivariate distributions. *Econometric Theory*, 20, 535–562.

Schweizer, B. & Sklar, A. (1983) *Probabilistic Metric Spaces.* New York, North Holland.

Sklar, A. (1959) Fonctions de répartition à n dimensions et leurs marges. *Publications de l'Institut de statistique de l'Université de Paris*, 8, 229–231.

CHAPTER 11

Joint porosity-permeability stochastic simulation by non-parametric copulas

V. Hernández-Maldonado, M. A. Díaz-Viera & A. Erdely-Ruiz

11.1 INTRODUCTION

On geological-petrophysical reservoir modeling exists the need to establish dependence models between petrophysical properties, such as porosity-permeability relationship. In practice, there is little information about permeability while the porosity is mostly sampled. By modeling the dependence structure between this two properties, it is possible to know the permeability profile through the available information on the porosity one.

In order to calculate permeability profiles in wells, typically is used a linear estimator in a regression form, see Balam (1995). However, to predict the average, or expected value of a property exists the disadvantage of not reproducing the the variability of the data and consequently, the estimated profiles do not reproduce the extreme values of the real information. This situation is critical when a reservoir is been modeled because this values may represent impermeable barriers or high permeability zones.

A competitive and much more systematic method was proposed by Deutsch & Cockerham (1994). It uses the application of multivariate stochastic simulations in order to predict the permeability or primary variable; where a proper specification of the petrophysical properties dependence structure is crucial. This approach involves the application of simulated annealing technique to model the joint distribution function of the porosity-permeability relation.

A modification of Deutsch's methodology was proposed by Díaz-Viera and Casar-González (2005). In that work, it was proposed the use of a bivariate t-copula to construct the joint distribution function rather than use the sampling one, also the dependence structure is specified by matching measures such as Kendall's τ and Spearman ρ. The above methodology was applied to simulate the permeability from the porosity profile in a double porosity carbonate systems restricted to one-dimensional case at well-log scale Díaz-Viera et al. (2006).

While Díaz-Casar methodology can reproduce adequately the observed data, and also their extreme values using the t-copula, the critical problem of this proposal is that the copula used is parametric type, i.e. it is based on a given distribution function, the student t. This causes the data not to be naturally represented, because the real data distribution is hard to fit properly with only one parametric distribution.

The methodology presented here makes co-dependent geostatistical simulations using Bernstein copulas, which have the ability to model the dependence structure between random variables, as petrophysical properties, without assume a given distribution function as t-copulas do.

11.2 NON-CONDITIONAL STOCHASTIC SIMULATION METHODOLOGY BY USING BERNSTEIN COPULAS

The proposed method is based on Díaz-Viera's modification to Deutsch and Cockerham methodology for geostatistical simulation. Broadly speaking the methodology can be described in two stages. The first one is to carry out a non-parametric permeability (K) simulation by Bernstein copulas, using the porosity (PHI) as secondary variable. At this stage we are reproducing the joint dependence pattern between these two petrophysical properties. In the second one, a geostatistical

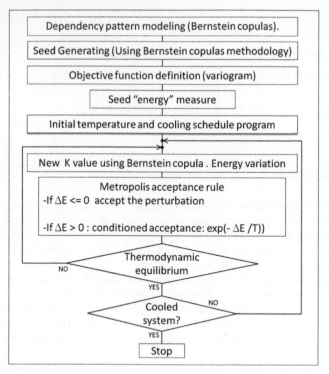

Figure 11.1   Methodology diagram. It shows the steps to produce a geostatistical co-simulation by simulated annealing, using Bernstein copulas.

simulation of permeability is performed by simulated annealing method, whose objective function is the variogram model, see Deutsch & Journel (1998), Figure 11.1.

A more detailed description of each step of the methodology is described below:

1. Modeling the petrophysical properties dependence pattern, using non-parametric copulas or Bernstein copulas.
2. Generating the seed or initial configuration for simulated annealing method, using the non-parametric simulation algorithm.
3. Defining the objective function.
4. Measuring the energy of the seed, according to the objective function.
5. Calculating the initial temperature and the simulated annealing schedule using the methodology proposed by Dreo et al. (2006).
6. Performing the simulation:
   • Accept or reject a new proposed permeability value using the non-parametric simulation algorithm.
   • The simulation finalizes when the objective function error (previously defined) is reached; an accumulation of 3 stages without change occurs; or when the maximum attempted perturbations is reached.

11.3   APPLICATION OF THE METHODOLOGY TO PERFORM A NON-CONDITIONAL SIMULATION WITH SIMULATED ANNEALING USING BIVARIATE BERNSTEIN COPULAS

We propose a copula-based non-parametric approach to model the relationship between the permeability and porosity of the double porosity carbonate formations of a South Florida Aquifer

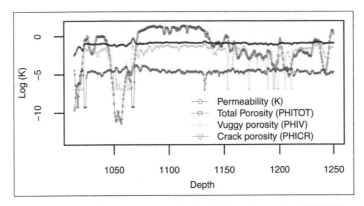

Figure 11.2   Crack porosity (PHICR), vuggy porosity (PHIV), total porosity (PHITOT) and permeability (K) derived from NMR Log (K).

in the western Hillsboro Basin of Palm Beach County, Florida, see Figure 11.2. After that, this methodology models the spatial distribution of the permeability in function of the depth.

The characterization of this aquifer for the borehole and field scales is given in Parra et al. (2001) and Parra & Hackert (2002), and a hydrogeological situation is described by Bennett et al. (2002). The interpretation of the borehole data and determination of the matrix and secondary porosity and secondary-pore types (shapes of spheroids approximating secondary pores) were presented by Kazatchenko et al. (2006a), where to determine the pore microstructure of aquifer carbonate formations the authors applied the petrophysical inversion technique that consists in minimizing a cost function that includes the sum of weighted square differences between the experimentally measured and theoretically calculated logs as in Kazatchenko et al. (2004).

In this paper we used the results of inversion obtained by Kazatchenko et al. (2006a) for carbonate formations of South Florida Aquifer that includes the following petrophysical characteristics: matrix porosity, secondary vuggy and crack porosities. It should be noted that the secondary-porosity system of this formation has complex microstructure and corresponds to a model with two types of pore shapes: cracks (flattered ellipsoids) with the overall porosity of 2% and vugs (close to sphere) with the porosity variations in the range of 10–30%. Such a secondary-porosity model can be interpreted as the interconnected by microfractures and channels vuggy formation.

The details of the statistical properties analysis of the sampled data were largely explained by Díaz-Viera et al. (2006), based on this study we determine that the highest observed dependence between the petrophysical properties corresponds to vuggy porosity (PHIV) and permeability (K).

11.3.1   *Modeling the petrophysical properties dependence pattern, using non-parametric copulas or Bernstein copulas*

According to Sklar's Theorem, see Sklar (1959), the underlying bivariate copula associated to a bivariate random vector (X, Y) represents a functional link between the joint probability distribution H and the univariate marginal distributions F and G, respectively:

$$H(x, y) = C(F(x), G(y)) \tag{11.1}$$

for all x, y in the extended real numbers system, where $C : [0, 1]^2 \rightarrow [0, 1]$ is unique whenever X and Y are continuous random variables. Therefore, all the information about the dependence between random variables is contained in their corresponding copula. Several properties may be derived for copulas, see Schweizer & Sklar (1983) and Nelsen (2006), and among them we have an immediate corollary from Sklar's Theorem: X and Y are independent continuous random variables if and only if their underlying copula is $\Pi(u, v) = uv$.

Let $S := \{(x_1, y_1), \ldots, (x_n, y_n)\}$ be observations of a random vector (X, Y). We may obtain empirical estimates for the marginal distributions of X and Y by means of

$$F_n(x) = \frac{1}{n} \sum_{k=1}^{n} \mathbb{I}\{x_k \leq x\}, \qquad G_n(y) = \frac{1}{n} \sum_{k=1}^{n} \mathbb{I}\{y_k \leq y\}, \tag{11.2}$$

where $\mathbb{I}$ stands for an indicator function which takes a value equal to 1 whenever its argument is true, and 0 otherwise. It is well-known, see Billingsley (1995), that the empirical distribution F_n is a consistent estimator of F, that is, $F_n(t)$ converges almost surely to $F(t)$ as $n \to \infty$, for all t.

Similarly, we have the empirical copula, Deheuvels (1979), a function C_n with domain $\{i/n : i = 0, 1, \ldots, n\}^2$ defined as

$$C_n \left(\frac{i}{n}, \frac{j}{n} \right) = \frac{1}{n} \sum_{k=1}^{n} \mathbb{I}\{rank(x_k) \leq i, rank(y_k) \leq j\} \tag{11.3}$$

and its convergence to the true copula C has also been proved, see Fermanian et al. (2004). The empirical copula is not a copula, since it is only defined on a finite grid, not in the whole unit square $[0, 1]^2$, but by Sklar's Theorem, see Sklar (1959), C_n may be extended to a copula.

We model vuggy porosities as an absolutely continuous random variable X with unknown marginal distribution function F, and permeability as an absolutely continuous random variable Y with unknown marginal distribution function G. We have bivariate observations from the random vector (X, Y). For simulation of continuous random variables, the use of the empirical distribution function estimates Eq. (11.2) is not appropriate since F_n is a step function, and therefore discontinuous, so a smoothing technique is needed. Since our main goal is simulation of porosity-permeability, it will be better to have a smooth estimation of the marginal quantile function $Q(u) = F^{-1}(u) = \inf\{x : F(x) \geq u\}$, $0 \leq u \leq 1$, which is possible by means of Bernstein polynomials as in Muñoz-Pérez & Fernández-Palacín (1987).

$$\tilde{Q}_n(u) = \sum_{k=1}^{n} \frac{1}{2}(x_k + x_{k+1}) \binom{n}{k} u^k (1 - u)^{n-k} \tag{11.4}$$

and the analogous case for marginal G in terms of values y_k. For a smooth estimation of the underlying copula we make use of the Bernstein copula as in Sancetta & Satchell (2004) and Sancetta (2007):

$$\tilde{C}_n(u, v) = \sum_{i=1}^{n} \sum_{j=1}^{n} C_n \left(\frac{i}{n}, \frac{j}{n} \right) \binom{n}{i} u^i (1 - u)^{n-i} \binom{n}{j} v^j (1 - v)^{n-j} \tag{11.5}$$

for every (u, v) in the unit square $[0, 1]^2$, and where C_n is as defined in Eq. (11.3).

11.3.2   *Generating the seed or initial configuration for simulated annealing method, using the non-parametric simulation algorithm*

In order to simulate replications from the random vector (X, Y) with the dependence structure inferred from the observed data $S := \{(x_1, y_1), \ldots, (x_n, y_n)\}$, accordingly to a result in Nelsen (2006), we have the following algorithm:

1. Generate two independent and continuous uniform $(0, 1)$ random variates u and t
2. Set $v = C_u^{-1}(t)$ where

$$C_u(v) = \frac{\partial \tilde{C}(u, v)}{\partial u} \tag{11.6}$$

and $\tilde{C}$ is obtained by (Eq. 11.5)

3. The desired pair is $(x, y) = (\tilde{Q}_n(u), \tilde{R}_n(v))$, where $\tilde{Q}_n(u)$ and $\tilde{R}_n(v)$, are the estimated and smoothed quantile functions of X and Y, respectively, according to (Eq. 11.4).

This non-parametric simulation methodology is used to generate the initial configuration or seed, which is used as a start point in the metropolis algorithm. The Figure 11.3 shows the resulting non-conditional simulation, note that the Bernstein copula reproduces very well the marginal distribution of each petrophysical property and do the same with the joint distribution function, therefore, the complex dependence structure existing between this two properties, vuggy porosity (PHIV) and permeability (K), is very well described by the non-parametric copulas.

Also note that the copula can reproduce the variability and the extreme values of the original data. This is verified, comparing some of most relevant statistics of the real and simulated permeability values using the Bernstein copulas, Figure 11.4.

Finally, Figure 11.5 shows the spatial distribution of permeability vs. a simple non-conditional Bernstein copulas simulation, using PHIV as secondary variable.

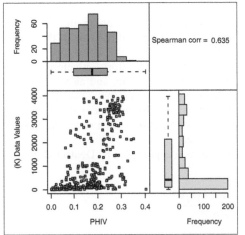 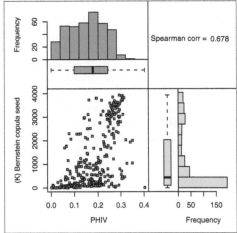

Figure 11.3   On the left, scatterplot of the original data, PHIV vs. K. On the right, the scatter plot of one single non-conditional simulation of PHIV vs. K, using Bernstein copulas, each scatter plot has their respective histograms.

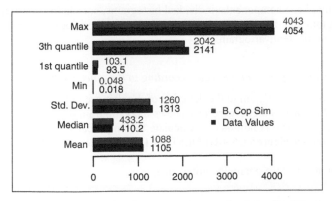

Figure 11.4   Comparative table and graph, of some statistics to real and simulated K values using Bernstein copulas.

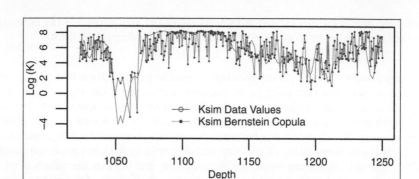

Figure 11.5   Spatial distribution of the real permeability and a simple non-conditional Bernstein copulas K simulation, using PHIV as secondary variable.

11.3.3   *Defining the objective function*

The simulated annealing technique runs many, often millions of perturbations in order to achieve an acceptable realization. A perturbation means modifying a permeability value of the current configuration, e.g. the seed we just calculated when the method just begins.

Executing millions of perturbations implies a high computational effort. Therefore, it is not recommended having too many components into the objective function, and also each one of them should be reasonably simple to compute, see Deutsch & Journel (1998). The advantage of model the dependence structure of petrophysical properties using Bernstein copulas is that they let us establish a simple objective function, i.e. an objective function with a single component, which does not affect the global computational performance.

The objective function is defined using the variogram, as it is proposed by Deutsch & Journel (1998).

$$FO = \sum_i \frac{[\gamma^*(h_i) - \gamma(h_i)]^2}{\gamma(h_i)^2} \tag{11.7}$$

where:

$$\gamma(h) = \frac{1}{2N(h)} \sum_{i=1}^{N(h)} [Z(x_i + h) - Z(x_i)]^2 \tag{11.8}$$

Deutsch and Cockerham methodology specifies the dependence structure through a multi-objective function, i.e. 5 components: individual histograms, correlation coefficient, variogram, indicator variogram and distribution conditional. In our methodology we propose the use of a single component, the semivariogram.

11.3.4   *Measuring the energy of the seed, according to the objective function*

It has been proposed a variogram in the direction of 0, tolerance of 90 (isotropic variogram), 50 intervals and a 2.27 meters lag. Figure 11.6 shows the variogram of the original data (objective function). Under the same conditions we calculate the empirical variogram and the model for seed configuration, see Figure 11.6. RGEOESTAD, Díaz-Viera et al. (2010), was used to analyze these results.

Using Equation (11.7) we obtain the initial energy:

$$E_i = 159.6313$$

In Table 11.1 they are shown the models to each configuration.

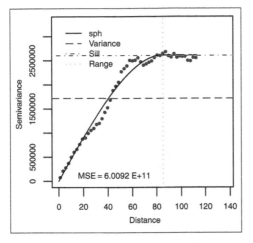 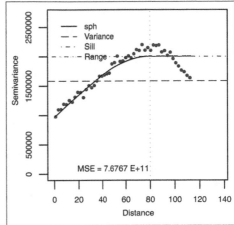

Figure 11.6   To the left, estimated variogram, 50 lag's for the data set. To the right, estimated variogram, 50 lag's for the seed.

Table 11.1   Variograms models to the dataset and seed.

	Nugget	Sill + Nugget	Range
Dataset	0.00	2611621.52	85.07
Seed	962645.81	2004671.28	80.00

11.3.5   Calculating the initial temperature, and the most suitable annealing schedule of simulated annealing method to carry out the simulation

To calculate the annealing schedule it is followed the procedure recommended by Dreo et al. (2006). The initial temperature is calculated based on Equation (11.9).

$$T_0 = \left(-\frac{\Delta E}{\log(\tau_0)} \right) \tag{11.9}$$

where ΔE ($E_{perturb} - E_{initial}$) is the mean of the energy differences between 100 perturbed configurations and the seed. The perturbation mechanism consists of uniformly and randomly chooses a value of porosity and generates a new permeability value using the simulation Bernstein copulas algorithm.

Each perturbation always starts in the same initial configuration. This is because our objective is to measure the system's energy change caused by a single perturbation. With this, we can get an idea of the acceptance rate (τ_0) we need to introduce. Let us remember that the rate of acceptance (τ_0) is a value that discriminates, within Boltzmann equation, the quality of a configuration that we will accept W. L. Coffee (1993). Therefore, an acceptance rate (τ_0) close to one will accept all perturbed configuration (low quality acceptance rate), while rates close to zero, only will accept perturbed configurations that reduce the overall system energy (high quality acceptance rate).

On Figure 11.7 they are shown 23 of the 100 perturbations of the initial configuration with an acceptance rate of 0.5 (medium quality). Also, it is made the calculation of each term of Equation (11.9) to show how metropolis criterion works to accept or reject the effect of a perturbation in the overall system energy.

Note that in the second column of this table that ΔE can have positive or negative differences, for example, 9.58 is a positive difference and corresponds to the energy of 169.21 (fifth row); or −2.69, which is a negative difference and corresponds to an energy of 156.93 (penultimate row). Positive differences indicate that the energy of the perturbed configuration is greater than the

100 perturbation energy vs. OF	ΔE	Metropolis acceptance (1)	$-\Delta E/T$	Metropolis acceptance (2) Boltzmann $r < \exp\left(-\{\Delta E\}/T\right)$
160.2733	0.6420 ✗		-0.4835	0.6166
159.6089	-0.0224 ✓			
159.6204	-0.0109 ✓			
163.5322	3.9009 ✗		-2.9381	0.0530
169.2158	9.5845 ✗		-7.2189	0.0007
164.3231	4.6918 ✗		-3.5338	0.0292
159.6924	0.0611 ✗		-0.0460	0.9550
165.3448	5.7135 ✗		-4.3033	0.0135
160.1211	0.4898 ✗		-0.3689	0.6915
159.2698	-0.3615 ✓			
161.3739	1.7426 ✗		-1.3125	0.2691
159.6402	0.0089 ✗		-0.0067	0.9933
159.6389	0.0076 ✗		-0.0057	0.9943
159.5629	-0.0684 ✓			
159.6319	0.0006 ✗		-0.0005	0.9995
159.647	0.0157 ✗		-0.0118	0.9882
160.4136	0.7823 ✗		-0.5892	0.5548
162.3441	2.7128 ✗		-2.0432	0.1296
159.6323	0.0010 ✗		-0.0008	0.9992
159.9063	0.2750 ✗		-0.2071	0.8129
156.9334	-2.6979 ✓			
159.4875	-0.1438 ✓			
159.3414	-0.2899 ✓			

Figure 11.7   23 perturbations of the initial configuration with an acceptance rate of 0.5, plus the calculation of each term of Equation (11.9).

energy of the initial configuration, negative differences indicate that the energy of the perturbed configuration is less than the energy of the initial configuration. Therefore, using the Metropolis criterion in its first phase means that only the negative energy differences will be accepted (green check mark, third column of Figure 11.7 and positive differences will be rejected (red check mark).

In its second phase the metropolis criterion tries to accept certain configurations which initially were rejected. In last column of Figure 11.7 are calculated, using the expression $\exp(\Delta E/T)$, the probability of acceptance for each initially rejected energy. Based on what is shown in Figure 11.7, those perturbed configurations, that have a great positive energetic difference will have little chance of being accepted, while those perturbed configurations that have a small positive energetic difference will have a high probability to be accepted.

In Figure 11.8, they are shown the same 23 perturbations of the initial configuration, with a rate of acceptance of 0.01. Note that the smaller rate to acceptance, the lower probability of acceptance.

In Figure 11.9 they are shown the same 23 perturbations of the initial configuration, now with an acceptance rate of 0.99. Note that most of the perturbations initially rejected have a high probability of being accepted.

Finally, we consider an acceptance rate of 0.5, because the Bernstein copula generates a "good quality" seed. Therefore, using Equation (11.9) the initial temperature is:

$$T_0 = 1.14$$

100 perturbation energy vs. OF	ΔE	Metropolis acceptance (1)	- ΔE/T	Metropolis acceptance (2) Boltzmann $r < \exp\left(-\langle \Delta E \rangle/T\right)$
160.2733	0.6420 ✘		-3.7248	0.0241
159.6089	-0.0224 ✔			
159.6204	-0.0109 ✔			
163.5322	3.9009 ✘		-22.6324	0.0000
169.2158	9.5845 ✘		-55.6077	0.0000
164.3231	4.6918 ✘		-27.2210	0.0000
159.6924	0.0611 ✘		-0.3545	0.7015
165.3448	5.7135 ✘		-33.1488	0.0000
160.1211	0.4898 ✘		-2.8417	0.0583
159.2698	-0.3615 ✔			
161.3739	1.7426 ✘		-10.1103	0.0000
159.6402	0.0089 ✘		-0.0516	0.9497
159.6389	0.0076 ✘		-0.0441	0.9569
159.5629	-0.0684 ✔			
159.6319	0.0006 ✘		-0.0035	0.9965
159.647	0.0157 ✘		-0.0911	0.9129
160.4136	0.7823 ✘		-4.5388	0.0107
162.3441	2.7128 ✘		-15.7392	0.0000
159.6323	0.0010 ✘		-0.0058	0.9942
159.9063	0.2750 ✘		-1.5955	0.2028
156.9334	-2.6979 ✔			
159.4875	-0.1438 ✔			
159.3414	-0.2899 ✔			

Figure 11.8   23 perturbations of the initial configuration with an acceptance rate of 0.01, plus the calculation of each term of Equation (11.9).

The change of stage is done by following the next conditions:

- $12 * N$ accepted perturbations $= 12 * 380 = 4500$
- $100 * N$ attempted perturbations $= 100 * 380 = 38000$

where N is the data number.

The decrease in temperature is calculated using geometric law Equation (11.10), see Figure 11.10.

$$T_{k+1} = 0.8 * T_k \tag{11.10}$$

11.3.6   *Performing the simulation*

The methodology proposed in this work is a simulation divided into two steps. First, we have to perturb the configuration system by generating a new permeability value, using bivariate Bernstein copulas simulation. Second, apply the simulated annealing method to generate a stochastic simulation accepting or rejecting the before perturbations, by using Metropolis criterion.

As it was mentioned in the previous section, each perturbation consists of generate a new permeability value (K), by the non-parametric and non-conditional simulation algorithm, using porosity (PHI) as a secondary variable. The Bernstein copulas suggest a new permeability value

100 perturbation energy vs. OF	ΔE	Metropolis acceptance (1)	- ΔE/T	Metropolis acceptance (2) Boltzmann $r < \exp\left(-\langle \Delta E\rangle/T\right)$
160.2733	0.6420	✗	-0.0081	0.9919
159.6089	-0.0224	✓		
159.6204	-0.0109	✓		
163.5322	3.9009	✗	-0.0494	0.9518
169.2158	9.5845	✗	-0.1214	0.8857
164.3231	4.6918	✗	-0.0594	0.9423
159.6924	0.0611	✗	-0.0008	0.9992
165.3448	5.7135	✗	-0.0723	0.9302
160.1211	0.4898	✗	-0.0062	0.9938
159.2698	-0.3615	✓		
161.3739	1.7426	✗	-0.0221	0.9782
159.6402	0.0089	✗	-0.0001	0.9999
159.6389	0.0076	✗	-0.0001	0.9999
159.5629	-0.0684	✓		
159.6319	0.0006	✗	0.0000	1.0000
159.647	0.0157	✗	-0.0002	0.9998
160.4136	0.7823	✗	-0.0099	0.9901
162.3441	2.7128	✗	-0.0343	0.9662
159.6323	0.0010	✗	0.0000	1.0000
159.9063	0.2750	✗	-0.0035	0.9965
156.9334	-2.6979	✓		
159.4875	-0.1438	✓		
159.3414	-0.2899	✓		

Figure 11.9   23 perturbations of the initial configuration with an acceptance rate of of 0.5, plus the calculation of each term of Equation (11.9).

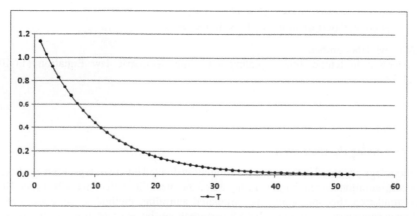

Figure 11.10   The geometric law is used to calculate the Decreasing program schedule of temperature, to perform the simulated annealing cooling.

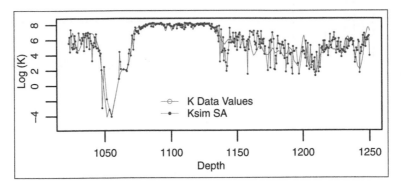

Figure 11.11   Spatial distribution of permeability in a single non-conditional SA simulation of K, with bivariate Bernstein copula, using PHIV as secondary variable.

which is well fitted within the dependence structure of these properties. This is what makes this approach attractive, because while traditional regression methods impose a linear dependence, Bernstein copulas allow modeling complex structures and nonlinear dependence that such methods do not. Therefore, each perturbation will always respect the relationship existing between the petrophysical properties.

The acceptance-rejection procedure also depends on the temperature of each stage when the simulation is in progress, so, it is necessary to allow the simulated annealing method to make a sufficient number of simulations for each temperature stage to reach its thermodynamic equilibrium. In other words, it is necessary to ensure that, by time of lowering the temperature, (phase change), the solution space has been sufficiently explored so that it contains the best solution.

Every time the temperature is decreased, the acceptance metropolis criterion becomes more restrictive, so that, at lower temperatures, only is accepted the best configurations, i.e. those with the lowest system energy.

The simulation finalizes when the objective function error is reached (previously defined); an accumulation of stages without change occurs; or when the maximum attempted perturbations is reached.

11.3.7   *Application of the methodology for stochastic simulation by bivariate Bernstein copulas to simulate a permeability (K) profile. A case of study*

We model the relationship between the permeability and porosity of the double porosity carbonate formations of a South Florida aquifer in the western Hillsboro Basin of Palm Beach County, Florida, see Figure 11.2. Until now, we have used bivariate Bernstein copulas to obtain an initial configuration (seed); also we have proposed an objective function and using it to measure the energy of the seed; finally we have calculated the initial temperature and the schedule program to perform the geostatistical simulation with simulated annealing. Figure 11.11, shows a single non-conditional and non-parametric simulated annealing simulation of K using PHIV (vuggy porosity) as a secondary variable. The simulation shows that the permeability values follow the spatially pattern as the original data, although, even now there is still small-scale variability on results, but it is smaller than the seed.

Figure 11.12 shows the scatterplot between K and PHIV from single non-conditional SA simulation using bivariated Bernstein copula and their respective histograms. As we could anticipate, the dependence structure between this two petrophysical properties is well represented. In the same Figure 11.12, it is shown the empirical variogram and its model (objective function) of the simulation.

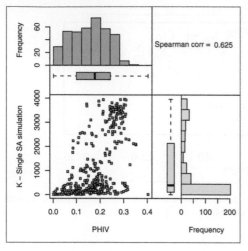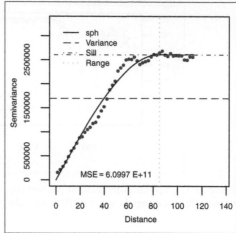

Figure 11.12　Scatter plot and variogram of a single non-conditional simulated annealing simulation for K using PHIV as a secondary variable.

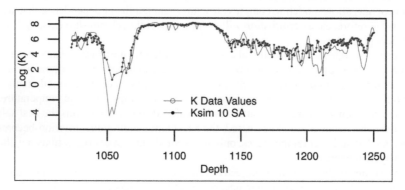

Figure 11.13　Spatial distribution of permeability in a median of 10 non-conditional SA simulations of K, with bivariate Bernstein copula, using PHIV as secondary variable.

Also it was realized a median of 10 non-conditional simulated annealing simulations in order to reduce small-scale variability, see Figure 11.13.

We put also the median scatter plot and its respective variogram, Figure 11.14. The reason for using a median rather than an average is because the median is not skewed as easily as the average.

As it can be seen in the median of 10 simulations, it is also very well represented the bivariate dependence structure of porosity-permeability relationship. Below is presented a comparative table of the variogram models, see Table 11.2.

As it can be seen, in the 10 median simulations is also very well represented the bivariate dependence structure of porosity-permeability relationship.

All simulations (including seed) have very similar statistics, Figure 11.15, this is because the dependence structure of the petrophysical properties is been modeled by Bernstein copulas. In the same figure, the bar labeled as K SA 1, which represents a simple simulation, is the best approaching to the original data, the bar labeled as Data; followed by the median of the 10 simulations, the bar labeled as K SA Median, this is because although the median smooths the variability of each simulation also accumulates each one of their mistakes; finally the seed, the bar labeled as K Seed.

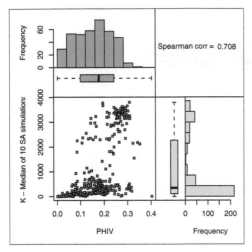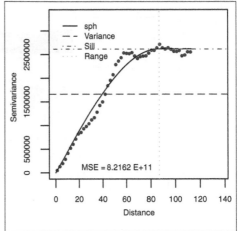

Figure 11.14   Scatter plot and variogram of a median of 10 non-conditional SA simulations of K using PHIV as a secondary variable.

Table 11.2   Variograms Models of the Dataset, seed, single non-conditional SA simulation and the median of 10 SA non-conditional simulations.

Configuration	Nugget	Sill + Nugget	Range
Dataset	0.00	2.61162×10^6	85.07
Seed	962645.81	2.00467×10^6	80.00
K SA 1	0.00	2.612×10^6	84.99
K SA Median	0.00	2.620×10^6	86.84

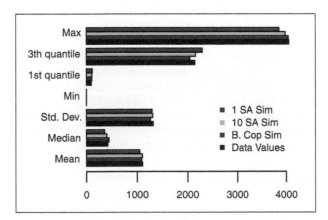

Figure 11.15   Statistical comparison of to the original data, initial configuration, a single SA simulation, and the median of 10 SA simulations.

Finally, Figure 11.16 shows a comparison chart of the mean square error (MSE). As we can see the values that have the greatest MSE are the data of the seed, 3.309. The simulated annealing simulations using bivariate Bernstein copulas, single and median shows MSE very reduced 1.415 and 0.921 respectively.

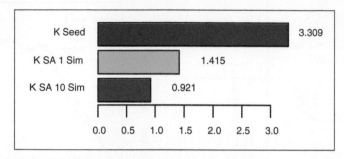

Figure 11.16   Mean Square Error (MSE) comparison between the initial configuration (seed), a single SA simulation, and the median of 10 SA simulations.

11.4   COMPARISON OF RESULTS USING THREE DIFFERENT METHODS

We propose the use of Bernstein copulas instead of establishing linear function using semi-logarithmic transformations, because they eventually end up giving results that are far away from the real data distribution. Also we say that a parametric copula based approach can represent very well a simple distribution structure but in data with complex relationship, the data may not to be naturally estimated, due the real data distribution is hard to fit properly with only one kind of distribution.

In this section we perform a comparative analysis between three methodologies used to simulate the spatial distribution of permeability (K), using its dependence structure with the petrophysical property: vuggy porosity (PHIV). The purpose of this analysis is to demonstrate that the Bernstein copulas is a better and sophisticated tool.

The comparison of permeability simulations is performed using the following methodologies:

1. Sasim of GSLIB Deutsch & Jouenel (1998). Stochastic simulation methodology by simulated annealing whose multiobjective function consists of two individual histograms, a correlation coefficient, a semivariogram moldel, and its conditional distribution, Equations (11.11, 11.12, 11.4 & 11.15).
2. *t*-copula Díaz-Viera et al. (2005). Stochastic spatial simulation methodology by simulated annealing using *t*-copulas to model the dependence struture of the petrophysical properties. Its multiobjective function is composed of the correlation coefficient, a semivariogram model, and conditional distribution, Equations (11.12, 11.14 & 11.15).
3. Bivariate Bernstein copulas. Stochastic spatial simulation methodology by simulated annealing using Bernstein copulas to model the dependence struture of the petrophysical properties. Its objective function consists only of a semivariogram model, Equation (11.12).

Here are the equations to calculate the objective function:

$$O_1 = \sum_z [F^*(z) - F(z)]^2 \tag{11.11}$$

$$O_2 = \sum_h \left[\frac{\gamma^*(h) - \gamma(h)]^2}{\gamma(h)^2} \right] \tag{11.12}$$

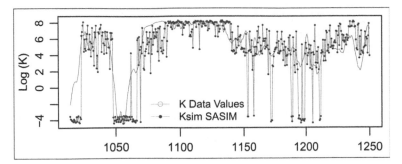

Figure 11.17 Spatial distribution in a single non-conditional simulation using SASIM methodology (Deutsch 1998), and using PHIV as a secondary variable.

where:

$$\gamma(h) = \frac{1}{2N(h)} \sum_{i=1}^{N(h)} [Z(x_i + h) - Z(x_i)]^2 \tag{11.13}$$

$$O_3 = [\rho^* - \rho]^2 \tag{11.14}$$

$$O_4 = \sum_{j=0}^{n_s} \sum_{i=0}^{n_p} [f_i^*(j) - f_i(j)]^2 \tag{11.15}$$

Before beginning, an overview of the performed comparisons using the three methodologies is presented:

- A single non-conditional simulation, and a median of 10 non-conditional simulations of permeability profile.
- A single 10% conditional simulation, and a median of 10, 10% conditional simulations of permeability profile.
- A single 50% conditional simulation, and a median of 10, 50% conditional simulations of permeability.
- A single 90% conditional simulation, and a median of 10, 90% conditional simulations of permeability profile.

11.4.1 *A single non-conditional simulation, and a median of 10 non-conditional simulations of permeability*

Figures 11.17, 11.18 & 11.19 show the spatial distribution of permeability in a single non-conditional simulation with simulated annealing, using as secondary variable PHIV. Simulation methodologies: SASIM. *t*-copula and Bivariate Bernstein copula.

A complete table of the MSE's differences between each method, in terms of percentage, is presented below (Table 1.3). Note that the methodology with the greater mean square error is SASIM with a value of 7.76, it is followed by the *t*-copula with 5.55 (which represents the 71% respecting SASIM-GSLib methodology) after the Bernstein copulas appear with a value of 1.85 (which represents the 23% respecting SASIM-GSLib methodology).

Figures 11.20, 11.21 & 11.22 show the comparison of the spatial distribution of the permeability of a median of 10 non-conditional simulations with simulated annealing, using PHIV as a secondary variable. Simulation methodologies: SASIM. *t*-copula and Bivariate Bernstein copula.

A complete table of the MSE's differences between each method, in terms of percentage, is presented (Table 11.4). Once again the methodology with a greater mean square error is the

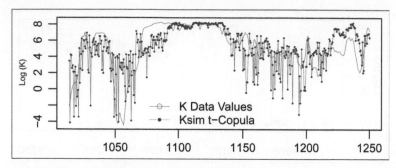

Figure 11.18   Spatial distribution in a single non-conditional simulation using *t*-copula methodology (Díaz et al. 2005), and using PHIV as a secondary variable.

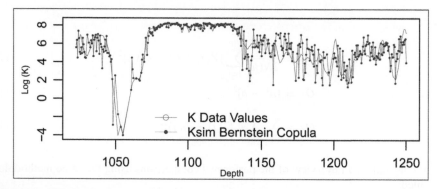

Figure 11.19   Spatial distribution in a single non-conditional simulation using Bivariate Bernstein copula methodology, and using PHIV as a secondary variable.

Table 11.3   Complete table of error differences between each method in terms of percentage. For a single simulation.

Method	MSE	vs. SASIM	vs. *t*-Copula
SASIM	7.76	100%	–
t-Copula	5.55	71.5%	100%
Bernstein Copula	1.85	23.8 %	33.3%

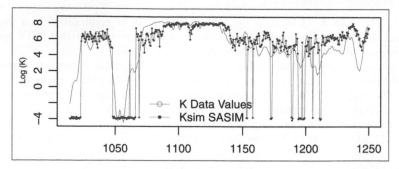

Figure 11.20   Spatial distribution in a median of 10 non-conditional simulations using SASIM methodology (Deutsch 1998), and using PHIV as a secondary variable.

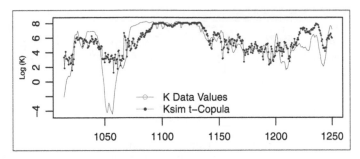

Figure 11.21   Spatial distribution in a median of 10 non-conditional simulations using *t*-copula methodology (Díaz et al. 2005), and using PHIV as a secondary variable.

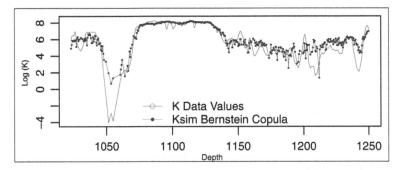

Figure 11.22   Spatial distribution in a median of 10 non-conditional simulations using Bivariate Bernstein copula methodology, and using PHIV as a secondary variable.

Table 11.4   Complete table of error differences between each method in terms of percentage. For a median of 10 nonconditional simulations.

Method	MSE	vs. SASIM	vs. *t*-Copula
SASIM	7.00	100%	–
t-Copula	3.82	54.6%	100%
Bernstein Copula	1.17	16.7%	30.6%

performed with SASIM with a value of 7.00 followed by the *t*-copula with 3.82 (which represents the 54% respecting SASIM-GSLib methodology) after copulas Bernstein appear with a value of 1.17 (which represents the 16% respecting SASIM-GSLib methodology).

Between 1 and 10 simulations, each methodology has a MSE reduction, 9% for SASIM (from 7.76 to 7.00); 31% for the t-copulas (from 5.55 to 3.82); 36% for Bernstein copulas (from 1.85 to 1.17).

11.4.2   *A single 10% conditional simulation, and a median of 10, 10% conditional simulations of permeability*

The Figure 11.23, 11.24 & 11.25 show the spatial distribution of permeability in a single 10% conditional simulation using as secondary variable PHIV. Simulation methodologies: SASIM. *t*-copula and Bivariate Bernstein copula.

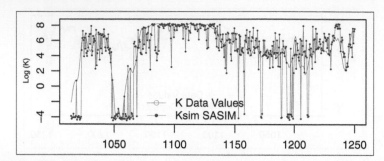

Figure 11.23   Spatial distribution in a single 10% conditional simulation using SASIM methodology (Deutsch 1998), and using PHIV as a secondary variable.

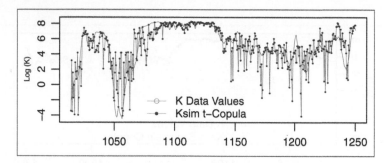

Figure 11.24   Spatial distribution in a single 10% simulation using *t*-copula methodology (Díaz et al. 2005), and using PHIV as a secondary variable.

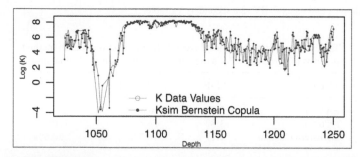

Figure 11.25   Spatial distribution in a single 10% simulation using Bivariate Bernstein copula methodology, and using PHIV as a secondary variable.

A complete table of the MSE's differences between each method, in terms of percentage, is presented (Table 1.5). Note that the methodology with a greater mean square error is the performed with SASIM with a value of 6.66 followed by the *t*-copula with 3.58 (which represents the 54% respecting SASIM-GSLib methodology) after copulas Bernstein appear with a value of 1.42 (which represents the 21% respecting SASIM-GSLib methodology).

Figures 11.26, 11.27 and 11.28 show the comparison of the spatial distribution of the permeability of a median of 10, 10% conditional simulations with simulated annealing, using PHIV as a secondary variable. Simulation methodologies: SASIM. *t*-copula and Bivariated Bernstein copula.

A complete table of the MSE's differences between each method, in terms of percentage, is presented (Table 1.6). Once again the methodology with a greater mean square error is the performed with SASIM with a value of 5.91 followed by the *t*-copula with 2.90 (which represents

Table 11.5   Complete table of error differences between each method in terms of percentage. For a single 10% conditional simulation.

Method	MSE	vs. SASIM	vs. t-Copula
SASIM	6.66	100%	–
t-Copula	3.58	53.8%	100%
Bernstein copula	1.42	21.3%	39.7%

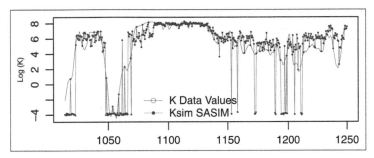

Figure 11.26   Spatial distribution in a median of 10, 10% conditional simulations using SASIM methodology (Deutsch 1998), and using PHIV as a secondary variable.

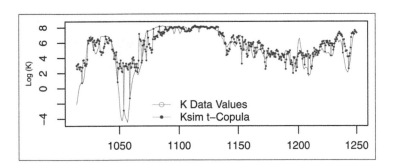

Figure 11.27   Spatial distribution in a median of 10, 10% conditional simulations using t-copula methodology (Díaz et al. 2005), and using PHIV as a secondary variable.

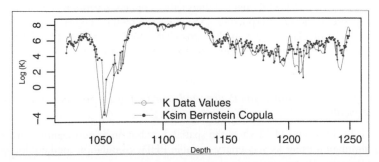

Figure 11.28   Spatial distribution in a median of 10, 10% conditional simulations using bivariate Bernstein copula methodology, and using PHIV as a secondary variable.

Table 11.6 Complete table of error differences between each method in terms of percentage. For a median of 10% conditional simulations.

Method	MSE	vs. SASIM	vs. *t*-Copula
SASIM	5.91	100%	–
t-Copula	2.90	49.1%	100%
Bernstein copula	1.02	17.3%	35.2%

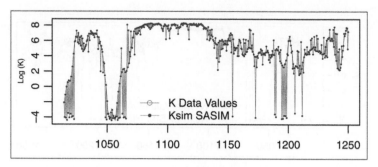

Figure 11.29 Spatial distribution in a single 10% conditional simulation using SASIM methodology (Deutsch 1998), and using PHIV as a secondary variable.

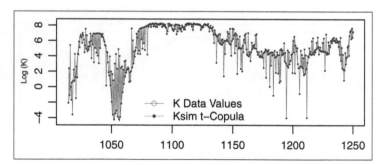

Figure 11.30 Spatial distribution in a single 10% simulation using *t*-copula methodology (Díaz et al. 2005), and using PHIV as a secondary variable.

the 49% respecting SASIM-GSLib methodology) after copulas Bernstein appear with a value of 1.02 (which represents the 17% respecting SASIM-GSLib methodology).

Between 1 and 10 simulation, each methodology has a MSE reduction, 11% for SASIM (from 6.66 to 5.91); 19% for the t-copulas (from 3.82 to 2.90); 36% for Bernstein copulas (from 1.42 to 1.02).

11.4.3 *A single 50% conditional simulation, and a median of 10, 50% conditional simulations of permeability*

The Figure 11.29, 11.30 and 11.31 show the spatial distribution of permeability in a single 50% conditional simulation using as secondary variable PHIV. Simulation methodologies: SASIM. *t*-copula and Bivariate Bernstein copula.

A complete table of the MSE's differences between each method, in terms of percentage, is presented below (Table 1.7). Note that the methodology with a greater mean square error is the

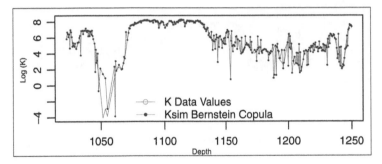

Figure 11.31 Spatial distribution in a single 10% simulation using Bivariate Bernstein copula methodology, and using PHIV as a secondary variable.

Table 11.7 Complete table of error differences between each method in terms of percentage. For a single 50% conditional simulation.

Method	MSE	vs. SASIM	vs. *t*-Copula
SASIM	3.22	100%	–
t-Copula	2.56	79.5%	100%
Bernstein copula	0.88	34.4%	34.4%

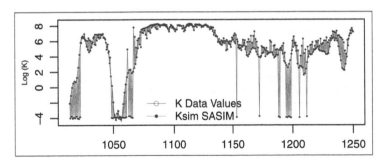

Figure 11.32 Spatial distribution in a median of 10, 50% conditional simulations using SASIM methodology (Deutsch 1998), and using PHIV as a secondary variable.

performed with SASIM with a value of 3.22 followed by the *t*-copula with 2.56 (which represents the 51% respecting SASIM-GSLib methodology) after Bernstein copulas appear with a value of 0.88 (which represents the 21% respecting SASIM-GSLib methodology).

Figures 11.32, 11.33 & 11.34 show the comparison of the spatial distribution of the permeability of a median of 10 50% conditional simulations with simulated annealing, using PHIV as a secondary variable. Simulation methodologies: SASIM. *t*-copula and Bivariated Bernstein copula.

A complete table of the MSE's differences between each method, in terms of percentage, is presented below (Table 1.8). Once again the methodology with a greater mean square error is the performed with SASIM with a value of 3.04 followed by the *t*-copula with 1.65 (which represents the 54% respecting SASIM-GSLib methodology) after Bernstein copulas appear with a value of 0.53 (which represents the 17% respecting SASIM-GSLib methodology).

Between 1 and 10 simulation, each methodology has a MSE reduction, 8% for SASIM (from 3.22 to 3.04); 35% for the t-copulas (from 2.56 to 1.65); 39% for Bernstein copulas (from 0.88 to 0.53).

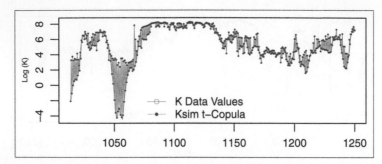

Figure 11.33 Spatial distribution in a median of 10, 50% conditional simulations using *t*-copula methodology (Díaz et al. 2005), and using PHIV as a secondary variable.

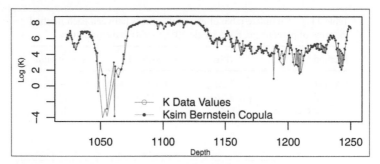

Figure 11.34 Spatial distribution in a median of 10, 50% conditional simulations using Bivariate Bernstein copula methodology, and using PHIV as a secondary variable.

Table 11.8 Complete table of error differences between each method in terms of percentage. For a median of 50% conditional simulations.

Method	MSE	vs. SASIM	vs. *t*-Copula
SASIM	3.04	100%	–
t-Copula	1.65	54.3%	100%
Bernstein copula	0.53	17.4%	32.1%

11.4.4 *A single 90% conditional simulation, and a median of 10, 90% conditional simulations of permeability*

The Figure 11.35, 11.36 & 11.37 show the spatial distribution of permeability in a single 90% conditional simulation using as secondary variable PHIV. Simulation methodologies: SASIM. *t*-copula and Bivariate Bernstein copula.

A complete table of the MSE's differences between each method, in terms of percentage, is presented below (Table 1.9). Note that the methodology with a greater mean square error is the performed with SASIM with a value of 0.92 followed by the *t*-copula with 0.51 (which represents the 44% respecting SASIM-GSLib methodology) after Bernstein copulas appear with a value of 0.40 (which represents the 55% respecting SASIM-GSLib methodology).

Figures 11.38, 11.39 and 11.40 show the comparison of the spatial distribution of the permeability of a median of 10 50% conditional simulations with simulated annealing, using PHIV as a secondary variable. Simulation methodologies: SASIM. *t*-copula and Bivariate Bernstein copula.

A complete table of the MSE's differences between each method, in terms of percentage, is presented below (Table 1.10). Once again the methodology with a greater mean square error is the

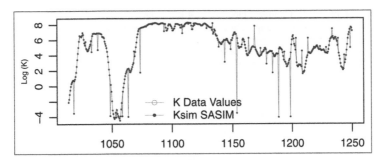

Figure 11.35   Spatial distribution in a single 90% conditional simulation using SASIM methodology (Deutsch 1998), and using PHIV as a secondary variable.

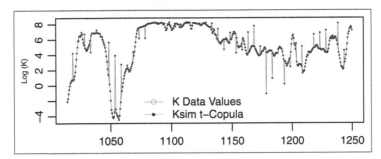

Figure 11.36   Spatial distribution in a single 90% simulation using *t*-copula methodology (Díaz-Viera et al. 2005), and using PHIV as a secondary variable.

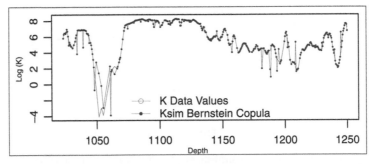

Figure 11.37   Spatial distribution in a single 90% simulation using Bivariate Bernstein copula methodology, and using PHIV as a secondary variable.

performed with SASIM with a value of 0.81 followed by the *t*-copula with 0.40 (which represents the 51% respecting SASIM-GSLib methodology) after copulas Bernstein appear with a value of 0.33 (which represents the 59% respecting SASIM-GSLib methodology).

Between 1 and 10 simulation, each methodology has a MSE reduction, 12% for SASIM (from 0.92 to 0.81); 21% for the t-copulas (from 0.51 to 0.40); 17% for Bernstein copulas (from 0.40 to 0.33).

11.5   CONCLUSIONS

The proposed method provides a very flexible tool to model the complex dependence relationships between pairs of petrophysical properties such as porosity and permeability. It can model bivariate

Table 11.9   Complete table of error differences between each method in terms of percentage. For a single 90% conditional simulation.

Method	MSE	vs. SASIM	vs. *t*-Copula
SASIM	0.92	100%	–
t-Copula	0.51	55.4%	100%
Bernstein copula	0.40	43.5%	78.4%

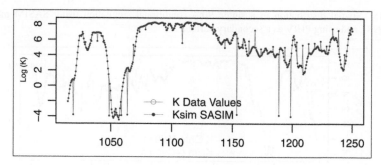

Figure 11.38   Spatial distribution in a median of 10, 90% conditional simulations using SASIM methodology (Deutsch 1998), and using PHIV as a secondary variable.

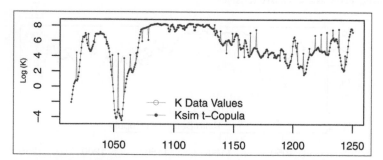

Figure 11.39   Spatial distribution in a median of 10, 90% conditional simulations using *t*-copula methodology (Díaz-Viera et al. 2005), and using PHIV as a secondary variable.

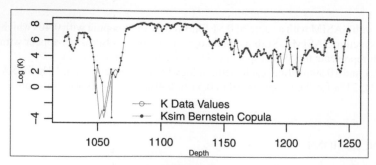

Figure 11.40   Spatial distribution in a median of 10, 90% conditional simulations using bivariate Bernstein copula methodology, and using PHIV as a secondary variable.

Table 11.10 Complete table of error differences between each method in terms of percentage. For a median of 90% conditional simulations.

Method	MSE	vs. SASIM	vs. *t*-Copula
SASIM	0.81	100%	–
t-Copula	0.40	49.4%	100%
Bernstein copula	0.33	40.7%	82.5%

dependencies in a much more efficient and accurate way. Hence, it serves as an alternative to traditional methods like linear regression, since it does not need the assumption of the existence of a linear dependency model between variables.

The methodology used in this work has three main advantages: First, an easy way to simulate bivariate data with the dependence structure and marginal behavior suggested by already observed data; second, a straightforward way to perform nonparametric quantile regression; and third, an easy way to implement a nonparametric copula into a stochastic geostatistical simulation.

In contrast to the parametric approach, the nonparametric one allows us to model non linear relationship between petrophysical properties without assume any distribution function as *t*-copula does, because the Bernstein copula is based on the empirical distribution function, and consequently, the may reproduce the data variability and the extreme values in a more natural way.

Since all the information about the dependence structure is contained in the underlying copula, in simulated annealing results, the histogram of permeability and porosity are automatically reproduced. Hence, the objective function is reduced and consequently its computational cost is lowered, we can take this advantage to introduce other components to objective function in order to get a better solution.

Another advantage about using the Bernstein copula is that there is no need to make logarithmic transformations of permeability, i.e. we do not have to make back transformations that can potentially bias the results. In fact, copulas are invariant under strictly increasing transformations of the variables.

In the case study, in contrast with the other two methods, SASIM of GSLIB and *t*-copula, the Bernstein copula has a mean squared error reduction about 83% (from 7.00 to 1.17) in non conditional simulations, consequently, it has more accurate results. It is necessary to say that there must be implemented efficient computational algorithms in order to speed up the computing time, because the Bernstein copula increases the computational effort in higher dimensions.

In this study case we noted that if we just performed a nonparametric quantile regression we obtain results that can compete (spatially speaking) with simulated annealing results (1.83 Bernstein copula median regression vs. 1.85 a single SA simulation).

The use of nonparametric copulas opens a promising line of research to model complex dependence structures between petrophysical properties and their intrinsic spatial dependence. In the geostatistical simulation framework, to make joint simulations we propose to use simulated annealing method, but we can use another optimization method which may give us more accurate results.

REFERENCES

Balan, B., Mohaghegh, S. & Ameri, S. (1995) State-of-the-art in permeability determination from well log data: Part 1. A comparative study, model development. *SPE 30978*.

Bennett, W.M., Linton, P.F. & Rectenwald, E.E. (2002) Hydrologic investigation of the Floridian aquifer system, western Hillsboro Basin, Palm Beach County, Florida. *Technical Publication WS-8*. South Florida Management District.

Billingsley, P. (1995) *Probability and Measure*. 3rd edition. New York, Wiley.

Deheuvels, P. (1979) La fonction de dépendance empirique et ses porpriétés. Un test nonparamétrique d'independance, *Académic Royale de belgique. Bulletin de la clane Sciences (5)*, 65(6): 274–292.

Deutsch, C.V. & Cockerham, P.W. (1994) Geostatistical modeling of permeability with annealing cosimulation (ACS). *SPE 28413.*

Deutsch, C.V. & Andre G. Journel. (1998) *GSLIB: Geostatistical Software Library and User's Guide.* 2nd edition. Oxford University Press, Oxford, U.K.: 369 pp.

Díaz-Viera, M., Barandela, A., Utset, R. & Fernandez, C. (1994) GEOESTAD: un sistema de computación para aplicaciones geoestadísticas. In: Barandela, R. (ed.). *Proceedings of GEOINFO*, 2nd Iberoamerican Workshop on Geomathematics, Havana.

Díaz-Viera, M. & Casar-González, R. (2005) Stochastic simulation of complex dependency patterns of petrophysical properties using t-copulas. *Proceedings IAMG'05: GIS and Spatial Analysis*, 2, 749–755.

Díaz-Viera, M., Anguiano-Rojas, P., Mousatov A., Kazatchenko E. & Markov M. (2006) Stochastic modeling of permeability in double porosity carbonates applying a Monte-Carlo simulation method with *t*-copulas. *SPWLA 47th Annual Logging Symposium*, Veracruz, Mexico, 4–7 June 2006.

Díaz-Viera, M., Hernández-Maldonado, V. & Mendez-Venegas, J. (2010) RGEOESTAD: Un programa de código abierto para aplicaciones geoestadísticas basado en R-Project, México. Available from: *http://mmc2.geofisica .unam.mx/gmee/paquetes.html.*

Dréo, J., Pétrowski, A., Siarry, P. & Taillard, T. (2006) *Metaheuristics for Hard Optimization.* Berlin, Heidelberg, Springer-Verlag.

Erdely, A. & Díaz-Viera, M.A. (2010) Nonparametric and semiparametric bivariate modeling of petrophysical porosity-permeability dependence from well log data. (Chapter 13). In. Jaworski, P., Durante, F., Hrdle, W.K. & Rychlik, T. (eds.). *Copula Theory and Its Applications, Lecture Notes in Statistics*, 198. Berlin Heidelberg, Springer-Verlag, pp. 267–278.

Fermanian, J-D. Radulović, D & Wegcamp M. (2004) Weak convergence of empirical copula processes. *Bernoulli*, 10, 847–860.

Kazatchenko, E., Markov, M. & Mousatov, A. (2004) Joint inversion of acoustic and resistivity data for carbonate microstructure evaluation. *Petrophysics*, 45, 130–140.

Kazatchenko, E., Markov, M., Mousatov, A. & Parra, J. (2006a) Carbonate microstructure determination by inversion of acoustic and electrical data: application to a South Florida Aquifer. *Journal of Applied Geophysics*, 59, 1–15.

Kazatchenko, E., Markov, M. & Mousatov, A. (2006b) Simulation of the acoustical velocities, electrical and thermal conductivities using unified pore structure model of double-porosity carbonate rocks. *Journal of Applied Geophysics*, 59, 16–35.

Muñoz-Pérez, J. & Fernández-Palacín, A. (1987) Estimating the quantile function by Bernstein polynomials. *Computational Statistics and Data Analysis*, 5, 391–397.

Nelsen, R. B. (2006) *An Introduction to Copulas.* 2nd edition. New York, Springer.

Parra, J.O., Hackert, C.L., Collier, H.A. & Bennett, M. (2001) A methodology to integrate magnetic resonance and acoustic measurements for reservoir characterization. *Report DOE/BC/ 15203-3.* Tulsa, Oklahoma, National Petroleum Technology Office, Department of Energy.

Parra, J.O. & Hackert, C.L. (2002) Permeability and porosity images based on crosswell reflection seismic measurements of a vuggy carbonate aquifer at the Hillsboro site, South Florida. *72nd Annual Meeting of the Society of Exploration Geophysicists*. Paper VCD P1.2.

Sancetta, A. (2007) Non-parametric estimation of distributions with given marginals via Bernstein–Kantorovic polynomials: L_1 and pointwise convergence theory. *Journal of Multivariate Analysis*, 98, 1376–1390.

Sancetta, A. & Satchell, S. (2004) The Bernstein copula and its applications to modeling and approximations of multivariate distributions. *Econometric Theory*, 20, 535–562.

Schweizer, B. & Sklar, A. (1983) *Probabilistic Metric Spaces.* New York, North Holland.

Sklar, A. (1959) Fonctions de répartition á *n* dimensions et leurs marges. *Publications de l'Institut de statistique de l'Université de Paris*, 8, 229–231.

CHAPTER 12

Stochastic simulation of a vuggy carbonate porous media

R. Casar-González & V. Suro-Pérez

12.1 INTRODUCTION

The porosity of a rock is defined as the percentage of empty spaces in relation to the total volume of the rock, as well, a pore is defined as the space not occupied by solid material. In carbonate sedimentary rocks, porosity formation involves complex processes in which intervene factors such as the depositional environment, diagenetic processes and the fracturing of the rock.

The system porous carbonate rocks consists of two main elements: the porosity associated to the matrix of the rock, called primary porosity, and open spaces associated with fractures and/or dissolution cavities, also known as vugs called secondary porosity (Choquette & Pray 1970). The porosity in a rock is a feature that distinguishes it as a system capable of storing fluids, but for a porosity system of this nature constitutes an oilfield, it requires that the pores are interconnected, allowing the fluids to have mobility, in others words the medium must be permeable.

This paper aims to describe a methodology to obtain a stochastic model of a vugular carbonate porous medium, as it is presented in an actually oil reservoir. To accomplish these objectives it was necessary to establish a formal, systematic and analytical procedure, which allowed to identify, quantify and model the porosity in carbonate rocks with vugular type porous system. The proposed methodology is based on geostatistical concepts such as spatial modeling and stochastic simulation (Journel, 1989; Deutsch & Journel, 1992; Chiles & Dolphins 1999, Deutsch 2002). Likewise, the methodology is based on the use of specialized technologies such as the X-ray computed tomography applied to rocks (Van Geet et al. 2000).

The information used in this work, and the methodologies and techniques are documented in the works of Casar-González (2003), Casar-González & Suro-Pérez (2001), Casar-González & Suro-Pérez (2000), and Suro-Pérez (1997).

12.2 X-RAY COMPUTED TOMOGRAPHY (CT)

One way to obtain information about the internal structure of porous rock and their system is using images from X-ray computed tomography (CT) (Van Geet et al. 2000). This paper proposes to use geostatistical techniques applied to the analysis of tomography information, which we can learn about and model three-dimensional architecture, geometry and connectivity of a porous media system.

We analyzed information from CT, which comes from a fragment of core (rock sample taken from an oil well) of carbonate rock with vugular porosity. The CT scanner was obtained with tomography equipment Technicare 2060. The fragment of core were divided by six fragments, each fragment is about 10 centimeters long by 9.4 centimeters in diameter. The density of sampling scanner was every 2 millimeters; so that, each CT image consists of a data matrix of 512×512 (262,144 data), with a resolution or pixel size of 0.2344 millimeters. The CT data were transformed to porosity according to the following linear relationship

$$\phi_i = \left[\frac{CT_{mx} - CT_i}{CT_{mx} - CT_{fluid}} \right] \times 1000 \qquad (12.1)$$

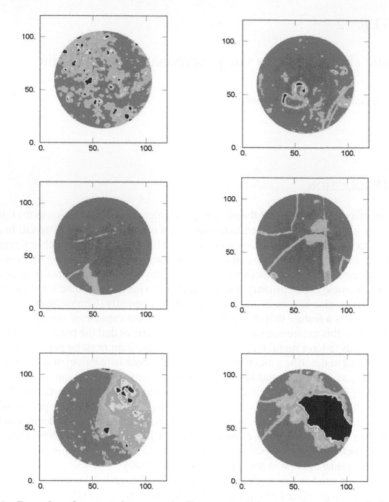

Figure 12.1   Examples of computed tomography X-ray converted to porosity of the studied core. Note the distribution of vugs, high porosity halos surrounding vugs areas and the presence of compact rock without vugs.

where ϕ_i is the porosity associated with pixel i; CT_{mx} is a constant dependent on the lithology, in this case, 3340 to dolomite, CT_{fluid} is the value of tomography for air (1000) and CT_i is the value of CT in the pixel i.

According to the above expression, the calculation of porosity is reduced to a linear scaling of observed CT values, where the critical variable is the constant CT_{mx}, which is dependent on the composition and characteristics of each rock.

The conversion of CT data to porosity data was carried out, then a visual analysis of images was completed. From this analysis, we concluded that vugular structures are the main component of the high porosity system. The presence of vugs is erratic, irregular and, therefore, highly heterogeneous. It is possible to differentiate areas which are vugs abundant and in contrast areas with absence of these, it meaning areas of compact rock. The rounded vugs have different sizes from 0.2 to 50 millimeters. The small vugs are usually rounded; the large ones tend to be elongated. According to the classification of Lucia (1995), this type of porosity corresponds to secondary porosity vugs connected. It is shown some examples of images of X-ray tomography converted to porosity, and three-dimensional reconstructions of some of the fragments analyzed (Figures 12.1 and 12.2).

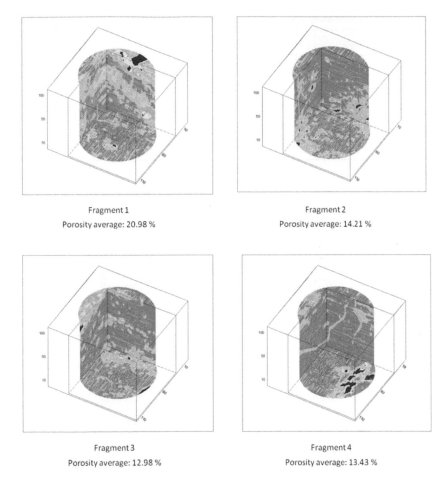

Figure 12.2 Three-dimensional reconstructions of core fragments by X-ray computed tomography converted to porosity values.

12.3   EXPLORATORY DATA ANALYSIS OF X-RAY COMPUTED TOMOGRAPHY

We performed a exploratory data analysis of porosity, the results are shown in Table 12.1. In the histograms of the different fractions of the core, we can see that the distributions for the first four fractions are strongly asymmetric and are biased toward low porosity, while for the last two fractions (5 and 6), histograms show better symmetry (Fig. 12.3).

12.4   TRANSFORMATION OF THE INFORMATION FROM POROSITY VALUES TO INDICATOR VARIABLE

The segmentation of the porosity information through the concept of indicator function has the objective to analyze and modeling the porous system architecture. The indicator function of the porosity was defined as follows:

$$I(x, z_c) = \begin{cases} 1, & \text{if } \phi(x) \leqslant z_c \\ 0, & \text{if } \phi(x) > z_c \end{cases} \tag{12.2}$$

Table 12.1 Exploratory data analysis of CT in the studied core. Statistics of porosity in the core fragments.

Fragment	Pixels	Min.	Max.	Mean	Variance
1	6,095,750	0.0	1.0	0.2098	0.0379
2	6,095,750	0.0	1.0	0.1421	0.0415
3	6,095,750	0.0	1.0	0.1298	0.0375
4	6,095,750	0.0	1.0	0.1343	0.0547
5	6,095,750	0.0	1.0	0.1755	0.0186
6	6,095,750	0.0	1.0	0.3402	0.0175
Total	36,574,500	0.0	1.0	0.1886	0.0268

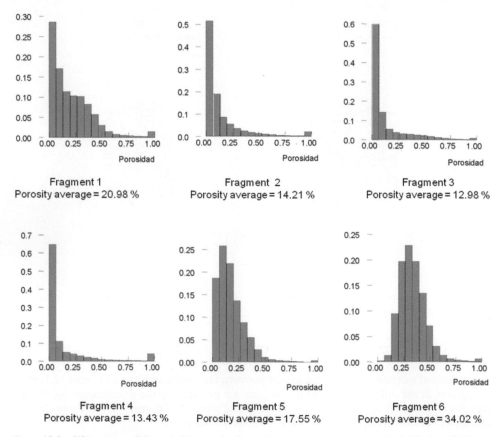

Figure 12.3 Histograms of the variable porosity for each fragment of the core studied. The length of each fragment is 10 centimeters.

where $I(x, z_c)$ is the local indicator function x, $\phi(x)$ is the porosity value of x, and z_c is a threshold value, in this case corresponds to a decile of the porosity probability distribution.

By this transformation indicator images were created, which are composed of ones and zeros, represented in the images as black and white. These images record which data are smaller than a certain threshold and which are larger. This segmentation information is intended to capture the geometry of the porous system. Figure 12.4 shows an image sequence indicator following the criteria described above and whose purpose is to determine the threshold that best defines the geometry of the porous system.

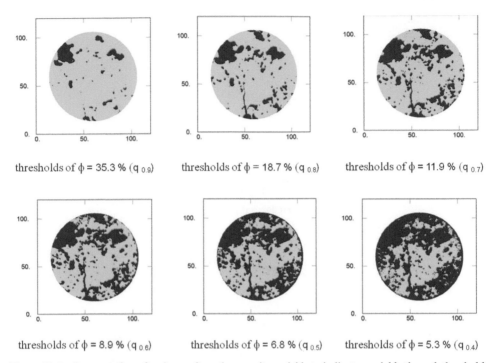

thresholds of $\phi = 35.3\%$ ($q_{0.9}$)   thresholds of $\phi = 18.7\%$ ($q_{0.8}$)   thresholds of $\phi = 11.9\%$ ($q_{0.7}$)

thresholds of $\phi = 8.9\%$ ($q_{0.6}$)   thresholds of $\phi = 6.8\%$ ($q_{0.5}$)   thresholds of $\phi = 5.3\%$ ($q_{0.4}$)

Figure 12.4   Segmentation of an image from the porosity variable to indicator variable through thresholds, which corresponding to the deciles values of porosity. In dark color observed porosity values are greater than or equal to the threshold. The sequence of images confirms the idea of dissolution as the process responsible for the formation of the vuggy porous media.

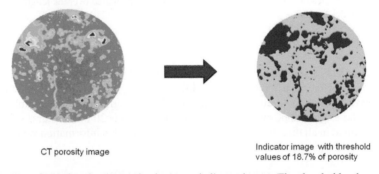

CT porosity image   Indicator image with threshold values of 18.7% of porosity

Figure 12.5   Transformation from porosity image to indicator image. The threshold value used for the transformation corresponds to the eighth deciles of the statistical distribution of porosity of the core studied, this values correspond to 18.7 percent of porosity.

A visual analysis of binary porosity images obtained using the threshold values corresponding to the deciles of the porosity distribution, suggests that the indicator image that best represents the geometry of the porous system is the image obtained taking the deciles value corresponding to 80 percent (18.7 percent of porosity) (Fig. 12.5).

12.5   SPATIAL CORRELATION MODELING OF THE POROUS MEDIA

Variogram function provides the spatial continuity of the porous medium through the analysis of data pairs separated by a distance h aligned in one direction. The spatial variable porosity,

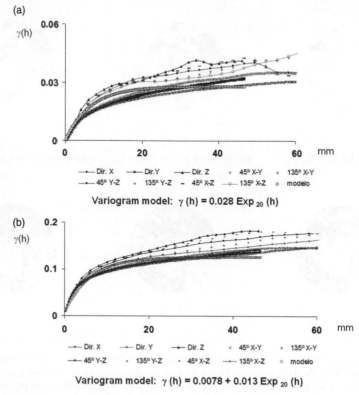

Figure 12.6 a) Variograms of porosity calculated over the length of the core analyzed by tomography and the fitted model. b) Indicator variable variograms of CT porosity calculated over the length of the examined core and fitted model.

associated with the CT core information, was investigated by directional variograms calculated for each fragment of the core in nine different directions along the X, Y and Z, and the directions of 45 and 135 degrees at the planes XY, YZ and XZ. The variograms for the variable porosity are continuous at the origin, scope or range correlation maximum is 20 millimeters, a similar spatial behavior was found in all directions and in all fragments, with this information we concluded that the spatial continuity porous media model is:

$$\gamma(h) = 0.028(1 - \exp(-3h/20)) \tag{12.3}$$

where the variogram of the form: $\gamma(h) = (1 - \exp(-3h/20)$ corresponds to a exponential variogram model, a is the range or correlation length and h is the distance between pairs of information aligned on the same direction. The number 0.028 is the value of the sill.

The same procedure was performed for the indicator variable information; variograms were calculated for each core fragment analyzed, along the main axes, X, Y and Z, and the directions of 45 and 135 degrees at the planes XY, YZ and XZ. The results are similar to those found for the porosity variable. Therefore, the vuggy porous medium characterized as an indicator variable can also be modeled as stationary and isotropic; the maximum correlation between data varies between 18 and 20 millimeters, reflecting a dependence on the size of the vugular structures. The variograms was models with an exponential model (Fig. 12.6).

$$\gamma(h) = 0.006 + 0.12(1 - \exp(-3h/20)) \tag{12.4}$$

12.6   STOCHASTIC SIMULATION OF A VUGGY CARBONATE POROUS MEDIA

A first step in creating a vuggy porous medium equivalent to that observed, is to reproduce the geometry of the vuggy structures according to the spatial relationship assessed by variograms. To this task we selected the algorithm known as Sequential Indicator Variable Simulation, according to Massonnat & Alabert (1990), Suro-Pérez & Journel (1990), in particular the program SISIMPDF was used (Deutsch & Journel 1992).

The basic information to run the program of Sequential Indicator Variable Simulation is:

1. The definition of the number of classes, in this case two: vugs and no vugs. This information is equivalent to the proportion of ones and zeros that has the image to reproduce (80-20 percent),
2. The variogram function that represents the model of spatial distribution of the indicator variable porosity, and
3. The definition of constraints data, in this case proceeded without constraints data, that means, the simulation perform is a non-conditional simulation.

The goal is to obtain a porous media equivalent to the real observed at the core. To have this medium, it has to reproduce the univariate statistics, retain the original proportions and reproduces the spatial variability model (variogram). It is no necessary that the resulting images reproduce a particular image, for this reason it is neither necessary to use of constraints data. Figures 12.7 and 12.8 show examples of realizations obtained in two and three dimensions: two dimensions: 512×512, and three dimensions: $512 \times 512 \times 25$, with a cell size of 0.2344 millimeters.

One way to corroborate the efficiency of the algorithm is by comparing statistics and spatial correlation of outputs with the original porous medium. To draw these comparisons were conducted statistical analysis and spatial distribution analysis of the simulation made. Through this analysis it was found that the selected algorithm adequately reproduces basic statistics that is, kept in an acceptable range of 80–20 percent ratio. Also, found a correspondence in terms of spatial variability of the phenomenon. The simulation with the selected algorithm is efficient and computing resources to generate each realization are minimal. Each realization is obtained from 2 to 3 minutes using a personal computer with a Pentium I at 166 megahertz, 96 megabytes of random access memory (RAM).

(a)

(b)

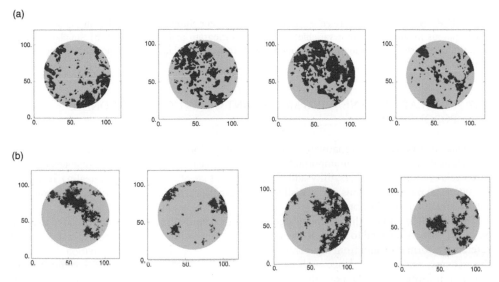

Figure 12.7   Simulations made with Sequential Indicators Simulation algorithm. a) Actually vuggy porous media represent by indicator variable. b) Simulated vuggy porous media.

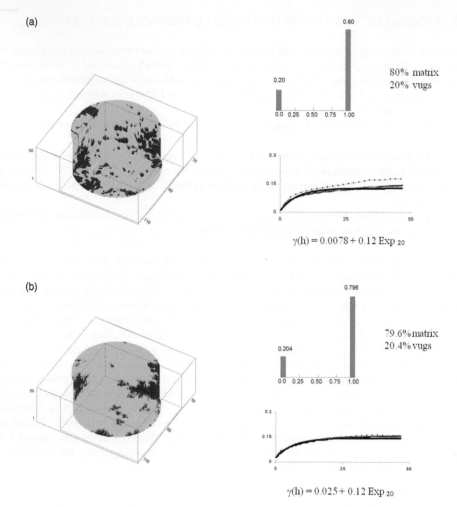

Figure 12.8   Simulations performed with Sequential Indicator Simulation algorithm in 3D. a) Actually vuggy porous medium (indicator variable), histogram, variogram and fitted model. b) Simulated vuggy porous media, histogram, variogram, fitted model. Note that the simulated media reproduces properly both statistics.

12.7   SIMULATION ANNEALING MULTIPOINT OF A VUGGY CARBONATE POROUS MEDIA

A stochastic simulation alternatively, can offer a simulation methodology known as annealing. This algorithm has become an important method of stochastic simulation in geosciences because of its relative simplicity, versatility and flexibility. Unlike other methods, annealing is able to include all types of functions and statistics or combinations of them in addition to its power to satisfy any conditional information.

The implementation of the annealing algorithm to the simulation of phenomena related to Earth Sciences may correspond to situations where certain information is intended to serve as a base or reference information and from it generate realizations that reproduce the information. Following this scheme, and in accordance with the aim of simulating the geometry of a vuggy porous medium, information of the porosity indicator variable is the actual information that represents the geometry of the porous medium, and will be considered as initial information or training

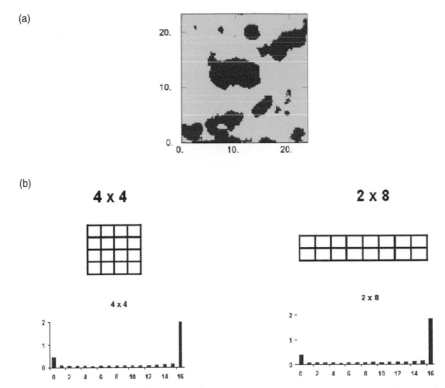

Figure 12.9   Implementation of Annealing Multipoint Simulation. a) Original porous medium categorized from CT taken at the core. Mesh 100 × 100, pixel size equal to 0.2344 millimeters. b) Two templates configuration 4 × 4 and 2 × 8, were designed. The statistics of two categories were calculated, matrix and vugs.

image. Srivastava (1994) presents an approach in which, instead of incorporating sophisticated statistics in the objective function, suggests the use of simple statistics, such as the histogram. The idea of this approach is to include major statistics than the only punctual points (variogram). This procedure not only reproduces the univariate statistics, but also the global histogram of the image and the spatial correlation model.

To apply the multipoint annealing approach in this work, we used a mesh in two dimensions of 100 × 100 extracted from the three-dimensional CT images. These images are now considered as the training images, represents the geometry of the original porous medium that we want to reproduce.

In a next step, it generates a completely random image with the same statistical proportion of 1's and 0's from the real images. This scheme ensures from the beginning that univariate statistics of the simulated image is identical to the actual image. In this case, and in order to simulate vugular structures observed in the CT, two multipoint statistics or templates were selected. The first is a template of 4 × 4 cells, with equal cell size dimension that the original mesh that contains the training image. The second template is an array of 2 × 8, with the same characteristics as the previous (Fig. 12.9).

Each template is superimposed on the upper left corner, to the original mesh that contains the image of training. In this first place it must be counting all those 1's and 0's contained inside the template. Then the template is moving to the next cell in the X direction and again it must be counting all those 1's and 0's. After touring the entire image in the X direction, the template is now moving in the Y direction, and the displacement process continues through the whole image. When finished reading the original mesh, we obtain a statistic of the two categories (1's and

0's). This statistic is representing in a histogram, one histogram for template, and multiplied by a constant to scale effects. This information is the initial statistical training image, which contains some univariate statistics of the actual image and in an explicit way, the spatial distribution model.

With these multipoint statistics (two templates), it was formulated an objective function which calculates the actual image histogram for the two templates.

$$E = \sum_{i=1}^{n} \sum_{j=1}^{n} (P_{ij}^T - P_{ij}^I)^2 \tag{12.5}$$

where P_{ij}^T and P_{ij}^I represent the statistics of the j-th category for the i-th template or configuration support for the training image and the initial image, respectively.

According to the methodology of simulated annealing, we begging with a image complete randomly, two samples are chosen, the exchanged and updated the statistics. If the new objective function value is lower, it accepts the change, if it is greater, the change is accepted under an acceptance probability, P acceptance which it is defined as:

$$P_{accep} = \exp\left(\frac{E_{new} - E_{last}}{t}\right) \tag{12.6}$$

It is recalled that E_{new} is the value of energy once they made the exchange of two samples. E_{last} is the energy value before the change. The process is repeated until P_{accep} converges to zero.

Through this algorithm, we obtain category images of the vuggy porous media equivalent to the actual images. The degree of accuracy of the algorithm to reproduce the univariate statistics in both templates is optimal. The results obtained by multipoint simulation annealing are fully equivalent to the real porous media. The algorithm reproduces well the histogram and the spatial variability (Figures 12.10 and 12.11).

One of the most important comparisons between the two simulations methods are described in the sense of computing resources it takes for each one of them. Tests were conducted with a 100×100 mesh, using a personal computer with a Pentium I at 166 MHz and 96 MB of random access memory. The simulations with the Sequential Indicator Variable Simulation algorithm, takes to 10 to 20 seconds. The Annealing Multipoint Simulations run from 45 to 90 minutes. With these results, it is clear that the Sequential Indicator Variable Simulation is significantly quicker and more practical than the Annealing Multipoint Simulation, in the particular case discussed here.

12.8   SIMULATION OF CONTINUOUS VALUES OF POROSITY IN A VUGGY CARBONATE POROUS MEDIUM

The characteristics of the porous medium discussed here are related to the geometry of vugular structures, their distribution pattern, size and shape. It should be noted that the description of the CT images mentioned the presence of high porosity halos surrounding the vugular structures. These halos have a less porosity than the vugs themselves, but significantly higher with respect to the matrix that makes up the rest of the porous media.

These porous areas are undoubtedly a significant influence on the storage and connectivity. So far it has developed a process for generating two categories porous media, matrix and vugs, however, and based on the above remarks, it is important to have high porosity halo component in the simulated media. The objective is to generated an appropriate porous media according to the actually observed, it means the porous media has to containing the characteristic of a gradual change in porosity from low values in the matrix up to high in the vugs.

To simulate porous media with three categories, a second simulation was applied to the created media in the first step. The same sequential indicator simulation algorithm was applied. In this second simulation, the matrix remained unchanged and only the vugular structures was simulated, subdivided it into two categories, vugs and high porosity halo. Based on the observation of CT images, it was determined that from the 100 percent of the vugular structure, 50 percent correspond

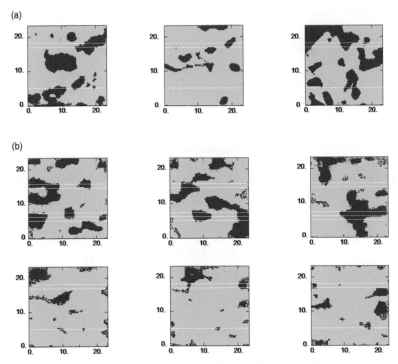

Figure 12.10   Application of annealing multipoint simulation. a) Examples of a vuggy porous media seg-mented by two categories matrix and vugs. b) Simulations by by annealing multipoint methodology of a porous medium formed by two categories vugs and matrix. Scale in millimeters.

to the high porosity halo and 50 percent correspond to vugs. In this way it was obtained a porous media with three categories: matrix, high porosity halo and vugs.

Another important aspect is the distribution of porosity values that make up the media and are associated with different present categories. The generation of porosity was carried out with Sequential Gaussian Simulation algorithm, in particular the program SGSIMM (Journel, 1988), (Deutsch & Journel 1992).

With regard to the allocation of porosity values, the approach to be adopted to assign values to each category is based on statistical analysis of the porosity observed in the core analyzed by CT. So for the porous matrix porosity values were considered value minor of the sixth decil ($q_{0.6}$), for halos of high porosity, porosities were allocated values between the sixth and eighth deciles ($q_{0.6}$–$q_{0.8}$) and for vugs the porosities values were greater than the eighth decile ($q_{0.8}$). Table 12.2 shows the approach taken to assign values of porosity for each of the categories listed in the range of values of porosity and the correspond variograms used in the simulation process.

Porosity values were generated in meshes of equal size to that containing the geometry of the porous medium in question with the sequential Gaussian simulation methodology without condi-tions, and were restricted to the area occupied by each category and in the range of values deter-mined for each category. Please note that the porosity information created by this way correspond to a Gaussian space, so at the end of the process a back transformation of the porosity data was performed, from a Gaussian space to the actually domain based on the statistics of the observations.

The last step is to combine the meshes that containing on one hand, vugular structure of the media and on the other hand, the porosity values appropriate to each category. Thus, the final product is a mesh that represents a vuggy porous media divided into three categories. Figure 12.12 shows the sequence of steps for created by simulation an equivalent vuggy porous media with the characteristics described above.

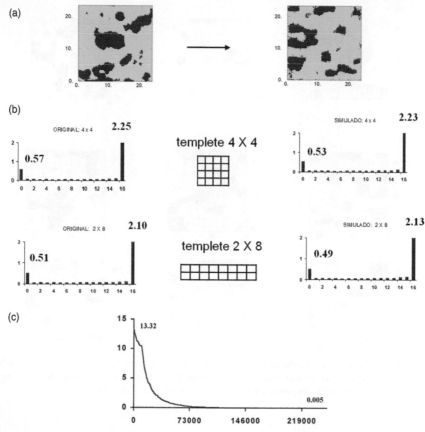

Figure 12.11 A simulated annealing multipoint realization. a) Actually vuggy porous media (left) and simulated vuggy porous media (right). b) Statistics on the proportion of matrix and vugs categories obtained by two templates, 4×4 and 2×8. c) Behavior of the objective function. The significant decrease occurs during the first 100,000 exchanges.

Table 12.2 Values of porosities observed in matrix, hight porosity halo and vugs. Also its variogram model.

Categories	Porosity observed	Variograms models
Matrix	Less or equal to $q_{0.6}$ ($\phi \leq 0.1126$)	$\gamma(h) = 0.0023 + 0.0023\,Exp_{4.55} + 0.0016\,Sph_{20}$
High porosity halo	Between $q_{0.6}$ and $q_{0.8}$ ($0.1126 < \phi \leq 0.2782$)	$\gamma(h) = 0.000185 + 0.00091\,Exp_{7.2}$
Vugs	Greater than $q_{0.8}$ ($\phi > 0.1126$)	$\gamma(h) = 0.006\,Exp_3$

12.9 ASSIGNING PERMEABILITY VALUES BASED ON POROSITY VALUES

The patterns of dependence between porosity and permeability are complex, may not always correlate with a linear correlation and with a model for ordinary least-squares fit. In the case of the core studied in this work we had information of porosity and permeability obtained in the laboratory, so it was possible to assign permeability values based in porosity values.

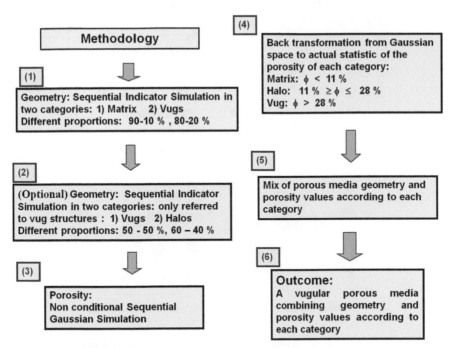

Figure 12.12 Process for obtaining vuggy porous media equivalent to actually porous media, which combines the geometry of vugular structures and their corresponding porosities values.

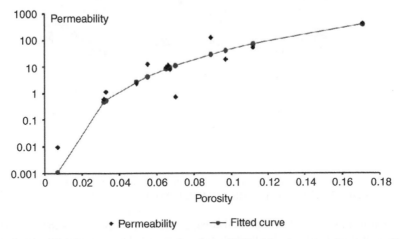

Figure 12.13 Porosity and permeability laboratory data calculated for the core, taken at a confining pressure of 400 psi. Based on the data presented in the table, it was calculated a transformation of porosity to permeability according to the fitted curve. Porosity in decimal fractions, permeability in milidarcies.

Based on the results of measurements of porosity and permeability (horizontal), taken at a confining pressure of 400 psi over the entire core, it was determined the following transformation involving the porosity with permeability. The curve fit to the observations is presented in Figure 12.13.

$$k = \begin{cases} (686.99\phi^2 + 0.1946)^2 & \text{for } \phi \leqslant 0.20 \\ 765.8 & \text{for } \phi > 0.20 \end{cases} \tag{12.7}$$

where ϕ corresponds to the porosity in decimal fraction, and k is the permeability in milidarcies.

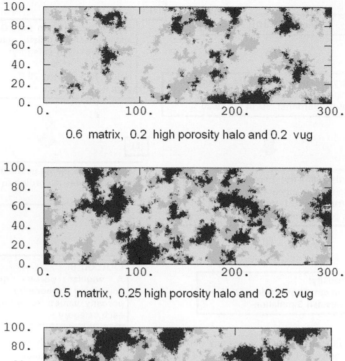

Figure 12.14   Example of three simulated vuggy porous media created by Sequential Indicator Variable Simulation and Sequential Gaussian Simulation. The media have different proportions of components: matrix, high porosity halo and vugs. The proportions shown are approximate. Scale in millimeters.

12.10   APPLICATION EXAMPLE: EFFECTIVE PERMEABILITY SCALING PROCEDURE IN VUGGY CARBONATE POROUS MEDIA

Following the methodology described in this paper, we created 24 porous media. Each media consists of a grid of 1280×427 cells with a cell size of 0.2344 millimeters, which is a core section of 300.03×100.08 millimeters (30×10 centimeters).

The 24 porous media represent four sets of six media each with different proportions of the components, matrix, high porosity halo and vugs, and the matrix percentages ranging from 90 to 40%. For the 24 porous media with porosity values, it were applied the transformation from porosity to permeability, creating a further 24 media with the same geometrical characteristics, but with permeability data (Fig. 12.14).

Using the commercial simulator software Eclipse (Schlumberger 2001), several flow simulation were conducted for the 24 simulated porous media (Fig. 12.15). The goal was, from the Darcy equation, to calculate trends for effective permeability of each medium. Flow was considered laminar, single phase and no chemical reaction between the fluid and the environment. The procedure was to saturate the media to 100 percent with water, then inject water,

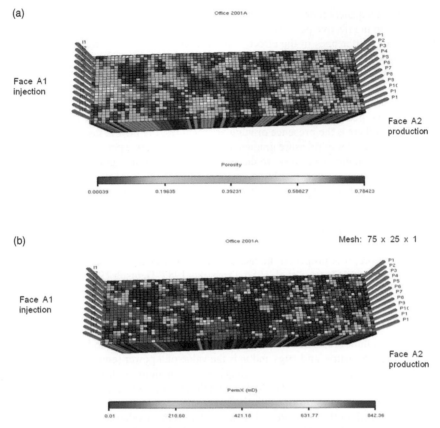

Figure 12.15   a) A mesh of porosity. b) A mesh of permeability. The meshes were loaded in the ECLIPSE flow simulation software. In the faces A1 and A2 are located 12 injection wells and 12 production wells respectively. Porous simulated media size is $30 \times 10 \times 1$ centimeters.

wait to observe the stabilization of the spending and the pressure and so obtain the necessary data to calculate effective permeabilities. Solving the permeability k from the Darcy equation, it results in:

$$k_e = \frac{\mu Q L}{A \Delta p} \tag{12.8}$$

where k_e is the effective permeability, μ is the viscosity of the fluid in centipoise, Q is the volume of fluid in square centimeters per second, L is the length of the medium in centimeters, A is the cross sectional area of the medium in square centimeters, Δp is the pressure difference between the faces of injection and production in atmospheres per centimeter.

For each media, 12 injection wells were considered in one of the faces (injection face) and 12 production wells on the opposite face (production face). It ran the process of moving fluid and analyzed the behavior of the rating curves and pressure with respect to time. With the rating data, pressure gradient and geometry of the media, effective permeability was calculated for each of the 24 simulated porous media.

The results of these calculated effective permeabilities are presented along with the average absolute permeability values held for each media. On the same graph shows the three most commonly used averages: the arithmetic, the geometric and the harmonic average. An analysis of the chart highlights some important facts: of the three most common methods to perform averages, none adequately represents the calculated effective permeabilities, moreover there is a strong dispersion in effective permeability data obtained by flow simulation (Fig. 12.16).

With regard to differences in the effective permeability values for different media studied, it can be explained by an increase in the value of the pressure gradient. The outlier in the pressure gradient occurs due to the spatial distribution of vuggy structures. In those porous media where it has compact zones in the face of injection, the fluid moves more slowly and this raises the pressure in the face of injection, resulting at the end a higher pressure gradient and thus a decrease in the value of effective permeability. By contrast, when the injected face has a high permeability values, its means there is the presence of vugular structures, the flow is faster and the pressure is smaller, resulting in a lower pressure gradient and a higher effective permeability. In other words, the effective permeability is sensitive to the spatial distribution of vugular structures. Moreover, knowing the exact matrix-vugs distribution present in the faces of injection, it would be possible to correlate this distribution and the resulting effective permeability values.

12.11   SCALING EFFECTIVE PERMEABILITY WITH AVERAGE POWER TECHNIQUE

As explained before, it is proposed the use of average power technique to model the effective permeability calculated (Journel et al. 1986, Desbarats 1987, Deutsch 1987, 1989). The effective permeability k_e is given by the equation:

$$k_e = [V_m k_m^\omega + (1 - V_m)k_v^\omega]^{1/\omega} \qquad (12.9)$$

where V_m is the volume or fraction of matrix in the porous medium, k_m and k_v are the permeability values for categories matrix and vugs and w is the value of a power (empirical value).

Thus the effective permeabilities for a vuggy porous medium with the characteristics of the media studied here, can be averaged or scaled with the average power methodology. The problem is resolved knowing the proportions or volume quantities of matrix and vugs, the value of the permeability for each of the two categories and to determine a value for the exponent w, which must be between -1 and $+1$. We present an option, in which it is fitting the experimental data and suggest values of permeability for the components matrix and vugs. For the w exponent it was assigned a value equal to 0.4, a value of 0.038 milidarcies was considered for the permeability at the matrix, which is the minimum value of permeability observed in the media, and 765.8 milidarcies for vugs permeability, which in turn is the maximum value of permeability observed in the porous medium (Figure 12.16).

12.12   SCALING EFFECTIVE PERMEABILITY WITH PERCOLATION MODEL

Another proposal for scaling the permeability is called percolation model (Kirkpatrick 1973, Hammersley & Welsh 1980, Deutsch 1989). The equation defining the effective permeability with the percolation model is given by:

$$k_e = ck_v(V_{mc} - V_m)^t \qquad (12.10)$$

where k_e is the effective permeability of the media, k_v is the vug permeability, V_{mc} is the critical volume of matrix (threshold at which the flow falls), V_m is the volume of matrix, t is an exponent, and c is a proportionality constant.

The application of this model involves knowing the V_{mc} variable, which is defined as critical volume fraction of the matrix from which the flow drops drastically. The percolation model is not established for cases in which this theoretical threshold value is not defined. In addition, this threshold cannot be defined with the experimental data that are available; however, when considering the effective permeability data in Figure 12.16, this threshold value should be between 0.90 and 1.0 of the matrix fraction, it meaning that, with the 90 percent of matrix, permeability

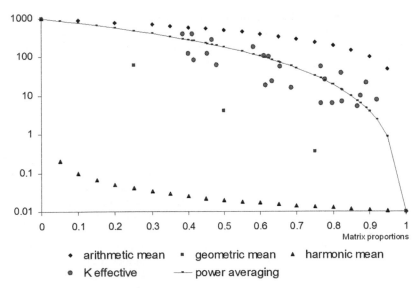

Figure 12.16   Effective permeability versus proportion of matrix related to the 24 vuggy simulated porous media. In the graph are shown as references, arithmetic, geometric and harmonic averages related to the data set. Permeabilities are given in milidarcies. Modeling effective permeability with average powers technique: $K_m = 0.01$, $K_v = 1000$, $\omega = 0.4$. Permeabilities are in milidarcies. For averaging permeabilities in vuggy porous media, it is necessary to know the proportion of matrix and vugs, the permeabilities of each category and the value of the exponent ω.

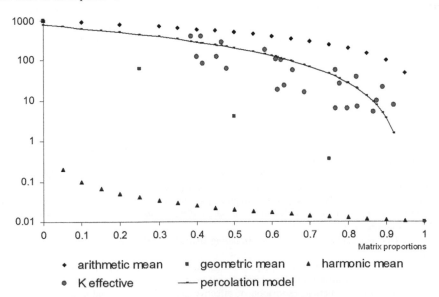

Figure 12.17   Modeling of effective permeability with percolation model: $K_v = 765.8$, $c = 1.1$ and $t = 1.8$. Permeabilities in milidarcies. Value of critical matrix volume V_{mc}, was set at 0.95.

values would be at the order of 5 to 9 milidarcies. However, it is likely that levels over 90 percent of matrix, the effective permeability decrease significantly.

Deutsch (1989), applies the percolation model to evaluate effective permeability in a sequence of sandstones and shales, and refers to a critical volume of shale (V_{mc}) of the order of 0.69 to 0.9%, an exponent t and a constant of proportionality c, between 1.5 and 2.0. In the context of vuggy media, and with these ideas, various tests were carried out to fit the experimental data to a percolation model, taking as a critical volume 0.95 percent of matrix. In one of the alternatives it proposed the permeability of the category vug equal to 765.8 milidarcies, the exponent t equal to 1.8 and the constant c equal to 1.1, the result is shown in Figure 12.17.

12.13　CONCLUSIONS AND REMARKS

The proposed stochastic approach has allowed quantitative description of such a vuggy porous medium present in the lot of carbonate reservoirs. This type of carbonate porous medium is of great importance in the Mexican oil industry, since much of the oil production comes from this type of porous media, such as the marine oil fields in the Sonda de Campeche.

The possibility of producing images of a carbonate porous medium in a controlled manner allows theirs application to the investigation of the behavior of different physical phenomena under various scenarios.

In conclusion, both approaches produce similar results in vuggy porous media. In both cases, it is required to know the proportions of matrix and vugs and its associated permeability values. For the average power technique, the exponent ω is estimated at 0.4, for values of matrix permeability on the order of 0.01 and vugs permeability of 1000 milidarcies.

In the case of percolation model, there is no clarity regarding the threshold of critical volume matrix from which the permeability decreases sharply. However, with consideration of a critical mass matrix of 95 percent, we find that the value of the constant of proportionality c varies between 1.0 and 1.1 and the exponent t between 1.8 and 2.0, considering vugs permeabilities in the range of 766 and 1000 milidarcies.

Several lines of investigation may follow from this proposal. One of them would be to adapt the methodology to model an environment in which the main element of high secondary porosity is fractures or the porous media that combine porous matrix, fractures and vugs.

REFERENCES

Alabert, F.G. & Massonnat, P. (1990) Heterogeneity in a complex turbidic reservoir: stochastic modeling of facies and petrophysical variability. SPE paper number 20604.

Casar-González, R. (2003) *Modelado estocástico de propiedades petrofísicas en yacimientos de alta porosi- dad secundaria*. Tesis Doctoral. Facultad de Ingeniería, División de Estudios de Posgrado, Universidad Nacional Autónoma de México, Mexico.

Casar-González, R. & Suro-Pérez, V. (2000) Stochastic imaging of vuggy formations. *2000 SPE International Petroleum Conference and Exhibition, Villahermosa, México*. SPE paper number 58998.

Casar-González, R. & Suro-Pérez, V. (2001) Two procedures for stochastic simulation of vuggy formations. *2001 SPE Latin American and Caribbean Petroleum Conference and Exhibition, Buenos Aires, Argentina*. SPE paper number 69663.

Chilès, J.P. & Delfiner, P. (1999) *Geostatistics Modeling Spatial Uncertainty*. John Wiley & Sons, New York, U.S.A.

Choquette, P.W. & Pray, L.C. (1970) Geologic nomenclature and classification of porosity in sedimentary carbonates. *The American Association of Petroleum Geologists Bulletin*, 54 (2), 207–250.

Desbarats, A.J. (1987) Numerical estimation of effective permeability in sand-shale formations. *Water Resources Research*, 23 (2), 273–286.

Deutsch, C.V. (1987) Estimating block effective permeability with geostatistics and powering averaging. Texas, USA. SPE paper number 15991.

Deutsch, C.V. (1989) Calculating effective absolute permeability in sandstone/shale sequences. *SPE Formation Evaluation Journal*, 343–348.

Deutsch, C.V. (2002) *Geostatistical Reservoir Modeling. Applied Geostatistics Series*. Oxford University Press.

Deutsch, C.V. & Journel, A.G. (1992) *GSLIB: Geostatistics Software Library and User's Guide*. Oxford University Press.

Hammersley, J.M. & Welsh, J.A. (1980) Percolation theory and its ramifications. *Contemporary Physics*, 21 (6), 593–605.

Journel, A.G. (1989) *Fundamentals of Geostatistics in Five Lessons. Short Course in Geology: Volume 8.* Washington, DC. American Geophysical Union.

Journel, A.G. & Alabert, F. (1988) Focusing on spatial connectivity of extreme-values attributes: Stochastic indicator models of reservoir heterogeneities. SPE paper number 18324.

Journel, A.G, Deutsch, C.V. & Desbarats, A.J. (1986) Power averaging for block effective permeability. *SPE California Regional Meeting, Oakland, USA*. SPE paper number 15128.

Kirkpatrick, S. (1973) Percolation and conduction. *Reviews of Modern Physics*, 45 (4), 574–588.

Lucia, F.J. (1995) Rock-fabric/petrophysical classification of carbonate pore space for reservoir characterization. *AAPG Bulletin*, 79 (9), 1275–1300.

Schlumberger (2001) *ECLIPSE: software de simulación numérica de flujos.*

Srivastava R.M. (1994) An annealing procedure for honoring change of support statistics in conditional simulation. *Geostatistics for the Next Century.* pp. 227–290. The Netherlands, Kluwer Academic Publishers.

Suro-Pérez, V. (1997) Reservoir description of a portion of the Abkatún field, México. PEMEX Exploración y Producción and Amoco Exploration and Production Sector. *Report No. S97-G-15*. Inedit. México.

Suro-Pérez, V. & Journel, A.G. (1990) Stochastic simulation of lithofacies: an improved sequential indicator approach. In: Guerrillot, D. & Gillon, O. *2nd European Conference on the Mathematics of Oil Recovery.* Technip. pp. 3–10.

Van Geet, M., Swennen, R. & Wevers, M. (2000) Quantitative analysis of reservoir rocks by microfocus X-ray computerised tomography. *Sedimentary Geology*, 132, 25–36.

CHAPTER 13

Stochastic modeling of spatial grain distribution in rock samples from terrigenous formations using the plurigaussian simulation method

J. Méndez-Venegas & M.A. Díaz-Viera

13.1 INTRODUCTION

The porous media characterization is a fundamental problem in areas such as soil sciences, hydrogeology, oil reservoir, etc. More precise knowledge one has about the porous media structure, the better is the accuracy in predicting effective petrophysical properties such as porosity, permeability, capillary pressure and relative permeabilities.

In particular, in the oil industry understanding the petrophysical properties concerning the rock formation, is a crucial element in reservoir management, since it allows us to accurately model the mechanisms that govern the recovery of hydrocarbons and consequently serve to propose and implement optimal secondary and enhanced recovery processes.

The aim in this work is to model the spatial grains distributions in rock samples from siliciclastic reservoir formations. As it is well known, the siliciclastic rocks are of sedimentary origin, usually formed in situ and were generated by erosion processes, transportation and deposition. Sedimentary rocks are formed by a packed grain structure that constitute the solid matrix and a pore system that is the space not occupied by the grains. The grains of the siliciclastic rocks are composed mainly of minerals such as quartz, clays, feldspars and other heavy minerals.

Usually the characterization of porous media is reduced to the study of just two categories: the solid matrix and the porous space (Okabe 2005, Oren 2003, Politis 2008). This approach possesses the disadvantage that it does not consider the mineralogical composition of the rock, therefore can not be taken in account the petrophysical property modifications because swelling phenomena when there are interactions between clay grains and water.

In this paper it is proposed a more general methodology applying the plurigaussian simulation method by which we can obtain the grain spatial distribution of siliciclastic rocks considering the rock mineralogical composition.

13.2 METHODOLOGY

The main goals of this work are to model the geometry of the pore space by simulating grain spatial distribution from images taken in siliciclastic rock samples, as well as, the spatial distribution of the minerals present in the solid matrix, using spatial stochastic simulations.

The procedure is applied in two successive stages. First are simulated two categories: matrix and the pore space, and subsequently is made the simulation of the mineralogy of interest.

The proposed approach has the advantage that, for simulating the spatial distribution of the clays, we can consider two cases: the first case is made under the assumption that the clays present in the rock come from another place (not part of the original rock) and it occupies part of the existing pore space, therefore, the clays are inserted into the pore space, while in the second case, the clays were formed together with the rock and is part of the matrix, so the clays are inserted in the matrix.

The methodology that was proposed in this work consists on the following steps:

1. Data image processing.
2. Geostatistical analysis.
   i. Exploratory data analysis.
   ii. Variographic analysis.
   iii. Stochastic simulation method.

In the following subsections each of the previous mentioned steps of the methodology will be described in details.

13.2.1 *Data image processing*

The images used as input data are obtained by scanning electron microscopy in backscatter electron mode. The sample preparation is similar to the one used for preparation of thin sections using light microscopic and consists of the following: the sample was impregnated with epoxy resin and polished on one side once the epoxy has hardened.

The process of extracting pores, clays and rock matrix from images of scanning electron microscopy is simpler, compared to the process for thin sections. A color thin section image contains three gray-level images in RGB space, while in scanning electron microscopy only one gray-level image is involved. In images of scanning electron microscopy, the rock matrix can be subdivided into ranges associated with different minerals using the atomic density contrast. For example, the pore space is associated with the darker gray tone because of the fact that the epoxy resin possesses smaller atomic density compared with the minerals contained in the matrix.

In the porous-media stochastic model are considered three categories: pore space, clay grains and rock matrix, where in the rock matrix category are included the rest of (no clay) mineralogies. A segmentation procedure developed by Fens (2000) enable automatic extraction of pore, clay and rock matrix categories. This procedure is based on fitting three Gaussian functions to the gray-level histogram (Fens 2000). In images of scanning electron microscopy the gray-values represent atomic density. Prior to processing and analyzing, these gray-values have to be calibrated which takes place using a set of standards with known gray-values. This calibration is essential to make quantitative use of the data provided by the analysis. The calibration standards used here consisted of artificial reservoir rock samples that contain only quartz and epoxy.

The total gray-value range in images of scanning electron microscopy can be divided in subranges. In Figure 13.1 the color bar below the histogram shows the division in these sub-ranges representing pores, clays, quartz/dolomite, feldspar/calcite and the heavy minerals. Two-level thresholding is used to extract the pixels in each range of gray-values. Thresholding is an image-to-image transformation, in this case a transformation from a gray-value image to a binary image (Fens 2000).

13.2.2 *Geostatistical analysis*

13.2.2.1 *Exploratory data analysis*
This step is essential in any practical statistical analysis. Usually, it is a combination of statistical and graphical techniques allowing us to obtain qualitative and quantitative information about the data sample probability distribution. In general, an exploratory data analysis pursue the verification of the basic assumptions that have to be fulfilled by the data sample to apply some statistical procedure.

In particular, in the context of a geostatistical analysis, it is required to verify if the data sample fulfills the following assumptions:

1. Its probability distribution is close to normal or almost symmetrical.
2. There is no significant trend, i.e. that at least meets the intrinsic hypothesis.
3. There are no outliers both distributional and spatial.

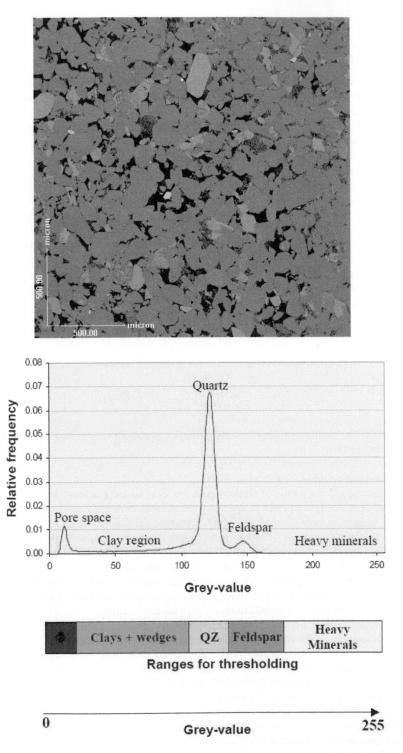

Figure 13.1   The gray-level histogram (top) calculated from an image of scanning electron microscopy (bottom) (Fens 2000).

Listed below are a set of techniques that are recommended to verify the previous assumptions.

1. Calculate the statistics basic (mean, median, variance, quartiles, skewness).
2. Graphics (histogram, box-plot, scatter plot, Q-Q plot).

Considering all the above, here the data sample statistical properties are explored in order to take into account or modify the features that do not meet the requirements (Armstrong 1980).

In this case, only the basic statistics were obtained and it was tested out the absence of trend in the data. The latter is carried out by estimating the variogram and verifying that it does not show a quadratic growth, as can be observed in the Figures 13.8, 13.9 and 13.10, where the estimated semivariance values show a bounded behavior.

13.2.2.2   *Variographic analysis*

The variographic analysis is one of the fundamental parts of any geostatistical analysis. Its aim is to model the underlying spatial structure in the data sample. For this purpose, depending on the stationarity degree of the data, you can use one of the following two functions: a variogram or a covariance function. In this paper it is used the variogram to determine the spatial dependence structure, since it is less restrictive from the point of view of the degree of stationarity. In short, a variographic analysis consists in estimate the sample variogram and find the variogram model which best fits to it.

The variogram function is defined as follows:

$$\gamma(h) = \frac{1}{2} Var(Z(x+h) - Z(x)) = \frac{1}{2} E[(Z(x+h) - Z(x))^2] \tag{13.1}$$

The most common variogram estimator $\hat{\gamma}(h)$ is given by:

$$\hat{\gamma}(h) = \frac{1}{2N(h)} \sum_{i=1}^{N(h)} [Z(x+h) - Z(x)]^2 \tag{13.2}$$

where $N(h)$ is the number of observations pairs ($Z(x)$ and $Z(x+h)$) and h is the separation distance between them.

13.2.2.3   *Stochastic simulation method*

A random function Z is a family of random variables $Z(x)$ where x belongs to $\mathbb{R}^d$ or some subset of it. In the case $d = 1$, we prefer to speak of a stochastic process. In this work, the resulting segmented image can be viewed as a discrete or categorical random function. There are a large variety of simulation methods of categorical random functions, grouped into two families: the object models and the cells models (Chilès 1999, Lantuejoul 2002).

In Object models each category is associated with a certain geometric shape (object) and are based on Poisson point processes, while in the cells models, a cell can take the value of one category and are based on the truncation of Gaussian random functions. Here we will use the second family of simulation methods to investigate the application.

The implementation of simulation of cells requires to characterize the discrete random function in terms of the spatial relationship of their categories, for which geostatistical analysis is done which consists of getting the proportions of occurrences of each category, the basic statistics and its variogram or the semivariance function, which is a dependence measure or spatial autocorrelation.

The proportions are calculated by dividing the sum of pixels of a given category between the total of pixels of the image, while the variogram is estimated by considering the value of the lag or interval equal to the size of the image pixel.

The proportions and the variogram obtained by categories are used as parameters in the spatial stochastic simulation method that is chosen.

As the stochastic simulation method is considered to apply the truncated plurigaussian method (Galli 1994, Le Loc'h 1996), which is described in the this section.

The truncated plurigaussian method is an extension of the monogaussian method, that is to say, while the monogaussian only use a Gaussian random function $Z(x)$ in the plurigaussian any number of Gaussian random functions may be used. However, in practice, to simplify the specification of the spatial relationships among the categories, the truncated plurigaussian method is restricted to the use of two Gaussian random functions $Z_1(x)$ and $Z_2(x)$; which may, or may not, be correlated:

$$Z_1(x) = Y_1(x) \tag{13.3}$$

$$Z_2(x) = \rho Y_1(x) + \sqrt{1 - \rho^2} Y_2(x) \tag{13.4}$$

where $Y_1(x)$, $Y_2(x)$, $Z_1(x)$ and $Z_2(x)$ are all standard Gaussian random functions, $Y_1(x)$ and $Y_2(x)$ are independent, and $Z_1(x)$ and $Z_2(x)$ have correlation coefficient ρ. If $\rho = 0$ then $Z_2(x) = Y_2(x)$ (Dowd 2003, Xu 2006).

In practice, $Y_1(x)$ and $Y_2(x)$ are generated using a simulation algorithm and $Z_1(x)$ and $Z_2(x)$ are calculated using Equations (13.3) and (13.4). The correlation coefficient ρ is used to introduce an additional degree of freedom in the modeling i.e. more flexibility. In general, a high correlation coefficient will introduce order among the facies while the contacts will be sharper and more disordered when the correlation coefficient is small (Le Loc'h 1996).

The spatial relationships and contacts allowed between units are defined by a family $(D_1, \ldots, D_n)$ subset $\mathbb{R}^m$, this partition is symbolized by a flag. The partition does not introduce any restriction on how many categories (n) and Gaussian random functions (m) should be used or which is the best configuration of the flag; these aspects must be established experimentally.

To carry out the work the plurigaussian and monogaussian method is considered. The monogaussian method was used in the first stage, for the second stage this method was discarded because for three or more classes, the method make a hierarchy of categories. This is best appreciated in Figure 13.2, where the facies 1 does not enter in contact with with the facies 3, it would therefore be inefficient in our case, where all the categories considered are in contact with each other simultaneously.

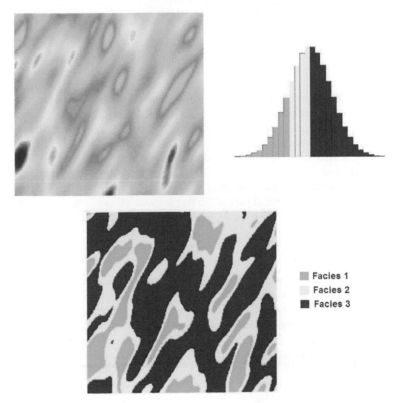

Facies 1
Facies 2
Facies 3

Figure 13.2   The monogaussian method.

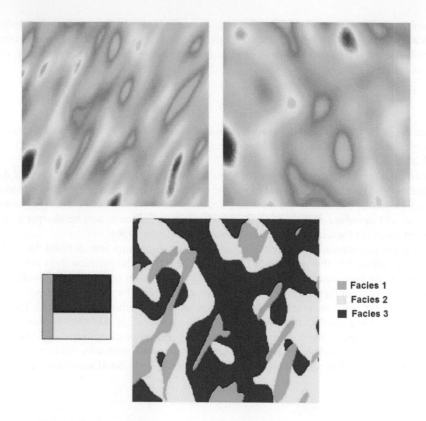

Facies 1
Facies 2
Facies 3

Figure 13.3 The plurigaussian method.

To perform the second stage, the plurigaussian simulation method was chosen because this method through the flag can control the contacts and proportions of more than two categories in a suitable way, which is the case of the present work. Figure 13.3 exemplifies most clearly this method, the two Gaussian random functions used are presented at the top; at the bottom (left) it is shown the flag, which shows that there are three facies, where it is observed three facies, that are in touch at the same time. The flag also tells us the proportion of each facies in the resulting simulation (bottom rigth). The final simulation is obtained by modeling the horizontal part of the flag by Gaussian 1, while the vertical part of the flag is modeled using the Gaussian 2.

13.3   DESCRIPTION OF THE DATA

In the case study presented, we used an image of a sandstone block obtained with the procedure described in the previous section (Fig. 13.4), which was taken of the PhD thesis of T. Fens (Fens 2000). The image size is 2×2 mm with a resolution of 256×256 pixels, with a pixel size equal to 0.0078 mm. The same five categories are clearly visible: quartz, clays, feldspars, heavy minerals and pore space.

According to the objectives of the work, the input image was segmented in three categories: matrix, clays and pore space (Fig. 13.5); quartz, heavy minerals and feldspars are grouped in a single category. This image has the same size as the original image and will help us compare the simulations obtained in the latter stages of the modeling procedure.

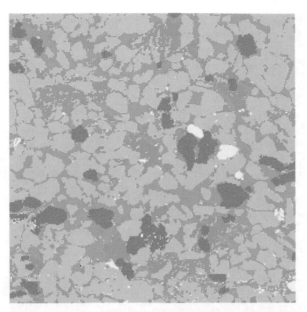

Figure 13.4   Reference image: quartz (orange), clays (green), pore space (blue), feldspars (dark green) and heavy mineralss (white) (Fens 2000).

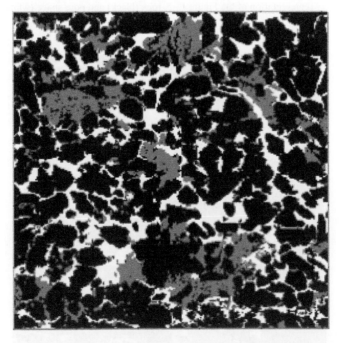

Figure 13.5   Segmentation of the Figure 13.4, rock (black), clays (gray) and pore space (white).

To compare the first stage of case 1, the input image is segmented as follows: rock matrix (quartz, feldspar and heavy minerals) and pore space (pore space and clay) (Fig. 13.6). The segmentation to compare stage 1 of case 2 is: rock matrix (quartz, heavy minerals, feldspar and clay) and pore space (Fig. 13.7).

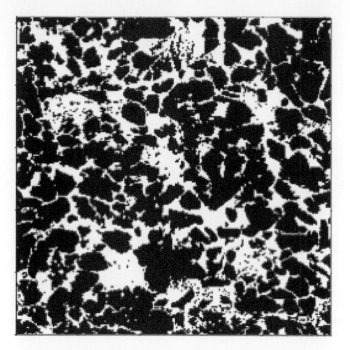

Figure 13.6   Binary representation of the Figure 13.5, rock (black) and pore space (white).

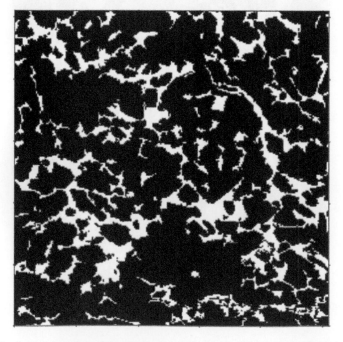

Figure 13.7   Binary representation of the Figure 13.5, rock (black) and pore space (white).

Table 13.1   Proportions of each category.

Category	Figure 13.6	Figure 13.7	Figure 13.5
Pore space	34.6%	19.9%	19.9%
Rock	65.4%	80.1%	65.4%
Clays			14.7%

Table 13.2   Basic statistics.

Statistics	Figure 13.6	Figure 13.7	Figure 13.5
Minimum	0.000	0.000	0.000
First quartile	0.000	1.000	1.000
Medium	1.000	1.000	2.000
Third quartile	1.000	1.000	2.000
Maximum	1.000	1.000	2.000
Mean	0.661	0.799	1.460
Variance	0.224	0.160	0.649
Standard Deviation	0.473	0.400	0.806

Table 13.1 presents the proportions to be used in each of the stages.

13.4   GEOSTATISTICAL ANALYSIS

13.4.1   *Exploratory data analysis*

Table 13.2 presents the statistical data; these will be used to see to what degree the simulations reproduce the statistics of the original information.

13.4.2   *Variographic analysis*

The variograms were calculated under the assumption that the information has no trend or anisotropy. These assumptions were corroborated by obtaining the variograms, because they do not have a quadratic growth and comparing the variograms in different directions, they do not show significant differences in sill, nor in range.

Figures 13.8, 13.9 and 13.10 show variograms for pore space (case 1 and case 2) and clays. To every variogram obtained a model was adjusted, the collection of which are presented in Table 13.3.

13.5   RESULTS

For simulating the first stage (case 1), we used the proportions of Figure 13.6 and the pore space model of Table 13.3. The simulation result is shown in Figure 13.11.

Figure 13.11 shows on the right the flag used in the simulation. The flag only have two divisions (pore space "white" and rock "black"), the division indicates the proportion of the category in the final simulation (left).

In the second stage, consider the proportions of Figure 13.5. The flag used is shown in Figure 13.13 (right). The model used in the first Gaussian random function is the same as that used in stage 1; for the second Gaussian random function, we used the clays model (Table 13.3). The simulation result of this stage is shown in Figure 13.13.

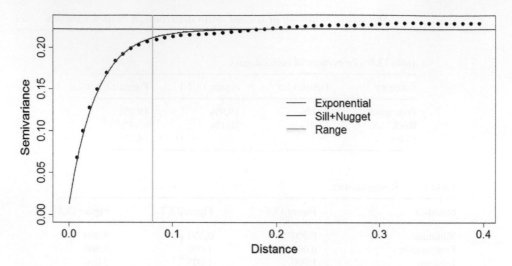

Figure 13.8 Estimated and fitted variogram model of pore space variable (case 1).

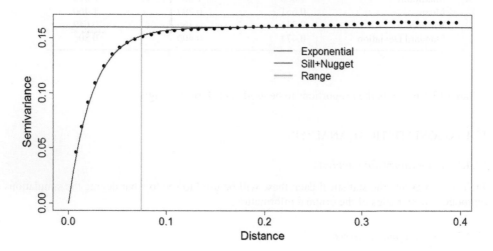

Figure 13.9 Estimated and fitted variogram model of pore space variable (case 2).

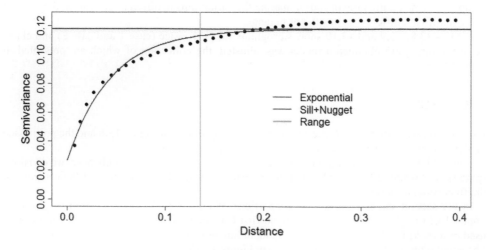

Figure 13.10 Estimated and fitted variogram model of clays variable (case 2).

Table 13.3   Fitted models variograms.

Variable	Model	Nugget	Sill	Practical range
Pore space (case 1)	Exponential	0.013	0.210	0.081
Pore space (case 2)	Exponential	0.000	0.160	0.075
Clays	Exponential	0.027	0.091	0.135

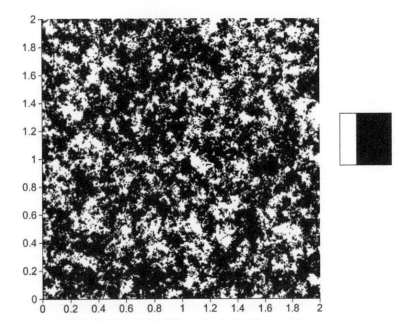

Figure 13.11   Pore space simulation (case 1).

The right of Figure 13.13 shows the flag, where according to the proposed case, first the rock pore and rock matrix is formed and then the clays, i.e. in the simulation of the first stage (Fig. 13.11), the clays are integrated within the pore space. This is done by including another category within the category that was occupied by the pore space in the flag of the first stage. The final simulation is shown at the left of Figure 13.13.

Figure 13.6 is the reference image of the simulation of the first stage (Fig. 13.11). Table 13.4 (column 2 and 3) shows the proportions of the reference image and the simulation. The simulation adequately reproduces the image proportions, basic statistics (Table 13.5) and the variogram models (Table 13.6 and Figure 13.12), when models are reproduced, indicating that the simulation reproduces properly the sizes of the structures.

The result of the second stage (Fig. 13.13) is compared with the reference image (Fig. 13.5). Table 13.4 (columns 4 and 5) and Table 13.5 show the proportions and basic statistics of the reference image and the simulation. At this stage it can be seen that the sizes of the clays are smaller compared to those in the reference image, i.e. the range of the simulation model is smaller than the reference image (Table 13.6 and Figure 13.14).

Table 13.4   Proportions of the reference image and the simulations of case 1.

Category	Figure 13.6	Figure 13.11	Figure 13.5	Figure 13.13
Pore space	34.6%	33.99%	19.9%	19.82%
Rock	65.4%	66.01%	65.4%	66.01%
Clays			14.7%	14.17%

Table 13.5   Basic statistics.

Category	Figure 13.6	Figure 13.11	Figure 13.5	Figure 13.13
Minimum	0.000	0.000	0.000	0.000
First quartile	0.000	0.000	1.000	1.000
Medium	1.000	1.000	2.000	2.000
Third quartile	1.000	1.000	2.000	2.000
Maximum	1.000	1.000	2.000	2.000
Mean	0.661	0.660	1.460	1.462
Variance	0.224	0.224	0.649	0.645
Standard Deviation	0.473	0.474	0.806	0.803

Table 13.6   Fitted variograms models.

Variable	Model	Nugget	Sill	Practical range
Pore space (case 1)	Exponential	0.071	0.152	0.075
Clays (case 1)	Exponential	0.062	0.059	0.078

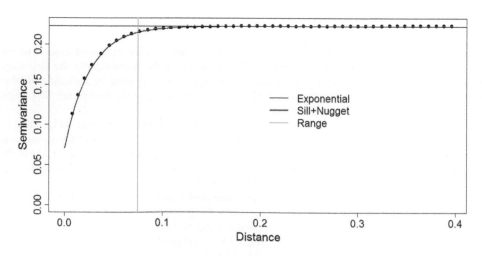

Figure 13.12   Estimated and fitted variogram model of the simulation of pore space (case 1).

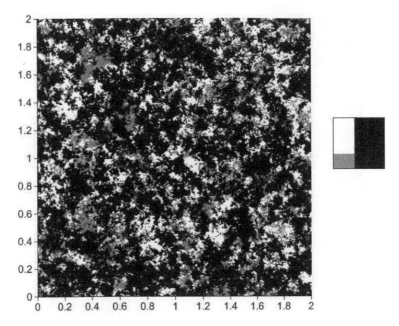

Figure 13.13   Clay distribution simulation (case 1).

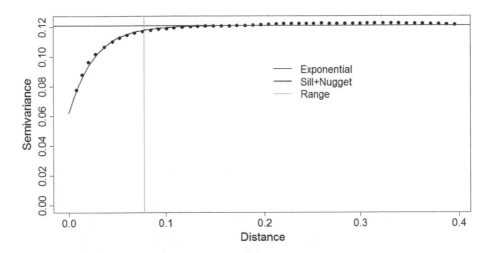

Figure 13.14   Estimated and fitted variogram model of the simulation of clays (case 1).

Figure 13.15 shows the simulation of the first stage (case 2), the ingredients of the simulation with the proportions of Figure 13.7 and the pore space model (Table 13.3).

Table 13.7 (columns 2 and 3) and Table 13.8 show respectively the proportions and basic statistics of the reference image and the simulation. Variogram models for the simulation are presented in Table 13.9 and Figures 13.16 and 13.18.

The simulation of the second stage (Fig. 13.17) is comparable to Figure 13.5. Table 13.7 (column 4 and 5) and Table 13.8 present the proportions and statistics of the image and the simulation. In both phases of this case the proportions, statistics and the sizes of the structures, present in the reference images, are well reproduced.

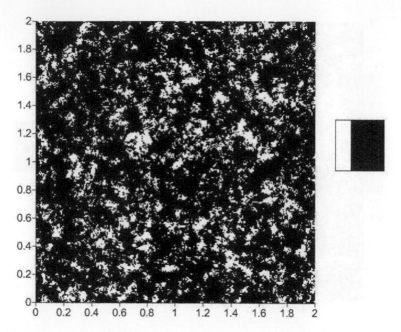

Figure 13.15   Pore space simulation (case 2).

Table 13.7   Proportions of the reference image and the simulations of case 2.

Category	Figure 13.7	Figure 13.15	Figure 13.5	Figure 13.17
Pore space	19.9%	19.99%	19.9%	19.99%
Rock	80.1%	80.01%	65.4%	65.37%
Clays			14.7%	14.64%

Table 13.8   Basic statistics.

Category	Figure 13.7	Figure 13.15	Figure 13.5	Figure 13.17
Minimum	0.000	0.000	0.000	0.000
First quartile	1.000	1.000	1.000	1.000
Medium	1.000	1.000	2.000	2.000
Third quartile	1.000	1.000	2.000	2.000
Maximum	1.000	1.000	2.000	2.000
Mean	0.799	0.800	1.460	1.454
Variance	0.160	0.160	0.649	0.647
Standard Deviation	0.400	0.474	0.806	0.805

13.6   CONCLUSIONS

This work is part of a line of research that attempts to investigate the impact of the interaction of reservoir fluids within themselves and/or injected chemicals on the petrophysical properties of the rock (porosity, permeability, relative permeability, etc.; and consequently in the patterns of fluid flow through the rock), and the changes in occupied volume by the clays at pore scale.

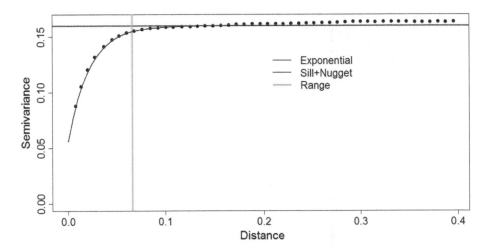

Figure 13.16 Estimated and fitted variogram model of the simulation of pore space (case 2).

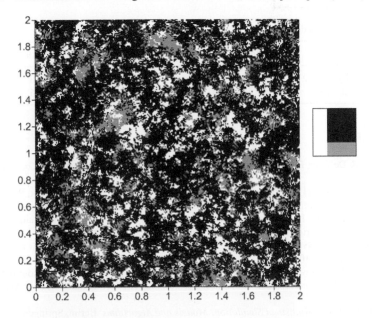

Figure 13.17 Clay distribution simulation (case 2).

The results presented are preliminary. However it has been found that the simulations using the plurigaussian method adequately reproduces the proportions, basic statistics and sizes of the structures present in the studied reference images, but does not reproduce properly the existing connectivity in them.

As a future work it should be considered to combine the plurigaussian method with other methods, as might be a multipoint geostatistical simulation method (Okabe 2007) or alternatively include information about the connectivity of the structures within the simulation process, through an indicator of the topology such as Euler's number (Wu 2006).

Although the work presented is restricted to 2D images, the methodology can be extended to 3D to achieve the reconstruction of the geometry of the porous medium, allowing a more adequate estimation of petrophysical properties.

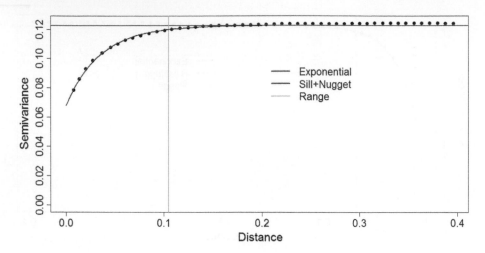

Figure 13.18   Estimated and fitted variogram model of the simulation of clays (case 2).

Table 13.9   Fitted models variograms.

Variable	Model	Nugget	Sill	Practical range
Pore space (case 2)	Exponential	0.056	0.104	0.066
Clays (case 2)	Exponential	0.068	0.055	0.105

REFERENCES

Armstrong, M. & Delfiner, P. (1980) *Towards a More Robust Variogram: Case Study on Coal*. Fontainebleau, Springer.

Chilès, J.P. & Delfiner, P. (1999) *Geostatistics: Modeling Spatial Uncertainty*. New York, Wiley.

Deutsch, C. (2002) *Geostatistical Reservoir Modelling*. New York, Oxford University Press.

Dowd, P., Pardo-Igzquiza, E. & Xu, C. (2003) Plurigau: a computer program for simulating spatial facies using the truncated plurigaussian method. Computers and Geosciences, 29, 123–141.

Fens, T. (2000) *Petrophysical Properties from Small Rock Samples Using Image Analysis Techniques*. Ph.D. thesis. Netherlands, Delft University Press.

Galli, A. Beucher. H., Le Loc'h, G., Doligez, B. & Heresim Group. (1994) The pros and cons of the truncated Gaussian method, *in Geostatistical Simulations*. Dordrecht, The Netherlands. Kluwer Academic Press. pp. 217–233.

Lantuejoul, C. (2002) *Geostatistical Simulation: Models and Algorithms*. Berlin, Springer.

Le Loc'h, G. & Galli, A. (1997). Truncated plurigaussian method: Theoretical and practical points of view. In *Geostatistics Wollongong'96*. Dordrecht, The Netherlands. Kluwer Academic Press, vol. 1, pp. 211–222.

Okabe, H. & Blunt, M. (2005) Pore space reconstruction using multiple point statistics. *Petroleum Science and Engineering*, 46, 121–137.

Okabe, H. & Blunt, M. (2007) Pore space reconstruction of vuggy carbonates using microtomography and multiple-point statistics. *Water Resources Research*, 43.

Oren, P. & Bakke, S. (2003) Reconstruction of Berea sandstone and pore-scale modeling of wettability effects. *Petroleum Science and Engineering*, 39 (3–4), 177–199.

Politis, M., Kikkinides, E., Kainourgiakis, M. & Stubos, A. (2008) Hybrid process-based and stochastic reconstruction method of porous media. *Microporous and Mesoporous Materials*, 110, 92–99.

Wu, K., Van Dijke, M., Couples, G., Jiang, Z. Ma, J., Sorbie, K., Crawford, J., Young, I. & Zhang, X. (2006) 3D stochastic modeling of heterogeneous porous media – Applications to reservoir rocks. *Transport in Porous Media*, 65, 443–467.

Xu, C., Dowd, P., Mardia, K. & Fowell, R. (2006) Flexible true plurigaussian code for spatial facies simulations. *Computers and Geosciences*, 32, 1629–1645.

CHAPTER 14

Metadistances in prime numbers applied to integral equations and some examples of their possible use in porous media problems

A. Ortiz-Tapia

14.1 INTRODUCTION

14.1.1 *Some reasons for choosing integral equation formulations*

Many phenomena which can be modeled through partial differential equations (PDEs), can be modeled as well with integral equations (IEs). Choosing IE over a PDE can be advantageous in many instances. One reason is that a PDE in s dimensions would require to be discretized into a set of simultaneous equations with N^s unknowns (N being related to the discretization); whereas its corresponding IE would require a set of N^{s-1} unknowns. For example Hunt and Isaacs (1981), describe a formulation with IE of ground-water flow, explaining that the original boundary-value problem required a set of simultaneous equations with N^2 unknowns, while its corresponding IE discretizes to only N unknowns.

14.1.2 *Discretization of an integral equation with regular grids*

Once we have chosen to use an IE, and we have it at hand, its solution can be achieved either analytically or numerically. To solve an IE analytically, it has to be in a form that can be treated with Laplace transforms, trigonometrical identities or some algebraic relations allowing to solve for the unknown function (see for example the solution of the Abel equation in Frank 1973). Since it is not always possible or convenient to solve an IE analytically, often a good alternative is to use a numerical method.

Solving an IE numerically implies to discretize it somehow, and there are several possibilities to do so. There are discretizations into regular grids. For example (Annunziato & Messina 2010) use heterogeneous maps for space and time. Consider the Volterra type integral equation

$$y(x,t) = f(x,t) + \int_0^t k(t-\eta)y(g(x,t,\eta),\eta)\,d\eta \tag{14.1}$$

Now suppose a domain S divided up by an uniform grid $\Pi_M := \{x_1, \ldots, x_M\}$ whose size is h. Let

$$P(\xi,\eta) = \sum_{l=l_1}^{l_2} L_l(\xi)y(x_l,\eta) \tag{14.2}$$

be the interpolating polynomial approximation, of degree $r = l_2 - l_1$, to the exact solution of Equation (14.1), where $L_l(x)$ denotes the l-th Lagrange fundamental polynomial associated with the $l_2 - l_1 + 1$ distinct nodes $x_{l_1}, \ldots, x_{l_2}$, then the semidiscretization in space obtained after the approximation of y by Eq. (14.2), leads to the following

$$y_m(t) = f_m(t) + \int_0^t k(t-\eta) \sum_{l=l_1}^{l_2} L_l(g(x_m,t,\eta))y_l(\eta)\,d\eta \tag{14.3}$$

$t \in [0, T], m = 1, \ldots, M$, where $f_m(t) = f(x_m, t)$ and the M functions $y_m(t)$ represent the numerical approximations to the exact solutions $y(x_m, t)$. Equations (14.3) represent a system of M linear Volterra IEs for the unknown continuous functions $y_1(t), y_2(t), \ldots, y_M(t)$. Let $\Pi_N := \{t_n : 0 = t_0 < t_1 < \cdots < t_N = T\}$ be a partition of the time interval $[0, T]$, with constant step size $\tau = t_{n+1} - t_n, n = 0, \ldots, N - 1$. For the discretization in time of the semidiscrete problem in Eq. (14.3), it can be chosen one of the numerical methods present in the literature of Volterra IEs (for example see Press et al. 1992). For this case, a classical s-step direct quadrature method has been chosen, yielding

$$y_{m,n} = f_m(t_n) + \tau \sum_{j=0}^{n} w_{n,j} k(t_n - t_j) \sum_{l=l_1}^{l_2} L_1(g(x_m, t_n, t_j)) y_{l,j} \quad m = 1, \ldots, M \qquad (14.4)$$

for $n = s, \ldots, N$, where the m-th component $y_{m,n}$ represents an approximation to the exact solution y of Eq. (14.1) at point (x_m, t_n) and $y_{m,0} = f_m(0) \, y_{m,1}, \ldots, y_{m,s-1}$ are given. Here, $w_{n,j}, n, j = 0, \ldots, N$, are the weights of the direct quadrature method and $l_1 = l_1(m, n, j)$ and $l_2 = l_2(m, n, j)$ are dynamically chosen for each triplet (x_m, t_n, t_j) in such a way that $g(x_m, t_n, t_j)$ belongs to the interval $[x_{l_1}, x_{l_2}]$. Thus, if we define the vector $Y_n = (y_{1,n}, \ldots, y_{M,n})^T$ of the numerical approximations in t_n, the matrix formulation of Equation (14.4) reads

$$Y_n = F(t_n) + \tau \sum_{j=0}^{n} w_{n,j} K(t_n, t_j) Y_j. \qquad (14.5)$$

Here it is assumed that τ is small enough to assure that $\det \left(I - \tau w_{n,n} K(t_n, t_n) \right) \neq 0$ and the uniqueness of the solution. Since the total error is dependent to the number of points $(x_m, t_n) \in \Pi_M \times \Pi_N$, and the tendency is to create a regular grid in the combined domain of space and time S, the accuracy of the solution depends in general to a grid which grows roughly like $\mathcal{O}(S^2)$. For complicated kernels, and/or boundaries which are functions, the number of points required might be large.

14.1.3 *Solving an integral equation with MC or LDS*

Monte Carlo methods (MC) can be used for the quadrature part of the discretization process, requiring for achieving certain accuracy a number of points whose convergence is roughly growing as $\mathcal{O}(1/\sqrt{S})$, i.e. quadrupling the number of sampled points will halve the error, regardless of the number of dimensions (Caflisch 1998). A refinement of this method is to somehow make the points random, but more likely to come from regions of high contribution to the integral than from regions of low contribution (Lepage 1978). In other words, the points should be drawn from a distribution similar in form to the integrand. Understandably, doing this precisely is just as difficult as solving the integral in the first place, but there are approximate methods available (Wikipedia 2010, Lepage 1978).

A similar approach involves using low-discrepancy sequences (LDS) instead: the quasi-Monte Carlo method. Quasi-Monte Carlo methods can often be more efficient at numerical integration, because the sequence fills the area better in the sense that points generated through LDS tend not to overlap, and samples more of the most important points that can make the simulation converge to the desired solution more quickly (Caflisch 1998, Morokoff & Caflisch 1994 and 1995). As an example of LDS, Halton sequences (Halton 1960, Kuipers & Niederreiter 1974), and metadistances of prime numbers will be used which are the core reason for this work, and which have been reported as concerns of quadratures in previous works (Ortiz-Tapia 2008a, 2008b, and references therein). In the next section details will be given for the algorithms. In here, it can be said that the most general idea of the algorithm for discretization using non-regular sequences, either pseudo or quasi-random numbers, which sample the kernel and the given functions of the integral equation. With that collection of numbers, we can form a linear system, which, depending on its nature, it can be solved by somehow inverting a matrix, or solving an eigensystem.

14.2 ALGORITHMS DESCRIPTION

14.2.1 *Low discrepancy sequences*

A sequence is an ordered list of numbers. A low-discrepancy sequence is a sequence with the property that for all N, the subsequence $x_1, \ldots, x_N$ is almost uniformly distributed in a given s dimensional space, and $x_1, \ldots, x_{N+1}$ is almost uniformly distributed as well (Kuipers & Niederreiter 1974). Let I^s be an integration domain in s dimensions. Let c_J be the characteristic function (integrand) to be evaluated at the points of the sequence. The sequence $\mathbf{x}_1, \mathbf{x}_2, \ldots$ (notice the multivariate notation) of points in I^s is called uniformly distributed in I^s if:

$$\lim_{N \to \infty} \frac{1}{N} \sum_{n=1}^{N} c_J \mathbf{x_n} = |J| \tag{14.6}$$

holds for all subintervals J of I^s, where $|J|$ denotes the s-dimensional Lebesgue measure, i.e. the (hyper-)volume of J. Intuitively, this means that the points $\mathbf{x}_1, \mathbf{x}_2, \ldots$ are spread out over the unit cube I^s according to the principle of proportional representation (Niederreiter 1978). The perceptive lector might notice that the definition just given is descriptive, but not constructive. In other words, all we know about LDSs is that they distribute uniformly, in such a way of being potentially useful for doing quadrature, in a sense sufficiently general though, since we are departing from Lebesgue measures. Having said that, in principle any sequence that intuitively can be thought to be good candidate to fulfill the criteria given through Equation (14.6), might be tested for quadratures. Certain numerical sequences so-called "metadistances", obtained in turn from the sequence of prime numbers, have been tested by trial and error, and the results have been promising enough to attempt to expand their applicability (Ortiz-Tapia 2008a, 2008b). In this work Halton LDSs are also tried for solving IEs.

14.2.2 *Halton LDSs*

The Halton sequence is constructed according to a deterministic method that uses a prime number as its base. As a simple example, to generate the sequence for 2, we start by dividing the interval $(0, 1)$ in half, then in fourths, eighths, etc, which generates

$$1/2, 1/4, 3/4, 1/8, 5/8, 3/8, 7/8, 1/16, 9/16, \ldots \tag{14.7}$$

The sequence in Eq. (14.7) can be used for one dimensional quadratures. One further example: to generate the sequence for 3, we divide the interval $(0, 1)$ in thirds, then ninths, twenty-sevenths, etc., which generates

$$1/3, 2/3, 1/9, 4/9, 7/9, 2/9, 5/9, 8/9, 1/27, \ldots \tag{14.8}$$

Both sequences in Eqs. (14.7) and (14.8) can be paired up with themselves, or with each other thus obtaining a sequence of points for the unit square:

$$(1/2, 1/3), (1/4, 2/3), (3/4, 1/9), (1/8, 4/9),$$
$$(5/8, 7/9), (3/8, 2/9), (7/8, 5/9), (1/16, 8/9), (9/16, 1/27) \tag{14.9}$$

For this work the procedure was to obtain the integer digits of several numbers, in base three. Then these collection of sequences of digits are reorganized, their length counted upon and finally a sum is made as follows, resulting in each Halton member (Eq. 14.10) of its corresponding LDS:

$$Halton_i = \sum_{k=1}^{Length SBT_i} (SBT_{i,k} \times 3^{-k}) \tag{14.10}$$

SBT_i stands for Sequence of digits Base Three for the ith Halton fraction.

Even though standard Halton sequences perform very well in low dimensions, correlation problems have been noted between sequences generated from higher primes. For example if we started with the primes 17 and 19, the first 17 pairs of points would have perfect linear correlation. To avoid this, it is common to drop the first 20 entries, or some other predetermined number depending on the primes chosen. In order to deal with this problem, various other methods have been proposed; one of the most prominent solutions is the scrambled Halton sequence, which uses permutations of the coefficients used in the construction of the standard sequence. (Wikipedia 2010).

14.2.3   *What is a "metadistance"*

Before actually describing the algorithms for integral equations, it is convenient to explain what is understood by a metadistance, and how is it generated. Suppose the infinite sequence of prime numbers

$$P_1, P_2, \ldots, P_{i-1}, P_i, P_{i+1}, \ldots, P_\infty \tag{14.11}$$

Among the efforts to study the patterns within prime numbers, it is to describe the "Jumping Champions" (Odlizko 1999) which are the most frequent differences between prime numbers, that is to say

$$\mathbb{J} = \max[P_{i+1} - P_i] \tag{14.12}$$

In Figure 14.1 it is shown a comparison between jumping champions coming from primes, and "random jumpers".

Since the most frequent "jumper" among primes is the number 6, at least until 1.74×10^{35} (Odlyzko 1999), this number was chosen to attempt to find deeper correlations between prime numbers (Ortiz-Tapia 2008a). Since no apparent patterns were found, the possibility that the sequence of prime numbers have some resemblance with uniformly distributed sequences was admitted, and from there trials on quadratures were made (Ortiz-Tapia 2008a). In particular, let the dyad be formed by a sequence of those prime numbers P_{i+2}, P_{i+1}, P_i such that

$$\{P_{i+2} - P_{i+1}, P_{i+1} - P_i\} = \{6, 6\} \tag{14.13}$$

Now, suppose that it is wanted to know what the distances between the dyads $\{6, 6\}$ are. To achieve such purpose, it can be noticed that prime numbers can be indexed from the first to the nth, i.e. knowing the index of the prime number is equivalent to pointing to a particular prime number. So, a difference between indexations is sought for in the following way. Let the choice of indices be $i, i + 2$, corresponding to enough information to generate a given dyad $\{6, 6\}$ and let the following such dyad be separated by k units of natural numbers, so that the two obtained dyads are found at the positions

$$\overbrace{\{i, i+2\}}^{j} \overbrace{\{i+k, i+k+2\}}^{j+1} \quad k \in \mathbb{N} \tag{14.14}$$

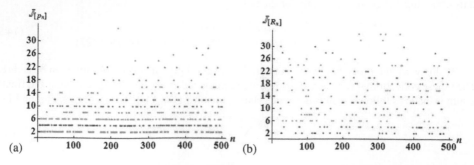

Figure 14.1   Jumping champions (a) compared with "jumpers" made out of integers taken at random (b).

Each k is unknown a priori, since the exact position of each couple is not known, unless of course the distribution of the prime numbers were to be known with full precision (Crandall & Pomerance 2005). A metadistance is defined as the distance between the chosen dyads $\{6, 6\}$, such that:

$$md = (i + k_{j+1}) - (i + k_j) \quad md \in \mathbf{Md} \tag{14.15}$$

where md stands as shorthand for meta-distances, implying distances embedded in the set of distances between prime numbers, $\mathbf{Md}$ is the set of all the meta-distances within a given range of prime numbers, and the way they are obtained is the difference between the first element of indices for a given set, minus the second element of the previous set. So, for example, suppose that we had the sequence of prime numbers

$$\{\{47_{15}, 53_{16}\}, \{53_{16}, 59_{17}\}\}, \tag{14.16}$$

where the indices 15, 16, 17 indicate which prime number it is, i.e. their position within the ordered sequence of prime numbers. Evidently $53 - 47 = 6$, $59 - 53 = 6$. This conforms the first dyad $\{6, 6\}$. The next such dyad is formed by the primes

$$\{\{151_{36}, 157_{37}\}, \{157_{37}, 163_{38}\}\} \tag{14.17}$$

Without loss of generality, let us take the last indices for each dyad and take the difference, which for this case it will be

$$38 - 17 = 23 = md_1 \tag{14.18}$$

We have thus obtained the first metadistance.

14.2.4 *Refinement of mds*

The observing reader might perhaps noticed that Halton LDS are non-repeating, whereas mds can repeat themselves. It has been suggested that mds performance can be improved by adjusting to this fact (Román-Mejía 2009), which was implemented by simply making a subset of mds, excluding those already present, while constructing a subset MD out of a given amount of prime numbers. For the present work, mds were chosen out of every million prime numbers, and from the resulting mds the aforementioned further filtering was applied. Nonetheless, results with similar low errors can be obtained by choosing $md < 32$.

14.3 NUMERICAL EXPERIMENTS

14.3.1 *Fredholm equations of the second kind in one integrable dimension*

A numerical solution for $f(t)$ is desired in Equation (14.19)

$$u(x) = f(x) + \lambda \int_a^b K(x,t)u(t)\,dt \tag{14.19}$$

without loss of generality, $\lambda = 1$, $[a, b] \in \mathbb{R}$, and the functions f and K (the so-called kernel) are given. Notice that the Right-Hand-Side (RHS) common of some other types of equations, is traditionally expressed very often as the Left-Hand-Side (LHS) in IE (Press et al. 1997). This equation is considered to be in one dimension, since the unknown $u(t)$ is a function of only one variable, or it can be thought that there is only one variable to be integrated. To approximate the function u on the interval $[a, b]$, a partition $x_0 = a < x_1 < \cdots < x_{m-1} < x_m = b$ is selected (for example) and the equations

$$u(x_i) = f(x_i) + \int_a^b K(x_i, t)u(t)\,dt, \quad \forall_i \in \{0, \ldots, m\} \tag{14.20}$$

are solved for $u(x_0), u(x_1), \ldots, u(x_m)$. Notice that solving the aforementioned linear system, and even though it was demanded that the partition remained within the interval $[a, b]$, no other condition is imposed, which means that there is no ordering within the sequence of elements belonging to the partition, implying that Monte-Carlo methods or LDSs may be used for selecting a useful partition. The integrals actually are approximated using quadrature formulas based on the nodes $x_0, \ldots, x_m \ \forall_t \in [a, b]$ and we can form the linear system in Equation (14.21)

$$\left(1 - \lambda \tilde{\mathbf{K}}\right) \cdot \mathbf{u} = \mathbf{f} \tag{14.21}$$

which can be solved using any suitable numerical algorithm for linear systems, although, of course, depending on the type of kernel, an sparse linear system might be obtained, or some other type of special linear system, for which special kinds of algorithms have been, and are being developed (chapter 2.7 of Press et al. 1997, and references therein).

14.3.2   *Results in one dimension*

Suppose that we have the IE

$$u(x) = 1 + \int_0^1 (1 + x + t + xt)^{1/2} u(t) \, dt \tag{14.22}$$

whose analytic solution is

$$u(x) = 1 + \frac{2(-1 + \sqrt{2})}{\sqrt{1 + x}(1 - \ln(2))} \tag{14.23}$$

Table 14.1 summarizes the central tendency measurements, resulting from MC, Halton and MDs. Results were rounded up to two decimal ciphers; "SD" stands for "Standard Deviation" (as opposed to σ, which will stand for "eingevalues" in this work throughout), "Max." is "Maximal"; "Min." is "Minimal" and $\langle \rangle$ is the average of all measurements.

Another example is

$$u(x) = x^2 + \int_0^1 \exp^{|x-t|} u(t) \, dt \tag{14.24}$$

No analytic solution is available for Equation (14.24), so the comparisons of accuracy were made with respect to the Riemann quadrature. Table 14.2 summarizes the central tendency measurements, resulting from MC, Halton and MDs, for this particular case.

Table 14.1   Summary of results from MC, Halton and MDs.

Method	Max. SD	Min. SD	$\langle SD \rangle$	Max. % error	Min. % error	$\langle$% error$\rangle$
MC	0.91	0.64	0.76	5.06	4.54	4.86
Halton	0.40	0.28	0.33	6.33	5.64	6.01
MDs	0.59	0.42	0.49	12.13	10.93	11.50

Table 14.2   Summary of results from MC, Halton and MDs.

Method	Max. SD	Min. SD	$\langle SD \rangle$	Max. % error	Min. % error	$\langle$% error$\rangle$
MC	1.74	0.73	1.09	137.69	49.02	73.56
Halton	0.36	0.18	0.24	2.13	0.55	1.64
MDs	0.26	0.12	0.18	12.96	1.81	7.46

Finally, a more elaborated example is

$$u(x) = 5\cos(x) - \frac{1}{4}\cos(x)\sin^2(1) + 0.1\int_0^1 \sin(t)\cos(x)u(t)\,dt \qquad (14.25)$$

whose analytic solution is

$$u(x) = 5\cos(x) \qquad (14.26)$$

Table 14.3 summarizes the central tendency measurements, resulting from MC, Halton and MDs, for this last example, Equation (14.25).

14.3.3   *Choosing a problem in two dimensions*

Extending the described methods to superior dimensions is convenient for those IEs which preferably have some features making them suitable for applying the aforementioned numerical methods. So, for example, an integral equation, where the multiple dimensions are those to be integrated, leaving only the dimension of the independent variable, collapse the whole system into a single-dimensional equation (Eq. 14.27); this is the most promissory case as could be seen from the quadratures described in (Ortiz-Tapia 2008a)

$$u(x) = f(x) + \int_D \cdots \int K(x,\mathbf{t})u(\mathbf{t})\,d\mathbf{t} \quad u(x) = \mathbb{R}^{n+1} \mapsto \mathbb{R} \qquad (14.27)$$

Another possibility is an integral equation where the integration is w.r.t. the unknown function, but where the kernel is in several dimensions. This will render a hypersurface (Eq. 14.28)

$$u(\mathbf{x}) = f(\mathbf{x}) + \int_a^b K(\mathbf{x},t)u(t)\,dt \qquad (14.28)$$

Actually for this work it was chosen an IE departing from the following problem, described in (Jiménez-Ángeles et al. 2006). A simple model for two like-charged parallel rods immersed in an electrolyte solution is considered. The model of the system is shown in Figure 14.2.

The IEs are based pretty much on the Ornstein-Zernike set of integral equations, and the resulting formulation is as depicted in Equation (14.29)

$$g_{\alpha j}(\mathbf{r}_3) \equiv g_{\alpha j}(\eta_3,\xi_3) = \exp\{*\},$$

$$* = \frac{4\pi R\sigma z_j e\beta}{\varepsilon}\left\{\ln\left[\left(\frac{\tau}{2}(\eta-\xi)\right)\right] + \ln\left[\left(\frac{\tau}{2}(\eta+\xi)\right)\right]\right\} - J(\eta_3,\xi_3) \qquad (14.29)$$

$$+ \int_{-1}^1 \int_{\eta_0(\xi_4)}^\infty \rho_{\alpha s}(\eta_4,\xi_4)K(\eta_3,\xi_3,\eta_4,\xi_4)\,d\eta_4\,d\xi_4$$

$$+ z_j \int_{-1}^1 \int_{\eta_0(\xi_4)}^\infty \rho_{\alpha s}(\eta_4,\xi_4)L(\eta_3,\xi_3,\eta_4,\xi_4)\,d\eta_4\,d\xi_4$$

$$+ z_j \int_{-1}^1 \int_{\eta_0(\xi_4)}^\infty \rho_{\alpha s}(\eta_4,\xi_4)F(\eta_3,\xi_3,\eta_4,\xi_4)\,d\eta_4\,d\xi_4,$$

Table 14.3   Summary of results from MC, Halton and MDs.

Method	Max. SD	Min. SD	⟨SD⟩	Max. % error	Min. % error	⟨% error⟩
MC	0.03	0.03	0.031	0.27	0.01	0.17
Halton	0.02	0.01	0.02	0.30	0.13	0.17
MDs	0.03	0.01	0.02	0.58	0.51	0.55

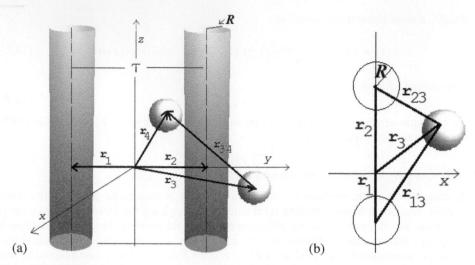

Figure 14.2   (a) Model and parameters for two rods in a restricted primitive model electrolyte. (b) Top view of the model.

where the function $(4\pi R\sigma z_j e\beta/\varepsilon)\{\ln[(\tau/2(\eta - \xi))] + \ln[(\tau/2(\eta + \xi))]\}$ is the Coulombian interaction between the rods and is not going to be taken in account (actually it is also ignored the fact that $\exp *$ affects all the calculation); $-J(\eta_3, \xi_3)$ is the spatial region excluded by the two rods, and is not taken in account in here; $L(\eta_3, \xi_3, \eta_4, \xi_4)$ as defined in Equation (14.30)

$$
L(\eta_3, \xi_3, \eta_4, \xi_4) = \frac{\tau^2(\eta_4^2 - \xi_4^2)}{4\sqrt{(\eta_4^2 - 1)(1 - \xi_4^2)}} \int_{-z_0}^{z_0} c^{sr}(r_{34})\, dz_4
$$

$$
= \frac{\tau^2(\eta_4^2 - \xi_4^2)}{\sqrt{(\eta_4^2 - 1)(1 - \xi_4^2)}} \frac{e^2\beta}{2\epsilon} \left(I_{-1} + \frac{2\Gamma I_0}{1 + \Gamma a} + \frac{\Gamma^2 I_1}{(1 + \Gamma a)^2} c_1 I_3 \right)
$$

(14.30)

serves for calculate the short-ranged ionic correlations, and it is not taken in account; $F(\eta_3, \xi_3, \eta_4, \xi_4)$, as defined in Equation (14.31)

$$
F(\eta_3, \xi_3, \eta_4, \xi_4) = \frac{\tau^2(\eta_4^2 - \xi_4^2)}{4\sqrt{(\eta_4^2 - 1)(1 - \xi_4^2)}} \frac{e^2\beta}{2\epsilon} \int_0^\infty \frac{dz}{r_{13}}
$$

$$
= \frac{\tau^2(\eta_4^2 - \xi_4^2)}{\sqrt{(\eta_4^2 - 1)(1 - \xi_4^2)}} \frac{e^2\beta}{2\epsilon} \ln A
$$

(14.31)

represent the long-ranged ionic correlations, and is not taken in account either; it remains the kernel described by Equation (14.32)

$$
K(\eta_3, \xi_3, \eta_4, \xi_4) = \frac{\tau^2(\eta_4^2 - \xi_4^2)}{4\sqrt{(\eta_4^2 - 1)(1 - \xi_4^2)}} \int_{-z_0}^{z_0} c^{hs}(r_{34})\, dz_4
$$

$$
= \frac{\tau^2(\eta_4^2 - \xi_4^2)}{2\sqrt{(\eta_4^2 - 1)(1 - \xi_4^2)}} \left(-c_1 I_0 + \frac{6\eta}{a} c_2 I_1 - \frac{\eta}{2a^3} c_1 I_3 \right)
$$

(14.32)

with $z_0 = \sqrt{a^2 - A^2}$, a being the particle diameter, τ is the separation between the centers of the rods, and

$$
A^2 = \frac{\tau^2}{4} \left\{ \left(\sqrt{(\eta_3^2 - 1)(1 - \xi_3^2)} - \sqrt{(\eta_4^2 - 1)(1 - \xi_4^2)} \right)^2 + (\eta_3\xi_3 - \eta_4\xi_4)^2 \right\}
$$

(14.33)

Finally, in the expressions for K, J, and L, it is defined

$$I_{-1} = 2 \int_0^{z_0} \frac{dz}{r_{13}} = 2 \ln \left[\frac{\sqrt{a^2 - A^2} + a}{A} \right]$$

(14.34)

$$I_0 = \int_0^{z_0} dz = 2\sqrt{a^2 - A^2}$$

(14.35)

$$I_1 = 2 \int_0^{z_0} r_{13} \, dz = a\sqrt{a^2 - A^2} + A^2 \ln \left[\frac{\sqrt{a^2 - A^2} + a}{A} \right]$$

(14.36)

$$I_3 = 2 \int_0^{z_0} r_{13}^3 \, dz = \frac{a^3}{2} \sqrt{a^2 - A^2} + \frac{3a}{4} A^2 \sqrt{a^2 - A^2} + \frac{3}{4} A^4 \ln \left[\frac{\sqrt{a^2 - A^2} + a}{A} \right]$$

(14.37)

For the Eqs. (14.32–14.37), the following parameters are defined

$$c^{hs}(r_{34}) = -c_1 + \frac{6\eta}{a} c_2 r_{34} - \frac{\eta}{2a^3} c_1 r_{34}^3,$$

$$c^{sr}(r_{34}) = \frac{e^2 \beta}{\epsilon} \left\{ \frac{1}{r_{34}} - \frac{2\Gamma}{1 + \Gamma a} + \frac{\Gamma^2 r_{34}}{(1 + \Gamma a)^2} \right\},$$

(14.38)

for $r_{34} \leq a$ and $c^{hs} r_{34} = c^{sr}(r_{34}) = 0$ for $r_{34} > a$, with

$$\eta = \frac{\pi a^3}{6} \sum_{i=+,-} \rho_i,$$

$$c_1 = \frac{(1 + 2\eta)^2}{(1 - \eta)^4},$$

$$c_2 = \frac{-(1 + 1/2\eta)^2}{(1 - \eta)^4},$$

(14.39)

$$\kappa = \frac{4\pi\beta e^2}{\epsilon} \sum_{i=+,-} z_i^2 \rho_i,$$

$$\Gamma a = -\frac{1}{2} + \frac{1}{2}\sqrt{1 + \kappa a}$$

14.3.4 *Transformation of the original problem*

Even after simplification of Equation (14.29), this remains as an improper integral, as can be seen in Equation (14.40), and is so considered in Press et al. (1997), since the upper limit of the inner integral is ∞

$$g_{\alpha j}(\eta_3, \xi_3) = \int_{-1}^{1} \int_{\eta_0(\xi_4)}^{\infty} \rho_{\alpha s}(\eta_4, \xi_4) K(\eta_3, \xi_3, \eta_4, \xi_4) \, d\eta_4 \, d\xi_4$$

(14.40)

The basic trick for improper integrals is to make a change of variables to eliminate the singularity, or to map an infinite range of integration to a finite one (Press et al. 1997). For example, the identity in Equation (14.41)

$$\int_a^b f(x) \, dx = \int_{1/b}^{1/a} \frac{1}{t^2} f\left(\frac{1}{t}\right) dt \qquad ab > 0$$

(14.41)

can be used with either $b \to \infty$ and a positive, or with $a \to -\infty$ and b negative, and works for any function which decreases towards infinity faster than $1/x^2$. The change of variable can be made analytically or numerically. Before actually trying to implement both the change of variable, and the application of the numerical methods described in this work, it is not altogether idle to create a workhorse artificial equation, and study over it the numerical behavior of the different methods. Without loss of generality, let us choose as workhorse IE Equation (14.42)

$$u(\eta_4, \xi_4, \eta_3, \xi_3) = \int_{-1}^{1} \int_{1}^{\infty} \left(\frac{1}{\eta_4 \xi_4} \frac{1}{\eta_4 \xi_4 (\eta_3 \xi_3)^2} \right) d\eta_4 \, d\xi_4 \tag{14.42}$$

whose kernel is Equation (14.43)

$$\frac{1}{\eta_4 \xi_4 (\eta_3 \xi_3)^2} \tag{14.43}$$

with an "unknown" function Equation (14.44)

$$\rho(\eta_4, \xi_4) = \frac{1}{\eta_4 \xi_4} \tag{14.44}$$

whose analytical solution is Equation (14.45)

$$-\frac{2}{\eta_3^2 \xi_3^2} \tag{14.45}$$

A plot of the analytic solution of Equation (14.45) is offered in Figure 14.3.

The change of variable as suggested in Equation (14.41) is made, so that Equation (14.46) is obtained

$$\int_{1}^{-1} \int_{0.0001}^{1} \left[\frac{1}{t^2} \frac{1}{tt^2} \rho \left(\frac{1}{tt}, \frac{1}{t} \right) \frac{1}{\frac{1}{tt} \frac{1}{t} (\eta_3 \xi_3)^2} \right] dtt \, dt \tag{14.46}$$

where $t = \xi_4$, and $tt = \eta_4$. Notice the change of domain of integration as well (including the fact that the lower limit of the inner integral is not set to 0 but to 0.0001, in order to avoid singularities) and that the accompanying function ρ has been changed into its abstract form, since it will be assumed "ignorance" over it, and this will be the form used to try and solve for numerically. The domains of η_3 and ξ_3 are as indicated in Equation (14.47)

$$\eta_3 \in \{0.1 \le \eta_3 \le 1\} \quad \text{(stepsize 0.1)}$$
$$\xi_3 \in \{0.1 \le \eta_3 \le 1\} \quad \text{(stepsize 0.1)} \tag{14.47}$$

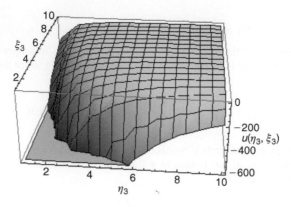

Figure 14.3 Analytic solution of the chosen problem in two dimensions, $u(\eta_3, \xi_3) = -2/(\eta_3^2 \xi_3^2)$.

14.3.5   *General numerical algorithm*

The kernel $K(\eta_3, \xi_3, \eta_4, \xi_4)$ is evaluated and points selected through either MC, i.e. random points within the domain of integration; or MDs or Halton points, as indicated by the primes over the parameters η_4' and ξ_4' in Equation (14.48)

$$K(\eta_3, \xi_3, \eta_4, \xi_4) \approx K(\eta_3, \xi_3, \eta_4', \xi_4'), \quad \eta_4' \in [1, -1] \quad \xi_4' \in [0.0001, 1] \qquad (14.48)$$

Let us recall that $f(x)$ is a given, i.e. known, in Equation (14.19); however in this case for two dimensions $f(x) = 0$ (as can be observed in Equation (14.42)). This would be the homogeneous case described in the theory of Fredholm IEs, and cannot be solved by matrix inversion; turning to the book Numerical Recipes, chapter 18-1 (Press et al. 1997), we find that the homogeneous case can be reexpressed, and then solved as an eigenvalue problem: Let λ be any constant multiplying the integral(s), and let the known function $f(x) = 0$ (as it was actually done in this case), so let $\lambda = 1/\sigma$ and $\mathbf{f} = 0$ in Equation (14.21), thus becoming a standard eigenvalue equation, as can be seen in Equation (14.49)

$$\tilde{\mathbf{K}} \cdot \mathbf{u} = \sigma \mathbf{u} \qquad (14.49)$$

which can be solved with any convenient matrix eigenvalue routine; for example those contained in the LAPACK routines (LAPACK 2010). Suppose further that the matrix $\tilde{\mathbf{K}}$ is symmetric. A symmetric matrix implies not only that all eigenvalues will be real, but also that the eigenproblem of Equation (14.49) can be more easily solved. So, once points have been chosen using MC, Halton or MDs, symmetry is achieved here by constructing only an upper-triangular matrix, (all the other entries were padded by zeros), and then the zero entries in the lower-triangular part are substituted, thus making the matrix symmetric, as shown in Equation (14.50)

$$\tilde{\mathbf{K}}_{i,j}^{\triangle} = \tilde{\mathbf{K}}_{j,i}^{\triangledown}, \qquad j > i \qquad (14.50)$$

where $\tilde{\mathbf{K}}^{\triangle}$ and $\tilde{\mathbf{K}}^{\triangledown}$ symbolize lower-triangular and upper-triangular matrices, respectively. Notice that according to Equation (14.42), this eigensystem must be solved for each value at the points defined by Equation (14.47), and this solution will give as many eigenvalues as the size of the original matrix. Fortunately, all that has to be done is to chose the largest eigenvalue found, and then associate each of these largest eigenvalues with the corresponding coordinate in Equation (14.47). Still, the solution is not obtained directly, since the collection of eigenvalues obtained must be rescaled approximately (Equation 14.51):

$$\sigma_i' = \frac{1}{\dfrac{\sigma}{10^{power_r}}} \times -k_{scale} \qquad (14.51)$$

where $power_r$ stands for rescaling power, and k_{scale} is a "refining" constant which both together would render a subset of local eigenvalues, which in turn, altogether, would render the error minimal. At the beginning both constants were obtained empirically and these are the results to be shown first, but clearly this is not a systematic procedure. There are several possibilities for obtaining these constants in a more systematic fashion. Assuming the most important constant should be the one taking the whole eigensystem to the proper order of magnitude, then the rescaling $power_r$ can be taken either as the logarithm of the maximal eigenvalue, the average of the logarithm of each one of them, or the logarithm of the average of just the maximal and the minimal eigenvalue, which in brief are described in Equation (14.52), indexed respectively as (a), (b), and (c).

$$power_r(a) = \log_{10}[\max(\sigma)]$$
$$power_r(b) = \langle \log_{10}(\sigma_i) \rangle \qquad (14.52)$$
$$power_r(c) = \langle \log_{10}(\max \sigma), \log_{10}(\min \sigma) \rangle$$

After several numerical experiments, it was chosen (c), i.e. Equation (14.53)

$$power_r(c) = \langle \log_{10}(\max \sigma), \log_{10}(\min \sigma) \rangle \tag{14.53}$$

as the systematic rescaling power, since it seemed the one rendering the lowest error. The multiplicative constant k_{scale} was also determined systematically as in Equation (14.54)

$$k_{scale} = \text{floor}_{10} \langle \log_{10}(\max \sigma), \log_{10}(\min \sigma) \rangle \tag{14.54}$$

where floor_{10} means that the closest power of 10 is taken, while applying the function floor. Figure 14.4 illustrates how the percentage of error may vary according to the change of $power_r$ and k_{scale}. This was the numerical experiment which lead to devising the possibilities for systematic rescaling.

14.3.6   *MC results, empirical rescaling*

In Figure 14.5 it is shown the collection of percentages of error, towards the minimal, departing from a matrix with 16 entries, corresponding to the number of points chosen with MC. In this case the average error is $\langle \epsilon \rangle \approx 55\%$, which was calculated classically, as shown in Equation (14.55)

$$\langle \epsilon \rangle = \frac{1}{n} \sum_{i=1}^{n} \frac{|u(\eta_{3,i}, \xi_{3,i}) - u'(\eta_{3,i}, \xi_{3,i})|}{|u(\eta_{3,i}, \xi_{3,i})|} \times 100 \tag{14.55}$$

Figure 14.6 depicts the approximated, and already interpolated function $u'(\eta_{3,i}, \xi_{3,i})$, using 4 and 16 points for generating the linear system. The interpolation was done with cubic splines. It is worth mentioning that, in the case of 16 points, regardless of the number of repetitions, i.e. changing the chosen points in MC, no effect was made on the minimal, average error. Such was not the case while using 4 points, and a wide dispersion of the percentage of error was found, depicted in Figure 14.7. Table 14.4 summarizes the results for MC. In this work it is used "SD" to stand for "Standard Deviation" instead of σ, since, as it can be seen it has been used to represent the eigenvalues.

14.3.7   *Halton results, empirical rescaling*

As it was mentioned before, Halton sequences are LDSs. Therefore, in order to obtain a more complete vision on the errors for the quadrature, several Halton LDSs were generated, and within

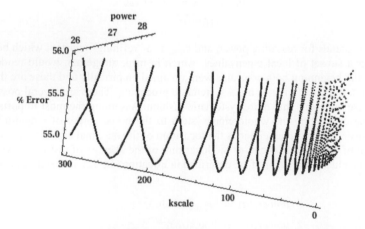

Figure 14.4   Extensive numerical search for minimal error in MDs, varying both $power_r$ and k_{scale}, ultimately leading to the formulas for systematic rescaling.

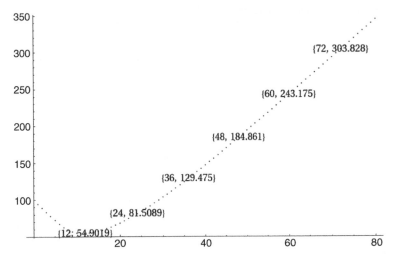

Figure 14.5   Finding the adjusting parameter k_{scale} empirically, which would minimize the percentage of error. The left number between braces is the sought for k_{scale}, while the number at the right corresponds to the percentage of error, which for the found $k_{scale} = 12$ it is $\approx 55\%$. In this case it is shown for MC, but similar explorations were done for Halton and MD LDSs.

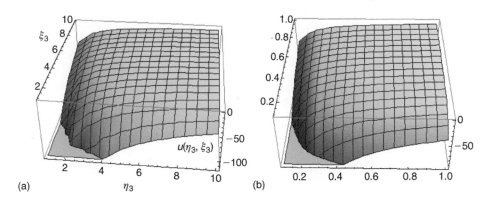

Figure 14.6   Approximated $u'(\eta_{3,i}, \xi_{3,i})$ (a) using 4 points in MC (b) using 16 points in MC.

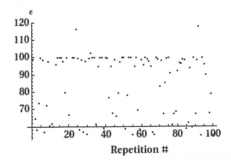

Figure 14.7   Collection of errors while approximating $u'(\eta_{3,i}, \xi_{3,i})$ using 4 points of MC.

Table 14.4   Summary of results from MC.

No. of points	No. of repetitions	$\langle \epsilon \rangle$	SD_ϵ	Max.%ϵ	Min.%ϵ
4	100	84.49	18.18	117.92	54.82
16	N.A.	54.90	0.00	54.90	54.90

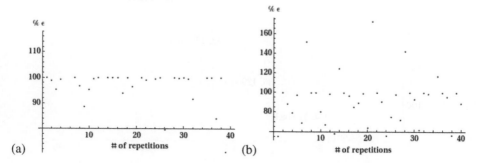

Figure 14.8   Collection of errors while approximating $u'(\eta_{3,i}, \xi_{3,i})$ (a) using 4 points of Halton LDSs (b) using 16 points of Halton LDSs.

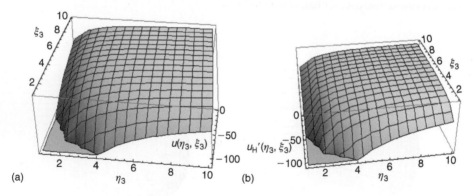

Figure 14.9   Approximated $u'(\eta_{3,i}, \xi_{3,i})$ (a) using 4 points of Halton LDSs (b) using 16 points of Halton LDSs.

each LDS a shift of ten units in their index was performed, in order to see the error of using different parts of the same LDS, and the error across all the Halton LDSs. Five LDSs were generated, each with seven thousand members. A total of 8 "repetitions" (actually, 8 shifts) was performed for each of those LDSs. So a total of 40 different trials was made for the total average and standard deviation of errors. Figure 14.8 shows the collection of errors for using 4 and 16 points, for Halton LDSs, while Figure 14.9 shows the approximating $u'(\eta_{3,i}, \xi_{3,i})$, with an error of $\approx 55\%$ when using 4 points, while there is an error of $\approx 55\%$ using 16 points. For clarity and reasons of scaling, in Figure 14.8(a) it is not shown the greatest errors.

Table 14.5 summarizes the results for Halton LDSs. Notice that augmenting the number of points in the quadrature did not substantially reduce the error, but it did reduce the standard deviation and the maximal error.

14.3.8   *MDs results, empirical rescaling*

The LDSs based on MDs were created out of the first 43 million prime numbers. From every million primes, it was collected the mds as described above (notice the usage of lower case "md"

Table 14.5   Summary of results from Halton LDSs.

No. of points	No. of repetitions	$\langle\epsilon\rangle$	SD_ϵ	Max.%ϵ	Min.%ϵ
4	40	113.67	66.64	375.15	55.37
16	40	92.01	25.22	171.81	54.96

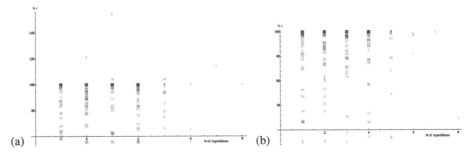

Figure 14.10   Collection of errors while approximating $u'(\eta_{3,i}, \xi_{3,i})$ (a) using 4 points of MD LDSs (b) using 16 points of MD LDSs.

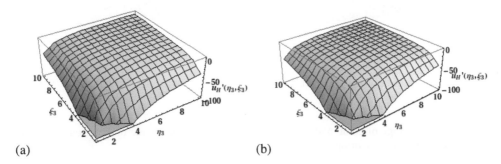

Figure 14.11   Approximated $u'(\eta_{3,i}, \xi_{3,i})$ (a) using 4 points of MD LDSs (b) using 16 points of MD LDSs.

for indicating the individual members of an MD "like this, in upper case", which stands for a whole LDS). Naturally, every set of one million primes would have a different amount of mds. Out of these different MDs a subset was taken, of only those mds smaller than 32, which helped minimize the error. Within each MD, it was performed a shift of one thousand units, in order to see the errors within each MD, and then accross all of the MDs. Figure 14.10 shows the collection of errors for 4 points and 16 points for MD LDSs, while Figure 14.11 shows the approximating $u'(\eta_{3,i}, \xi_{3,i})$ with an error of $\approx 55\%$ in using 4 points, and $\approx 55\%$ in using 16 points. Notice that there are two types of clustering: the first concerns the amount of repetitions per MD subset, but this is because of the diverse amounts of primes and of *mds* < 32. The second clustering is much more telling; Figure 14.10(a) shows that there are many repetitions with errors above 100%, while Figure 14.10(b) clearly shows that even with a clustering around 100% of error, none is above this cipher, and certainly many more repetitions are below that boundary.

Table 14.6 summarizes the results for MD LDSs. In this table, $\langle\langle\epsilon\rangle_{MD_i}\rangle$ stands for the average over all MD_i, of the average of errors, this last average is over those errors coming from the different repetitions (shiftings) within the same MD_i; $SD\langle\epsilon\rangle_{MD_i}$ is the Standard Deviation over the averages for all repetitions pertaining to the same MD_i; $\langle SD_i\rangle$ is the average over the Standard Deviations belonging to all MD_i (each one, in turn, composed of several shiftings, or "repetitions"); SD of $\langle SD_i\rangle$ is the Standard Deviation for all those belonging to all MD_i; Max.%ϵ indicates the

Table 14.6   Summary of results from MD LDSs.

No. of points	$\langle\langle\epsilon\rangle_{MD_i}\rangle$	$SD\langle\epsilon\rangle_{MD_i}$	$\langle SD_i\rangle$	SD of $\langle SD_i\rangle$	Max.%ϵ	Min.%ϵ
4	90.67	6.45	13.12	7.18	154.56	54.85
16	92.89	5.71	8.80	6.65	100.00	54.81

Table 14.7   Summary of results from MC, systematic rescaling.

No. of points	No. of repetitions	$\langle\epsilon\rangle$	SD_ϵ	Max.%ϵ	Min.%ϵ
4	100	84.27	2.98	84.88	69.75
16	100	84.88	3.95×10^{-14}	84.88	84.88

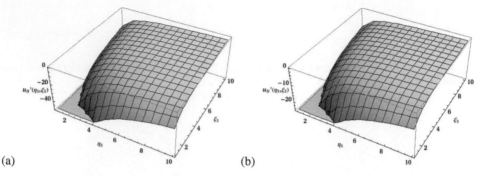

(a)   (b)

Figure 14.12   Approximated $u'(\eta_{3,i}, \xi_{3,i})$ (a) using 4 points in MC (b) using 16 points in MC, systematic rescaling.

maximum percentage of error, and Min.%ϵ is the minimum percentage of error, out of all MD_i and every shifting or repetition.

14.3.9   *MC results, systematic rescaling*

Using Equations (14.53) and (14.54), the found results are illustrated in Table 14.7, and the best interpolated function can be seen in Figure 14.12. Notice that in this case, an increase in the number of points considered in the matrix of eigenvalues, does not imply an improvement in the accuracy. It would seem the contrary, actually.

14.3.10   *Halton results, systematic rescaling*

The results for Halton LDSs are shown both in Figure 14.13 and in Table 14.8, as concerns of using a systematic rescaling for eigenvalues.

14.3.11   *MDs results, systematic rescaling*

The results of the systematic rescaling of eigenvalues, for MDs, are shown both in Figure 14.14 and Table 14.9. For these numerical experiments, a non-repeating set of mds was obtained, in lieu of those *mds* < 32.

14.3.12   *Accuracy goals*

There is not enough knowledge to predict with rigor how many points are needed from each method in order to achieve a given accuracy. However, if one would like to estimate how many

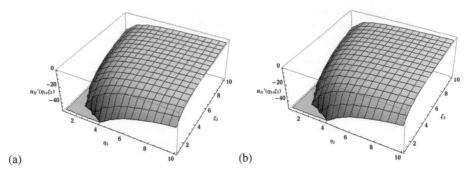

(a)   (b)

Figure 14.13   Approximated $u'(\eta_{3,i}, \xi_{3,i})$ (a) using 4 points of Halton LDSs (b) using 16 points of Halton LDSs, systematic rescaling.

Table 14.8   Summary of results from Halton LDSs, systematic rescaling.

No. of points	No. of repetitions	$\langle\epsilon\rangle$	SD_ϵ	Max.%ϵ	Min.%ϵ
4	355	84.02	3.49	84.88	69.75
16	355	72.14	5.52	84.88	69.75

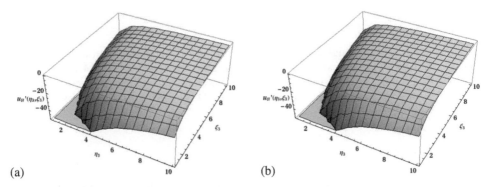

(a)   (b)

Figure 14.14   Approximated $u'(\eta_{3,i}, \xi_{3,i})$ (a) using 4 points of MD LDSs (b) using 16 points of MD LDSs, systematic rescaling.

Table 14.9   Summary of results from MD LDSs, systematic rescaling.

No. of points	$\langle\langle\epsilon\rangle_{MD_i}\rangle$	$SD\langle\epsilon\rangle_{MD_i}$	$\langle SD_i\rangle$	SD of $\langle SD_i\rangle$	Max.%ϵ	Min.%ϵ
4	75.92	2.01	7.96	0.63	84.88	69.75
16	69.75	$<10^{-13}$	$<10^{-13}$	$<10^{-13}$	69.75	69.75

points from each method would be needed to achieve an accuracy of say 1%, and assuming a linear behavior, then it can be chosen any of the measures of dispersion, and attempt to extrapolate up to the desired accuracy. It can be seen that this method would render meaningless results for some measures, since in some the slope is positive, rendering at the end a negative result. For the sake of comparison, it was taken the average of errors. Table 14.10 summarize the possible number of points needed for achieving an accuracy of 1%, taking for the extrapolation different dispersion measures.

Notice that even though MDs would need more points (based on the average error), the SD is much lower than using Halton LDSs, as can be seen from Tables 14.8 and 14.9.

Table 14.10　Estimated number of points needed to achieve ≈1% of accuracy for MC, Halton and MD LDSs.

MC	Halton	MDs
N.A.	88	150

Table 14.11　Estimated rate of convergence for MC, Halton and MD LDSs.

MC	Halton	MDs
1.01	0.86	0.92

14.3.13　*Rate of convergence*

The rate of convergence is estimated taking as a basis the percentage of error, as the number of used points change from 4 to 16, from the results of the aforementioned numerical experiments (systematic rescaling), and Equation (14.56) (Schatzman 2002), where it is said that a sequence convergence linearly to L, if there exists a number $\mu \in (0, 1)$ such that

$$\lim_{k \to \infty} \frac{|x_{k+1} - L|}{|x_k - L|} = \mu \tag{14.56}$$

Admittedly there are not enough data to anywhere claim that $k \to \infty$. However, with the available data, and taking $L = 1\%$, the estimated rate of convergence is shown in Table 14.11.

Notice that Halton and MD LDSs convergence linearly, and MC convergences sublinearly, since $\mu \gtrsim 1$. It is not idle to repeat that even if Halton LDSs do seem to converge faster, MDs do so with a SD much smaller.

14.4　CONCLUSIONS

The method which has shown the least standard deviation accross repetitions, while augmenting the number of points, so far seems to be the MD; the minimal error is comparable to MC and Halton LDSs, and both Halton and MD LDSs preserved consistency in tendency towards convergence, while augmenting the number of points to form the matrix of eigenvalues, perhaps more marked in MDs than in Halton LDSs. It remains to be studied as well the preservation of certain properties of the numerical solution such as monotonicity and conservativity.

REFERENCES

Caflisch, R.E. (1998) Monte Carlo and quasi-Monte Carlo methods. *Acta Numerica*, 7, 1–49.

Crandall, R. & Pomerance, C. (2005) *Prime numbers: A computational perspective*. New York, Springer.

Frank, L. (1973) *Matematika*. Praha, Nakladatelství technické literatury.

Halton, J.H. (1960) On the efficiency of certain quasi-random sequences of points in evaluating multi-dimensional integrals. *Numerische Mathematik*, 2 (1), 84–90.

Hunt, B. & Isaacs L.T. (1981) Integral equation formulation for ground-water flow. *ASCE*, 107 (10), 1197–1209.

Jiménez-Ángeles, F. & Odriozola, G. & Lozada-Cassou, M. (2006) Electrolyte distribution around two like-charged rods: Their effective attractive interaction and angular dependent charge reversal. *Journal of Chemical Physics*, 124 (13), 1349021–13490218.

Kuipers, L. & Niederreiter, H. (1974) *Uniform Distribution of Sequences*. New York, Wiley Interscience.

LAPACK. Available from: http://www.netlib.org/lapack/. [Accesed in 2010].

Lepage, D.P. (1978) A New Algorithm for Adaptive Multidimensional Integration. *Journal of Computational Physics*, 27 (2), 192–203.

Morokoff, W.J. & Caflisch, R.E. (1994) Quasi-Random Sequences and Their Discrepancies. *SIAM Journal of Scientific Computing*, 15 (6), 1251–1279.

Morokoff, W.J. & Caflisch, R.E. (1995) Quasi-Monte Carlo Integration. *Journal of Computational Physics*, 122 (2), 218–230.

Niederreiter, H. (1978) Quasi-Monte Carlo methods and pseudo-random numbers. *Bulletin of American Mathematical Society*, 84 (6), 957–1041.

Odlyzko, A. & Rubinstein, M. & Wolf, M. (1999) Jumping champions. *Experimental Mathematics*, 8 (2), 107–118.

Ortiz-Tapia, A. (2008a) Some patterns in primes and their possible applications as quasi-Monte Carlo methods in multivariable integration. In: Suárez-Arriaga, M.C. et al. (eds.) *Numerical Modeling of Coupled Phenomena in Science and Engineering. Practical Uses and Examples.* Amsterdam, Taylor & Francis Balkema. pp. 55–69.

Ortiz-Tapia, A. (2008b) On the functional convergence of a probability density function from some patterns in primes indices, and some perspectives of usage as a quasi-Monte Carlo method for multivariate integration and Monte Carlo simulation of chemical potentials common in the oil industry. In: Suárez-Arriaga, M.C. et al. (eds.) *Numerical Modeling of Coupled Phenomena in Science and Engineering. Practical Uses and Examples.* Amsterdam, Taylor & Francis Balkema. pp. 243–252.

Press, W.H., Teukolsky, S.A., Vetterling, W.T. & Flannery, B.P. (1997) *NUMERICAL RECIPES in Fortran 77he Art of Scientific Computing (Vol. 1 of Fortran Numerical Recipes).* Cambridge, Cambridge University Press.

Román-Mejía, J.A. (2009) Personal communication.

Schatzman, M. (2002) *Numerical Analysis: A Mathematical Introduction.* Oxford, Clarendon Press.

Wikipedia de web Encyclopaedia. Available from: *http://en.wikipedia.org/wiki/Monte_Carlo_method*. [Accessed in 2010].

Wikipedia de web Encyclopaedia. Available from: *http://en.wikipedia.org/wiki/Halton_sequence*. [Accessed in 2010].

Section 4:
Waves

CHAPTER 15

On the physical meaning of slow shear waves within the viscosity-extended Biot framework

T.M. Müller & P.N. Sahay

15.1 INTRODUCTION

Studies of wave attenuation and dispersion in heterogeneous porous materials caused by fluid-solid interactions are often based on Biot's equations of poroelasticity (Biot 1956a,b, 1962). This theory has been further developed by Spanos & de la Cruz (1985), Sahay et al. (2001) and Spanos (2002). These authors used the volume-averaging technique to upscale the pore-scale equations. More recently, Sahay (2008) analyzed the consequences of incorporating the fluid strain rate tensor into Biot's constitutive relation as is suggested within the volume-averaging framework. This guarantees that by tending porosity to unity, the constitutive relation render a Newtonian fluid. We refer to this framework as the viscosity-extended Biot framework. A particularity of this extended framework is the prediction of an additional shear mode with non-vanishing velocity, which essentially describes the out-of-phase shear motions of the two phases. In the classical Biot theory, it is a non-propagating process as a result of the neglect of viscous stresses in the fluid phase. In analogy to the second or slow compressional wave of the Biot theory we refer to this second shear wave as slow shear wave (see also Mavko et al. 2009).

In heterogeneous media each wave can be converted into another through conversion scattering at heterogeneities. In heterogeneous elastic continua the conversion scattering from compressional to shear waves is of importance for the interpretation of seismic signals (Sato & Fehler 1998). In heterogeneous poroelastic continua the conversion scattering mechanism from fast compressional into the slow compressional wave may occur. The latter process has been thoroughly analyzed and it is well-known that it results into compressional wave attenuation for frequencies much smaller than critical Biot frequency Ω_b, i.e., if the slow Biot wave is a diffusion wave (Johnson 2001, Pride et al. 2004, Müller & Gurevich 2005). Therefore, the question arises of how the slow shear wave mode interacts with propagating compressional and shear waves in the presence of inhomogeneities.

In order to quantify the conversion scattering we analyze the coherent wave field in randomly heterogeneous media. A powerful tool for this purpose is the method of statistical smoothing (Karal & Keller 1964, Rytov et al. 1989). This method has been previously employed to analyze scattering in wave-energy conservative systems by analyzing the amount of wave energy transferred from the coherent wave into the incoherent wave field. For example, Rytov et al. (1989) obtained an expression for the effective wave number for the scalar scattering problem. These results were generalized for the case of randomly heterogeneous isotropic and anisotropic elastic media (Turner 1999, Turner & Anugonda 2001, Gold et al. 2000). More recently, Müller and Gurevich (2005a) showed that the method of statistical smoothing can be also used to describe conversion scattering in systems where the wave field energy is not conserved, i.e., where a portion of the coherent wave field energy is transfered to a dissipative wave mode. In particular, they derived an effective wave number accounting for the conversion scattering into a diffusion wave mode and showed that this results into significant attenuation and dispersion.

Müller and Sahay (2009, 2011a,b) use the method of statistical smoothing in order the compute an effective, dynamic equivalent wave number of a fast compressional wave accounting exclusively for the conversion scattering process into the slow shear wave. The slow shear wave can be regarded as a generalized diffusion wave as it obeys the standard diffusion equation or a diffusion equation with a damping term, depending on the frequency regime (Sahay 2008). In this paper we focus on the physical interpretation of the slow shear wave process. In Sections 15.2 and 15.3, we provide an outline of the theoretical framework of the slow shear wave conversion scattering mechanism. Many technical details are left out for the sake of clarity and the interested reader is referred to Müller and Sahay (2011a). In Section 15.4 we give a physical interpretation of the slow shear wave conversion process. It is argued that the the conversion scattering process is related to the amount of dissipation within the viscous boundary layer. The role of the conversion scattering process in the transition from the viscosity- to inertia-dominated regime is also discussed.

15.2 REVIEW OF THE VISCOSITY-EXTENDED BIOT FRAMEWORK

15.2.1 *Constitutive relations, complex phase velocities, and characteristic frequencies*

The viscosity-extended Biot constitutive relations are developed in Sahay (2008) and reproduced in the following. In an homogeneous and isotropic poroelastic medium the solid stress tensor and the fluid-stress tensor, upon incorporating for the viscous stress term, are given by

$$\tau^s_{jk} = K_0 e^s_{ll}\delta_{jk} + 2\mu_0 \breve{e}^s_{jk} - (\alpha - \eta)p^f\delta_{jk} \tag{15.1}$$

$$\tau^f_{jk} = -\eta p^f\delta_{jk} - \eta\frac{\xi_f}{K_f}\partial_t p^f\delta_{jk} + 2\mu_f(\eta\partial_t\breve{e}^f_{jk} + (\alpha_\mu - \eta)\partial_t\breve{e}^s_{jk}), \tag{15.2}$$

where p^f is the fluid pressure

$$p^f = -\alpha M e^s_{ll} - M\eta(e^f_{ll} - e^s_{ll}). \tag{15.3}$$

In Equations (15.1)–(15.3) e^s is the solid-frame strain tensor $e^s_{jk} = 1/2(u^s_{j,k} + u^s_{k,j})$ with u^s_i being the solid displacement vector. The trace-free part of the strain tensor is denoted as $\breve{e}^s$, i.e., $\breve{e}^s_{jk} = e^s_{jk} - 1/3 e^s_{ll}\delta_{jk}$, where δ_{jk} stands for a Kronecker delta and summation over repeated indices is assumed. In Equation (15.2) $\partial_t\breve{e}^s_{jk}$ and $\partial_t\breve{e}^f_{jk}$ denote the deviatoric part of the strain rate tensors of the solid and fluid phases, respectively.

The constitutive relations involve the following material constants. The Biot-Willis coefficient, α, and the fluid storage coefficient, M, are linked to the bulk moduli of the constituent solid, K_s, the drained frame, K_0, and the constituent fluid, K_f, as

$$\alpha = 1 - \frac{K_0}{K_s} \tag{15.4}$$

$$M = \left(\frac{\eta}{K_f} + \frac{\alpha - \eta}{K_s}\right)^{-1}, \tag{15.5}$$

where η denotes porosity. The shear modulus of drained frame is denoted as μ_0. The drained-frame and saturated compressional wave modulus are defined as $H_0 = K_0 + 4/3\mu_0$ and $H = P_0 + \alpha^2 M$, respectively. The 2nd and 3rd term on the right hand side of Equation (15.2) is the incorporated fluid viscous stress tensor term. This modification is based upon the assumption of Newtonian rheology for the pore fluid. It introduces the pore fluid bulk (ξ_f) and shear (μ_f) viscosities and the parameter α_μ into the constitutive relation. The parameter α_μ links drained frame (μ_0) and mineral (μ_s) shear moduli as

$$\alpha_\mu = 1 - \frac{\mu_0}{\mu_s}. \tag{15.6}$$

Neglecting this viscous stress tensor term results exactly into Biot's constitutive relations (Biot 1962):

$$\tau^s_{jk} = K_0 e^s_{ll}\delta_{jk} + 2\mu_0 \breve{e}^s_{jk} - (\alpha - \eta)p^f\delta_{jk} \tag{15.7}$$

$$\tau^f_{jk} = -\eta p^f\delta_{jk} = [\alpha M\eta e^s_{ll} + M\eta^2(e^f_{ll} - e^s_{ll})]\delta_{jk}. \tag{15.8}$$

In the limiting case of a pure fluid, i.e., porosity is 1, the constitutive relation of an isotropic Newtonian fluid is obtained, i.e.

$$\tau_{jk}^{f} = -p^{f}\delta_{jk} - \frac{\xi_{f}}{K_{f}}\partial_{t}p^{f}\delta_{jk} + 2\mu_{f}\partial_{t}\breve{e}_{jk}^{f}, \tag{15.9}$$

as detailed in Landau and Lifshitz (1987).

In the frequency domain, the equations of motion for the viscosity-corrected Biot framework can be represented by a 6×6 matrix differential operator, $\mathbf{L}(\partial, \omega)$ (explicit expressions can be found in Müller and Sahay 2011a). Its corresponding Green's function entails a total of four wave numbers indicating the existence of four wave processes. As shown in Sahay (2008) one pair, labeled k_1 and k_2, can be understood as wave numbers of the fast and the slow compressional wave, respectively. Likewise, the second pair, k_3 and k_4, can be understood as wave numbers of the fast and the slow shear wave, respectively. The wave numbers can be represented as $k_n = \omega/c_n$ involving circular frequency ω and the complex and frequency-dependent phase velocities

$$c_1, c_2 = \sqrt{\frac{\text{Tr}_{\alpha}}{2}\left(1 \pm \sqrt{1 - 4\frac{\Delta_{\alpha}}{\text{Tr}_{\alpha}^2}}\right)} \tag{15.10}$$

$$c_3, c_4 = \sqrt{\frac{\text{Tr}_{\beta}}{2}\left(1 \pm \sqrt{1 - 4\frac{\Delta_{\beta}}{\text{Tr}_{\beta}^2}}\right)}, \tag{15.11}$$

where Δ and Tr stands for the determinant and trace of the matrix appearing as its subindex. Simplified expressions of these matrices are given in Appendix 15.A.

Apart from the critical Biot frequency

$$\Omega_{b} = \frac{\eta\mu_{f}}{\kappa\rho_{f}} \tag{15.12}$$

(with permeability κ), the equations of motion entail additional relaxation frequencies associated with saturated-frame compressional and shear stress relaxation,

$$\Omega_{P} = \frac{H}{\eta M/K_{f}\xi_{f} + 4/3\mu_{f}} \tag{15.13}$$

$$\Omega_{S} = \frac{\mu_{0}}{\mu_{f}}, \tag{15.14}$$

as well as the compressional and shear pore fluid relaxation (in a deformable frame),

$$\Omega_{flP} = \frac{\eta M}{\eta M/K_{f}\xi_{f} + 4/3\mu_{f}} \tag{15.15}$$

$$\Omega_{flS} = \frac{\eta M}{\mu_{f}}, \tag{15.16}$$

respectively. For this viscosity-extended framework to be applicable, the frame relaxation frequency needs to lie above the peak frequency associated with Biot relaxation

$$\Omega_{i} = d_{f}\Omega_{b}, \tag{15.17}$$

(where $d_{f} = \eta\rho_{f}/\rho_{i}$ involves the modified reduced density $\rho_{i} = (S - m_{f})\eta\rho_{f}$, where S is the tortuosity) the frequency beyond which the fluid is free to move in the pores (apart from a thin boundary layer attached to the pore walls), i.e.,

$$\Omega_{S} \gg \Omega_{i}. \tag{15.18}$$

Moreover, the relation

$$\Omega_{P} \approx \Omega_{S} > \Omega_{flS} > \Omega_{flP} \gg \Omega_{b} > \Omega_{i} \tag{15.19}$$

will hold for a large class of poroelastic solids.

15.2.2 *Properties of the slow shear wave*

As we are particularly interested in the slow shear wave process, it is useful to examine basic properties of the slow shear wave already entailed in its complex wave number. For frequencies much lower than Biot's critical frequency Ω_b the slow shear wave mode obeys a diffusion equation with a damping term (Equation 27 in Sahay 2008) and, therefore, is a spatially extremely localized process. For frequencies larger than Ω_b this wave mode becomes a diffusion wave with the wave number

$$k_4 \approx \sqrt{\frac{i\omega S \rho_f}{\mu_f}}. \tag{15.20}$$

Expressions in Eqs. (15.11) and (15.20) for the slow shear wave number are plotted in Figure 15.1 using parameters typical for a consolidated rock. For frequencies higher than Ω_b (shown as vertical line) the exact expression in Eq. (15.11) (symbols) approaches the diffusion limit (curves labelled HF). Though the real part of k_4 scales linearly with frequency in the low-frequency band (like an ordinary propagative wave mode the corresponding asymptotic behavior is depicted by the curve labelled LF in Figure 15.1) the finite imaginary part of k_4 means that the slow shear wave is then spatially localized.

Ordinary diffusion waves are characterized through the equality of real and imaginary parts of their wave number (Mandelis 2000). As Figure 15.1 illustrates, there is a frequency band $\omega \leq \Omega_b$ in which the real and imaginary part of the slow shear wave are different. We refer to such a wave mode as generalized diffusion wave. The process of conversion scattering from a propagating wave mode into this generalized diffusion wave is analyzed in the following section.

15.3 CONVERSION SCATTERING IN RANDOMLY INHOMOGENEOUS MEDIA

15.3.1 *Effective wave number approach*

In order to gain further insight into the role of the slow shear wave process in heterogeneous porous media we analyze the conversion scattering mechanism from a fast compressional wave into the slow shear wave. Details of the analysis can be found in Müller and Sahay (2011a). Their approach is based on the framework of scattering in random media (e.g. Rytov et al. 1987). In the present context it is assumed that any material parameter X involved in the viscosity-extendend Biot framework can be represented as sum of a constant background value, $\bar{X}$, and a randomly fluctuating part, $\tilde{X}$, so that $X = \bar{X} + \tilde{X}$. Let the relative fluctuations $\epsilon_X = \tilde{X}/\bar{X}$ be vanishing after ensemble averaging, $\langle \epsilon_x \rangle = 0$, and their spatial correlation function be $B_{XX}(\delta r) = \langle \epsilon_X(r + \delta r)\epsilon_X(r) \rangle$ with the variance $\sigma_{XX}^2 = B_{XX}(0)$. The correlation function is assumed to approach zero with increasing argument.

In order to simplify, we take $\alpha_\mu = \alpha$, i.e. frame bulk and shear moduli are related to respective grain moduli in a likewise manner, and set the fluid bulk viscosity vanishing so that the effect of shear drag within the fluid can be examined exclusively. Further simplifications are introduced by restricting the analysis to the frequency band defined by

$$\Omega_i \leq \omega \leq \sqrt{\Omega_i \Omega_S}. \tag{15.21}$$

Note that the first inequality does not impose strong restrictions because of the strongly damped slow shear wave behavior for $\omega < \Omega_b$. The second inequality is merely introduced for clarity as it ensures that none of the fluid or solid relaxations mechanisms are operative (defined by the relaxation frequencies in Eqs. (15.13)–(15.16)). The background medium properties are assumed to be those obtained in the high-frequency, inertial regime $\Omega_i \ll \omega \leq \sqrt{\Omega_i \Omega_S}$. Then the slow compressional wave is a propagating wave mode and the fast compressional and shear waves propagate with constant velocities.

The random medium scattering theory suggests that analogous to the decomposition of the material parameters, the differential operators involved in the equation of motion can be also decomposed into constant background and fluctuating parts. Hence a scattering series can be

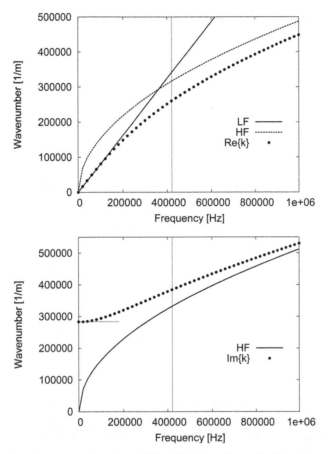

Figure 15.1   The slow shear wave dispersion relation. Shown are the real and imaginary parts of the slow shear wave number as a function of frequency (top and bottom, respectively). For frequencies higher than Ω_b (vertical solid line) the slow shear wave behaves like an ordinary diffusion wave with equal real and imaginary parts of the wave number. Conversely, for $\omega < \Omega_b$ its behavior is fundamentally different. The finite imaginary part of the wave number indicates the strong spatial localization of this wave mode.

formulated. Analytical methods for waves in random media only allow to compute the statistical moments of wave fields and therefore require ensemble averaging of the field equations (Frisch 1968, McCoy 1973, Rytov et al. 1989). Averaging the scattering series results into a matrix with the averaged Green tensors. This integral equation can be viewed as a matrix generalization of the so-called Dyson equation. It cannot be solved analytically unless an approximation is introduced. We apply the method of statistical smoothing (Karal & Keller 1964). This approximation is valid when the absolute value of the relative fluctuations is small, that is, $|\epsilon_X| < 1$. As a result an effective compressional wave number is obtained

$$k_1^* = k_1^\infty [1 + \Delta_1 \xi(\omega)], \tag{15.22}$$

where the frequency-dependent function $\xi(\omega)$ is given by

$$\xi(\omega) = 1 + k_4^2 \int_0^\infty r B(r) e^{ik_4 r} dr \tag{15.23}$$

containing an integral over the variance-normalized correlation function $B(r)$. The coefficient Δ_1 is given as

$$\Delta_1 = \Gamma_{HH} \sigma_{HH}^2 - \Gamma_{HC} \sigma_{HC}^2 + \Gamma_{CC} \sigma_{CC}^2 \tag{15.24}$$

entailing the variances of the random fluctuations of the moduli H and C. The dimensionless coefficients Γ_{XY} are given by

$$\Gamma_{HH} = \frac{(c_1^0)^4}{(c_1^\infty)^2 (c_3^0)^2} \tag{15.25}$$

$$\Gamma_{HC} = \frac{2 m_f \alpha}{S \eta} \frac{(c_1^0)^2 c_{fl}^2}{(c_1^\infty)^2 (c_3^0)^2} \tag{15.26}$$

$$\Gamma_{CC} = \frac{m_f \alpha^2}{S \eta^2} \frac{c_{fl}^4}{(c_1^\infty)^2 (c_3^0)^2}, \tag{15.27}$$

where the superscripts 0 and ∞ in the compressional and shear velocities $c_{1,3}$ denote Gassmann and inertial limit, respectively. c_{fl} is the sound velocity in the fluid in the presence of deformable frame. Expressions for these velocities are given in the Appendix. In the absence of random fluctuations $\Delta_1 = 0$ and the effective wave number reduces to the compressional wave number of the background medium, $k_1^\infty = \omega/c_1^\infty$. However, in the presence of random fluctuation medium parameters, the dynamic-equivalent wave number in Eq. (15.22) accounts for the conversion scattering process into the slow shear wave. Other mode conversions are neglected in this approximation. The analytic structure of k_1^* provides insight into the slow shear wave conversion process and is analyzed next.

15.3.2 Attenuation and dispersion due to conversion scattering in the slow shear wave

The imaginary part of k_1^* yields the attenuation coefficient $\gamma(\omega) = \Im\{k_1^*\}$. By inspection, γ is found to be positive resulting into an exponentially damped coherent wave. This demonstrates that the slow shear wave conversion scattering mechanism draws energy from the propagating compressional wave. The low and high frequency asymptotic scaling for attenuation are straightforward to obtain provided that the correlation function $B(r)$ tends rapidly to zero with increasing argument. Defining the reciprocal quality factor as $Q^{-1} = 2\Im\{k_1^*\}/\Re\{k_1^*\}$ we find

$$Q^{-1}(\omega \to 0) \propto \omega \tag{15.28}$$

and

$$Q^{-1}(\omega \to \infty) \propto \omega^{-1/2}. \tag{15.29}$$

Maximum attenuation is obtained when the resonance condition $k_4 \Lambda \approx 1$ is met, where Λ denotes a characteristic length scale of the inhomogeneities. Using Eq. (15.20) we obtain the cross-over frequency

$$\Omega_c = \frac{\mu_f}{S \rho_f \Lambda^2}. \tag{15.30}$$

Scaling laws in Eqs. (15.28) and (15.29) are typical for a relaxation type phenomenon in the context of wave attenuation in porous media (Müller & Rothert 2006). In the present case it can be understood as a shear stress relaxation process within the fluid phase. Figure 15.2 shows the reciprocal quality factor as a function of the normalized frequency f/f_c [where f_c denotes the critical Biot frequency] using the autocovariance model $B(r) = e^{-|r/\Lambda|}$ and $\Delta_1 = 0.06$. Variations in the correlation length Λ are indicated by the ratio $R = \Lambda/\sqrt{\kappa}$.

Attenuation is accompanied by phase velocity dispersion. The real part of k_1^* is related to the phase velocity of the compressional wave through $c_1(\omega) = \omega/\Re\{k_1^*\}$. The structure of Eq. (15.22) shows that the phase velocity of the coherent wave in the dynamic-equivalent medium is smaller than in the background medium. Indeed, from Eq. (15.23) we obtain $\Re\{\xi(0)\} = \xi_0$ with $0 < \xi_0 \leq 1$ so that

$$c_1(\omega = 0) = c_1^\infty [1 - \xi_0 \Delta_1] \tag{15.31}$$

and $\Re\{\xi(\infty)\} = 0$ yielding $c_1(\infty) = c_1^\infty$.

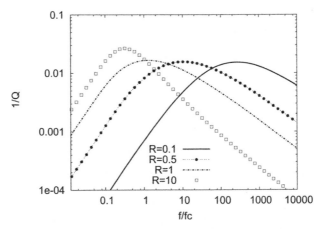

Figure 15.2   Attenuation (in terms of reciprocal quality factor) versus normalized frequency for varying heterogeneity scale. The normalization frequency is $fc = \Omega_b/2\pi$.

As the theory of conversion scattering is restricted to the frequency range in Eq. (15.21) a direct comparison between Ω_b and Ω_c shows that the characteristic scale of the inhomogeneities is confined to

$$\kappa \geq \Lambda^2, \tag{15.32}$$

i.e., on the order of, or smaller than the square root of the flow permeability. Approximating the pore space of the porous material by an ensemble of thin, parallel tubes, where the permeability is proportional to the squared radius of the tubes (R), the latter relation results into $R \geq \Lambda$. This means that the conversion scattering mechanism under consideration is only active in the presence of heterogeneities on the pore-scale.

15.4   PHYSICAL INTERPRETATION OF THE SLOW SHEAR WAVE CONVERSION SCATTERING PROCESS

15.4.1   *Slow shear conversion mechanism as a proxy for attenuation due to vorticity diffusion within the viscous boundary layer*

Let us compare the present conversion scattering into the slow shear wave with the conversion scattering into Biot's slow compressional wave in the presence of mesoscopic heterogeneities. Mesoscopic refers to a heterogeneity length scale which is much larger than any pore scale descriptor, such as the grain diameter. In fact, Müller and Gurevich (2005a) used also the method of statistical smoothing to derive a dynamic-equivalent wave number for the conversion scattering into the slow compressional wave. It is therefore not surprising that the expressions for the effective wave numbers have similar analytic structure (Equation (15.22) here, and Eq. (49) in Müller & Gurevich 2005a). However, the frequency band where these conversion scattering occurs and the underlying physics are in both cases substantially different.

The so-called mechanism of wave-induced flow at mesoscopic heterogeneities is operative if

$$\lambda_1 \gg \lambda_2 <> a \gg \Lambda \tag{15.33}$$

where λ_1 and λ_2 are the wavelengths of the fast and slow compressional waves, respectively, and a is the characteristic mesoscopic length scale. The symbol $<>$ means less or greater than. Conversely, for conversion scattering into the slow shear wave we have

$$\lambda_1 <> a \gg \lambda_4 <> \Lambda. \tag{15.34}$$

As the characteristic frequencies of both mechanisms scale with the square of the reciprocal heterogeneity size, these mechanisms occur in separated frequency bands. Qualitatively, Equation (15.33) means that centimeter-scale heterogeneities will affect waves with several meter or larger

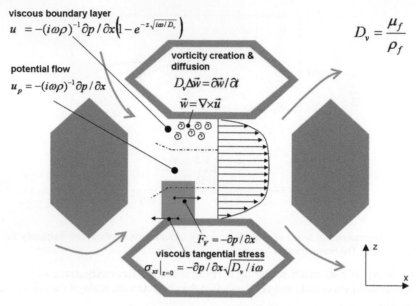

Figure 15.3   Viscous boundary layer and potential flow region of fluid flow in the pore space of a porous rock. The equations refer to the flow of a Newtonian fluid in straight cylindrical channel as outlined in Lighthill (1978).

wavelengths. Such a scenario is most important for seismic waves and mesoscopic heterogeneities have been observed and characterized in laboratory experiments (Toms et al. 2009). Accordingly, Equation (15.34) means that micro-meter heterogeneities will affect waves with millimeter-to-centimeter-scale wavelengths such as ultrasound.

As the physical mechanism is concerned, the conversion scattering into slow compressional waves can be interpreted as fluid pressure equilibration process governed by diffusion of fluid pressure wherein the fluid pressure gradients are induced during compressional and dilatational cycles of the wave. Müller and Rothert (2006) showed that the effect of this mesocsopic flow on propagating waves can be revealed by analyzing fluid pressure diffusion with pressure diffusivity

$$D_p = \frac{\kappa N}{\mu_f} \tag{15.35}$$

in conjunction with Darcy's law of filtration through porous media (with $N = MH_0/H$). However, with increasing frequency this mechanism becomes negligibly small as slow compressional waves become propagating wave modes and hence cannot be related to a dissipation process any longer. On the other hand it is not plausible that the dissipation of elastic wave energy will vanish as frequency increases and the relation in Eq. (15.33) is violated. This apparent contradiction can be resolved if the slow shear wave conversion mechanism is considered. As much as the slow compressional wave looses its diffusion character when approaching Biot's critical frequency, the slow shear wave becomes a diffusion wave mode. At the same time, it is very well known that viscous boundary layers in the pore channels develop as wave frequency approaches Biot's critical frequency. It seems therefore logical to relate the mechanism of slow shear wave conversion scattering with the dissipation occurring within the viscous boundary layer.

Viscous boundary layers develop in order to guarantee the zero particle velocity condition at the pore walls (Schlichting 2000). Within the viscous boundary layer the velocity of the potential flow region is gradually reduced to zero. This is achieved by momentum transfer between adjacent fluid layers. This is schematically illustrated in Figure 15.3. As a consequence rotational components of the fluid velocity field emerge:

$$\mathbf{w} = \nabla \times \mathbf{u} \neq 0 \tag{15.36}$$

while $\mathbf{w} = 0$ holds outside the viscous boundary layer. This so-called vorticity is generated at the pore walls and diffuses away obeying a diffusion equation

$$D_v \Delta \mathbf{w} = \frac{\partial \mathbf{w}}{\partial t} \tag{15.37}$$

with vorticity diffusivity

$$D_v = \frac{\mu_f}{\rho_f}. \tag{15.38}$$

We now argue that conversion scattering into the slow shear wave resembles a process of shear stress equilibration wherein the diffusing quantity is the vorticity generated at the fluid-solid interface.

The slow compressional and slow shear conversion mechanism can be interpreted as dissipation mechanisms a propagating wave mode experiences in heterogeneous fluid-saturated porous media. Both wave attenuation mechanisms are examples where dissipation is controlled through a diffusion process. The difference is the physical quantity that diffuses. For the slow compressional wave conversion mechanism it is fluid pressure diffusion, whereas for the slow shear wave mechanism it is vorticity diffusion. This difference manifests in the reciprocal manner the dissipation of wave field energy is controlled by the shear viscosity of the fluid (Equations 15.35 and 15.38).

Though, at first glance, both attenuation mechanisms seem to be unrelated, it is not difficult to see that they are complementary in nature. To this end it we consider the descriptions of the relative fluid-solid displacement at macro- and micro-scale. Darcy's equation defines a macroscopic relation between the fluid pressure gradient and the relative fluid-solid velocity (filtration velocity). Its pore-scale representation can be thought of as a sum of pressure drag and viscous shear drag contributions. If Δp denotes the pressure drop over the length L of a pore channel then

$$-\frac{\Delta p}{L} = \frac{1}{AL} \left[\int_{A_{fs}} p n_x \, ds - \int_{A_{fs}} \tau \, ds \right], \tag{15.39}$$

where n_x is a unit vector component in flow direction. The shear stress in flow direction is denoted as τ and is proportional to the shear viscosity of the fluid. The area over which needs to be integrated is the fluid-solid surface A_{fs}. A denotes the surface area at which the fluid enters the pore channel. Further, the viscous skin depth is defined as

$$\delta = \sqrt{\frac{2D_v}{\omega}}. \tag{15.40}$$

If the viscous skin depth is smaller than the typical pore channel half width viscous boundary layers will develop and the corresponding pore-scale flow equations need to be used (i.e., the right hand side of Equation 15.39 applies). Dissipation then occurs exclusively within the viscous boundary layer domain and is controlled by the shear stress. On the other hand, if the viscous skin depth exceeds the typical pore channel widths then the viscous boundary layer cannot develop and the upscaled transport equation is given by Darcy's equation and is controlled by pressure diffusion. So, depending on wave frequency either the pore-scale fluid flow equations or their upscaled counterparts should be employed to calculate the amount of wave energy dissipation.

In summary, the above considerations suggest that the description of wave energy dissipation involves a transition from pressure diffusion to vorticity diffusion while frequency increases. We model pressure diffusion controlled dissipation as conversion into the slow compressional wave. Analogously, we model vorticity diffusion controlled dissipation as conversion into the slow shear wave. This complementary nature of compressional wave attenuation is also illustrated in Figure 15.4. The characteristic frequency is Biot's critical frequency.

15.4.2   *The slow shear wave conversion mechanism versus the dynamic permeability concept*

It is particularly interesting to understand the role of the slow shear wave conversion process in the transition from viscosity-dominated to inertial regime in porous media. In fact, there

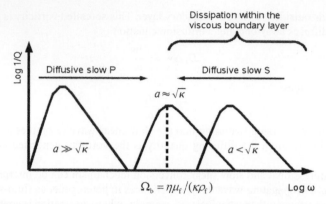

Figure 15.4   Role of the slow shear wave conversion scatterng mechanism for attenuation of fast compressional waves. At low frequencies [compared to the Biot critical frequency] fast compressional waves are attenuated through conversion scattering into the slow compressional wave (so-called "mesoscopic wave-induced flow" mechanism, Müller et al. 2010). In this frequency band the slow shear wave conversion scattering mechanism is not operative. At frequencies comparable or higher than Biot's crtitical frequency fast compressional wave attenuation is due to conversion into slow shear waves. This slow shear wave conversion scattering mechanism accounts for the presence of a viscous boundary layer.

are two processes involved in this transition regime. One is the wavelength-scale fluid pressure equilibration process between the peaks and troughs of a wave cycle (governed by fluid pressure diffusion as for the mesoscopic mechanism described above). Its characteristic frequency is very close to Biot's critical frequency (Pride 2005). This relaxation process is responsible for (slight) attenuation and dispersion of the fast compressional and shear waves and is captured by Biot's theory rendering the wave numbers k_1 and k_3 complex quantities in a macroscopically homogeneous medium. However, as Biot himself pointed out, if the viscous skin depth is smaller than the pore channel thickness the effect of the viscous boundary layer on the fluid flow behavior cannot be neglected. He included this boundary layer effect through the introduction of a complex viscosity resulting from the analysis of oscillatory fluid flow in elastically rigid cylinders (and slits) (Biot 1956). Later, Johnson et al. (1987) modeled this transition behavior in a more general way by developing a theory for the dynamic permeability (and tortuosity).

The present results may provide an alternative recipe to model the transition regime behavior. In fact, the constraint on the characteristic heterogeneity scale together with the proportionality to the fluid kinematic shear viscosity, $\nu_f = \mu_f/\rho_f$, of both Ω_c and Ω_b (Equations 15.12 and 15.30) indicates that the mechanism of conversion scattering into slow shear mode describes the process of shear stress relaxation in the viscous boundary layer. This means that in the viscosity-extended Biot framework it may be possible to describe the transition from the viscosity- to intertia-dominated regime without invoking the concepts of viscodynamic operator, or dynamic permeability, or memory variables (Biot 1956, Bedford & Stern 1984, Norris 1986, Gist 1994, Johnson et al. 1987, Carcione 2007). For example, Müller and Sahay (2011b) show that the attenuation of compressional and shear waves associated with the transition from the viscosity- to intertia-dominated regime can be modeled with the slow shear wave conversion scattering mechanism.

15.5   CONCLUSIONS

We draw the following conclusions.

1. The mechanism of conversion scattering into the slow shear wave is a process that draws energy from the propagating compressional wave and thus introduces attenuation and dispersion.

2. This attenuation mechanism is operative for frequencies comparable or larger than the critical Biot frequency, i.e., for many porous material including rocks at ultrasonic frequencies. Due to the strong damping of the slow shear wave at frequencies less than the Biot critical frequency, this conversion scattering process is negligible.
3. Typical scales of heterogeneities at which the conversion scattering takes place are on the order of pore-scale features such as the diameter of pore throats.
4. The conversion scattering process and the resulting compressional wave attenuation show similarities with the so-called dynamic permeability effect earlier introduced to Biot's theory of dynamic poroelasticity in order to account for the presence of viscous boundary layers in the transition from the viscosity- to inertia-dominated regime.

Based on the above arguments, we think that the incorporation of the strain rate tensor in the constitutive relation has a distinct physical meaning and should not be neglected a priori as has been suggested previously. Only in the low-frequency regime $\omega \ll \Omega_b$ of an infinite homogeneous poroelastic solid, this term can be neglected without incurring error. However, in heterogeneous poroelastic solids Ω_b can vary locally and the conversion into the slow shear wave should not be ignored.

The conversion scattering theory discussed here is based on a number of simplifying assumptions, such as weak scattering, neglect of bulk viscosity effects, and restriction to a certain frequency band. However, we believe that these assumptions will not alter the main conclusions drawn here. In this paper we have deliberately neglected all other wave field interactions, that is to say, the conversion scattering into other wave modes than the slow shear wave. In principle it is possible to include these interactions with the same formalism as presented here. For example, Müller et al. (2008) analyze the interplay of elastic scattering with the mechanism of wave-induced flow at mesoscopic heterogeneities (conversion scattering into Biot's slow compressional wave).

15.A APPENDIX

15.A.1 α *and* β *matrices*

The matrices $\boldsymbol{\alpha}$ and $\boldsymbol{\beta}$ are 2×2 matrices associated with compressional and shear waves, respectively, whose elements are dimensionally equal to velocity squared. They are

$$\boldsymbol{\alpha} = \boldsymbol{\Omega}^{-1}(\mathbf{C}_\alpha - i\omega \mathbf{N}_\alpha) \equiv \begin{pmatrix} \alpha^{mm} & \alpha^{mi} \\ \alpha^{im} & \alpha^{ii} \end{pmatrix}, \tag{15.A.1}$$

$$\boldsymbol{\beta} = \boldsymbol{\Omega}^{-1}(\mathbf{C}_\beta - i\omega \mathbf{N}_\beta) \equiv \begin{pmatrix} \beta^{mm} & \beta^{mi} \\ \beta^{im} & \beta^{ii} \end{pmatrix}. \tag{15.A.2}$$

$\boldsymbol{\Omega} = \boldsymbol{I} + i(\Omega_i/\omega)\mathbf{I}_0$, is the 2×2 diagonal matrix associated with the Biot relaxation frequency. Here $\mathbf{I}$ is the 2×2 identity matrix, $\mathbf{I}_0$ is the diagonal matrix $Diag(0, 1)$. The elements of the second-order $\mathbf{C}_\alpha$ and $\mathbf{C}_\beta$ matrices have dimensions of velocity squared, whereas the elements of the second-order $\mathbf{N}_\alpha$ and $\mathbf{N}_\beta$ matrices have dimensions of kinematic viscosity. These matrices are defined as

$$\mathbf{C}_\alpha = \rho^{-1}\left(K + \frac{4}{3}\mu\right), \tag{15.A.3}$$

$$\mathbf{N}_\alpha = \rho^{-1}\left(\xi + \frac{4}{3}\nu\right), \tag{15.A.4}$$

$$\mathbf{C}_\beta = \rho^{-1}\mu, \tag{15.A.5}$$

$$\mathbf{N}_\beta = \rho^{-1}\nu.$$

In the these equations $\rho = Diag(\rho_m, \rho_i)$ is the density matrix, $\mathbf{K} = \mathbf{M}^T \mathbf{K}_b \mathbf{M}$, $\boldsymbol{\xi} = \mathbf{M}^T \boldsymbol{\xi}_b \mathbf{M}$, $\boldsymbol{\mu} = \mathbf{M}^T \boldsymbol{\mu}_b \mathbf{M}$, $\boldsymbol{v} = \mathbf{M}^T \boldsymbol{v}_b \mathbf{M}$, with the transformation matrix

$$\mathbf{M} = \begin{pmatrix} 1 & m_f \\ 1 & m_s \end{pmatrix} \tag{15.A.6}$$

involving the fluid and solid mass fractions m_f and

$$\mathbf{K}_b = \begin{pmatrix} \dfrac{K_0}{M} + (\alpha - \eta)^2 & (\alpha - \eta)\eta \\ (\alpha - \eta)\eta & \eta^2 \end{pmatrix} M, \tag{15.A.7}$$

$$\boldsymbol{\mu}_b = \begin{pmatrix} 1 & 0 \\ 0 & 0 \end{pmatrix} \mu_0, \tag{15.A.8}$$

$$\boldsymbol{\xi}_b = \begin{pmatrix} 0 & 0 \\ (\alpha - \eta)\eta & \eta^2 \end{pmatrix} \dfrac{M}{K_f} \xi_f, \tag{15.A.9}$$

$$\boldsymbol{v}_b = \begin{pmatrix} 0 & 0 \\ \alpha_\mu - \eta & \eta \end{pmatrix} \mu_f. \tag{15.A.10}$$

$\boldsymbol{\xi}_b$ and $\boldsymbol{v}_b$ exist due to incorporated fluid viscous relaxation terms which render the viscosity-corrected Biot constitutive relation to be a 2×2 matrix generalization of viscoelastic constitutive relation.

15.A.2   *Inertial regime*

In the frequency band $\Omega_i \leq \omega \leq \sqrt{\Omega_i \Omega_S}$ simplified expression for some of the matrices involved in the conversion scattering analysis are obtained. In particular, the matrices $\boldsymbol{\alpha}$ and $\boldsymbol{\beta}$ defined in Equations (15.A.1) and (15.A.2) are then given as

$$\boldsymbol{\alpha} = (c_1^0)^2 \begin{pmatrix} 1 - i\Psi_1 \dfrac{\omega}{\Omega_P} & m_f \Psi_2 + i\Psi_3 \dfrac{\omega}{\Omega_P} \\ d_f \Psi_2 \left(1 - i\dfrac{\Omega_i}{\omega}\right) + i d_s \Psi_1 \dfrac{\omega}{\Omega_P} & d_f (m_f + \Psi_4)\left(1 - i\dfrac{\Omega_i}{\omega}\right) - i d_s \Psi_3 \dfrac{\omega}{\Omega_P} \end{pmatrix} \tag{15.A.11}$$

$$\boldsymbol{\beta} = (c_3^0)^2 \begin{pmatrix} 1 - i\alpha_\mu \dfrac{\omega}{\Omega_S} & m_f + i(\eta - m_f\alpha_\mu)\dfrac{\omega}{\Omega_S} \\ d_f \left(1 - i\dfrac{\Omega_i}{\omega}\right) + i d_s \alpha_\mu \dfrac{\omega}{\Omega_S} & d_f m_f \left(1 - i\dfrac{\Omega_i}{\omega}\right) - i d_s (\eta - m_f\alpha_\mu)\dfrac{\omega}{\Omega_S} \end{pmatrix}, \tag{15.A.12}$$

where $d_f = 1/(S - m_f)$, $d_s = m_s d_f/m_f$ and

$$\Psi_1 = \alpha + (\alpha_\mu - \alpha)\dfrac{4}{3}\dfrac{\Omega_{flP}}{\Omega_{flS}} \tag{15.A.13}$$

$$\Psi_2 = 1 - \dfrac{\alpha}{\eta}\dfrac{c_{fl}^2}{(c_1^0)^2} \tag{15.A.14}$$

$$\Psi_3 = \eta - m_f \Psi_1 \tag{15.A.15}$$

$$\Psi_4 = \left(\dfrac{1 - 2\alpha m_f}{\eta}\right)\dfrac{c_{fl}^2}{(c_1^0)^2}. \tag{15.A.16}$$

In these matrices

$$c_1^0 = \sqrt{\dfrac{H}{\rho_m}} \qquad c_3^0 = \sqrt{\dfrac{\mu_0}{\rho_m}} \tag{15.A.17}$$

are the compressional and shear wave velocities in the Gassmann limit (Mavko et al. 2009). In Equation (15A.12) c_{fl} denotes the velocity of sound in the fluid in the presence of a deformable solid-frame

$$c_{\mathrm{fl}} = \sqrt{\frac{\eta M}{\rho_{\mathrm{f}}}}, \tag{15.A.18}$$

which in the limit of a rigid frame reduces to $c_{\mathrm{fl}} = \sqrt{K_{\mathrm{f}}/\rho_{\mathrm{f}}}$ as $M \to K_{\mathrm{f}}/\eta$.

In the inertial regime $\Omega_{\mathrm{i}} \ll \omega \ll \{\Omega_{\mathrm{P}}, \Omega_{\mathrm{S}}\}$ the imaginary parts of these matrices can be neglected and their trace and determinant are simply given as

$$\Delta_\alpha^\infty = \eta^2 \frac{M H_0}{\rho_{\mathrm{m}} \rho_{\mathrm{i}}} \tag{15.A.19}$$

$$\mathrm{Tr}_\alpha^\infty = (c_1^0)^2 \left(1 + d_{\mathrm{f}}(m_{\mathrm{f}} + \Psi_4)\right) \tag{15.A.20}$$

$$\Delta_\beta^\infty = 0 \tag{15.A.21}$$

$$\mathrm{Tr}_\beta^\infty = (c_3^0)^2 (1 + m_{\mathrm{f}} d_{\mathrm{f}}). \tag{15.A.22}$$

The wave numbers of the fast and slow compressional and fast shear wave in the frequency band of Eq. (15.21) are real with $k_n = \omega/c_n^\infty$. The c_n^∞ are defined by Equations (15.10) and (15.11) with trace and determinant given by Eqs. (15.A.19)–(15.A.22), respectively.

REFERENCES

Bedford A., Costley D. & Stern M. (1984) On the drag virtual mass coefficients in Biot's equations. *Journal of the Acoustical Society of America*, 76, 1804–1809.

Biot M.A. (1956) Theory of propagation of elastic waves in a fluid-saturated porous solid. I. Low frequency range. *Journal of the Acoustical Society of America*, 28, 179–191 .

Biot M.A. (1956) Theory of propagation of elastic waves in a fluid-saturated porous solid. II. Higher frequency range. *Journal of the Acoustical Society of America*, 28, 179–191.

Biot M.A. (1962) Mechanics of deformation and acoustic propagation in porous media. *Journal of Applied Physics*, 33, 1482–1498.

Carcione J.M. (2007) *Wave Fields in Real Media, Handbook of Geophysical Exploration.* Amsterdam, Elsevier.

Frisch U. (1968) Wave propagation in random media. In: Bharucha-Reid, A.T. (ed.). *Probabilistic Methods in Applied Mathematics* New York, Academic Press, pp. 75–198.

Gist G.A. (1994) Fluid effects on velocity and attenuation in sandstones. *Journal of the Acoustial Society of America*, 96, 1158–1172.

Gold N., Shapiro S.A., Bojinski S. & Müller T.M. (2000) An approach to upscaling for seismic waves in statistically isotropic heterogeneous elastic media. *Geophysics*, 65, 1837–1850.

Johnson D.L., Koplik J. & Dashen R. (1987) Theory of dynamic permeability and tortuosity in fluid-saturated porous media. *Journal of Fluid Mechanics*, 176, 379–402.

Johnson D.L. (2001) Theory of frequency dependent acoustics in patchy saturated porous media. *Journal of the Acoustical Society of America*, 110, 682–694.

Karal F.C. & Keller J.B. (1964) Elastic, electromagnetic and other waves in random media. *Journal of Mathematical Physics*, 5, 537–547.

Liu Q.R. & Katsube N. (1990) The discovery of a second kind of rotational wave in a fluid-filled porous material. *Journal of the Acoustical Society of America*, 88, 1045–1053.

Landau L.D. & Lifshitz E.M. (1987) *Fluid Mechanics.* Oxford, Pergamon Press.

Lighthill J. (1978) *Waves in Fluids.* Cambridge, Cambridge University Press.

Mandelis A. (2000) Diffusion waves and their uses. *Physics Today*, 53, 29–34.

Mavko G., Mukerji T. & Dvorkin J. (2009) *The Rock Physics Handbook, Tools for Seismic Analysis of Porous Media. 2nd edition.* Cambridge, UK, Cambridge University Press.

McCoy J.J. (1973) On the dynamic response of disordered composites. *Journal of Applied Mechanics*, 28, 511–517.

Müller T.M. & Gurevich B. (2005a) A first-order statistical smoothing approximation for the coherent wave field in random porous media. *Journal of the Acoustical Society of America*, 117, 4, 1796–1805.

Müller T.M. & Gurevich B. (2005b) Wave-induced fluid flow in random porous media: Attenuation and dispersion of elastic waves. *Journal of the Acoustical Society of America*, 117, 5, 2732–2741.

Müller T.M. & Rothert E. (2006) Seismic attenuation due to wave-induced flow: Why Q in random structures scales differently. *Geophysics Research Letters*, 33, L16305.

Müller T.M. Gurevich B. & Shapiro S.A. (2008) Attenuation of seismic waves due to wave-induced flow and scattering in random porous media. In: Sato, H. & Fehler, M. (eds.). *Earth Heterogeneity and Scattering Effects on Seismic Waves, Advances in Geophysics Vol. 50*. Amsterdam, Elsevier. pp. 123–166.

Müller T.M. & Sahay, P.N. (2009) Attenuation due to slow S wave conversion. In: Ling, H., Smyth, A. & Betti, R. (eds.). *Proceedings of the Fourth Biot Conference on Poromechanics, New York, Columbia University, New York*. Lancaster, Destech Publications. pp. 733–739.

Müller T.M. Gurevich B. & Lebedev M. (2010) Seismic wave attenuation and dispersion resulting from wave-induced flow in porous rocks – A review: *Geophysics*, 75, 75A147–75A164.

Müller T.M. & Sahay P.N. (2011a) Fast compressional wave attenuation and dispersion due to conversion scattering into slow shear waves in randomly heterogeneous porous media. *Journal of the Acoustical Society of America*, 129, 5, 2785–2796.

Müller T.M. & Sahay P.N. (2011b) Porous medium acoustics of wave-induced vorticity diffusion. *Applied Physics Letters*, 98, 084101.

Norris A.N. (1986) On the viscodynamic operator in Biot's equation of poroelasticity. *Journal of Wave Material Interaction*, 1, 365–380.

Pride S.R., Berryman J.G. & Harris J.M. (2004) Seismic attenuation due to wave-induced flow, *Journal of Geophysical Research*, 109, B1, B01201 .

Pride S.R. (2005) Relationships between seismic and hydrological properties. Hydrogeophysics, 50, 253–290.

Rytov S., Kravtsov Y.A. & Tatarskii V.I. (1989) *Wave Propagation Through Random Media: Volume 4 of Principles of Statistical Radiophysics*, Berlin, Springer Verlag.

Sahay P.N. (1996) Elastodynamics of deformable porous media. *Proceedings of the Royal Society A*, 452, 1517–1529.

Sahay P.N. (2001) Dynamic Green's function for homogeneous and isotropic porous media. *Geophysical Journal International*, 147, 622–629.

Sahay P.N., Spanos T.J.T. & de la Cruz V. (2001) Seismic wave propagation in inhomogeneous and anisotropic porous media. *Geophysical Journal International*, 145, 209–222.

Sahay P.N. (2008) On the Biot slow S-wave. *Geophysics*, 73, N19–N33.

Sato H. & Fehler M. (1998) *Wave Propagation and Scattering in the heterogenous Earth*. New York, AIP-press.

Schlichting H. & Gersten K. (2000) *Boundary Layer Theory. 8th edition*. Berlin, Springer Verlag.

Spanos T.J.T. (2002) *The Thermophysics of Porous Media*. New York, Chapman and Hall/CRC.

Toms-Stewart J., Müller T.M., Gurevich B. & Paterson L. (2009) Statistical characterization of gas-patch distributions in partially saturated rocks. *Geophysics*, 74, WA51–WA64.

Turner J.A. (1999) Elastic wave propagation and scattering in heterogeneous anisotropic media: Textured polycrystalline materials. *Journal of the Acoustical Society of America*, 106, 541–552.

Turner J.A. & Anugonda P. (2001) Scattering of elastic waves in heterogeneous media with local isotropy. *Journal of the Acoustical Society of America*, 109, 1787–1795.

CHAPTER 16

Coupled porosity and saturation waves in porous media

N. Udey

16.1 INTRODUCTION

The porosity wave is an incompressible fluid flow process where a dynamic change in the porosity of a porous medium is coupled to a dynamic pressure change in the fluid, which in turn induces the fluid to flow. For this wave to occur, the solid matrix of the porous medium must be deformable, and the pressure changes must be slow enough for the fluid to behave incompressibly, i.e. the fluid is forced to move instead of just compressing. Therefore, in this process one obtains a damped travelling wave of porosity, pressure, and fluid. The theoretical description of porosity waves was first presented by Spanos (2002). In the diffusional limit, the inertial terms appearing in the theory are negligible, and the porosity wave becomes a porosity diffusion process. This diffusion process was first recognized and analysed by Geilikman et al. (1993).

The saturation wave is a fluid flow process where a dynamic change in the proportion of two incompressible fluids in a porous medium, i.e. the saturation, is is coupled to the pressure in each fluid. Since the fluid pressure in each fluid changes, so does their difference, namely the macroscopic capillary pressure. Saturation waves are inherently dispersional. Dispersion is a pore scale process where a displacing fluid tends to bypass a displaced fluid. It is a fundamental characteristic of fluid flow in porous media. Many of the essential features of dispersion have been presented in the classic book by Bear (1972). A theory of saturation waves and their dispersional nature was recently presented by Udey (2009).

Porosity waves and saturation waves are often coupled. Each wave can induce the other. Laboratory studies (Wang et al. 1998, Zschuppe 2001) have demonstrated that porosity waves can enhance fluid flow and displacement; and furthermore, they can generate saturation waves which are highly dispersional and can suppress viscous fingering. Consequently, when oil is being displaced by water using a porosity wave, more oil can be accessed and swept out by the saturation wave. At the field scale, these processes have generated an observed increase in total fluid volumes and increased oil production and reserves (Dusseault et al. 2002, Groenenboom et al. 2003, Spanos et al. 2003).

In order to implement reservoir simulations and enhanced recovery estimation tools, a deep understanding of the physics of porosity waves, saturation waves, and the coupling between them must be attained. The theory of porosity waves and the theory of saturation waves have been developed separately. In the research presented here, those theories will be used to guide the construction and analysis of the wave equations for coupled porosity and saturation waves in porous media.

16.2 THE GOVERNING EQUATIONS

16.2.1 *Variables and definitions*

Consider a porous medium consisting of a solid and two immiscible and incompressible fluids. The properties of these materials are labelled by the subscripts $A = s, 1, 2$. The densities of the solid and two fluids are ρ_A ($A = s, 1, 2$) respectively. The bulk moduli are K_A ($A = s, 1, 2$).

The solid's shear modulus is μ_s and the fluids' shear and bulk viscosities are μ_A ($A = 1, 2$) and ξ_A ($A = 1, 2$). The solid's displacement is $\mathbf{u}_s$ while the solid's velocity is $\mathbf{v}_s = \partial_t \mathbf{u}_s$. The fluid velocities are $\mathbf{v}_A$ ($A = 1, 2$).

The proportion of the volume occupied by the two fluids is the porosity η while the proportion occupied by the solid is the solidosity $\eta_s = 1 - \eta$. The proportion of each fluid is its fractional porosity η_A ($A = 1, 2$). The saturation of each fluid S_A ($A = 1, 2$) is defined by writing $\eta_A = \eta S_A$ ($A = 1, 2$). Since $\eta = \eta_1 + \eta_2$ then one has $S_1 + S_2 = 1$. The irreducible saturations of fluid 1 and fluid 2 are denoted by S_{1r} and S_{2r} respectively. The permeability of the porous medium is denoted by K.

The governing equations and physical quantities at the megascopic scale of the porous medium are obtained from their corresponding pore scale equations and quantities via the technique of volume averaging (Whitaker 1967, Slattery 1967, Spanos 1988). At the megascopic scale, the governing equations are the the equations of continuity for the fluids, the equations of motion for the solid and the two fluids, the porosity equation, and the saturation equation. Since temperature is not being incorporated into the description, the system of equations is isothermal. Also, the effects of gravity are being excluded in the analysis and one lets $\mathbf{g} = 0$.

At the megascopic scale the temporal and spatial variations of volume averaged quantities are considered to be of first order. For example, the porosity may be split into a zeroth and first order part by writing $\eta = \eta^o + \delta\eta$. Then one has $\partial_t \eta = \partial_t \delta\eta$, so no distinction is necessary between a derivative of η and its first order part $\delta\eta$. The velocities are first order since they represent a deviation from a background state of no flow. As an example, one can write

$$\nabla \cdot (\eta_1 \rho_1 \mathbf{v}_1) = \nabla(\eta_1^o \rho_1^o \mathbf{v}_1) = \eta_1^o \rho_1^o \nabla \cdot \mathbf{v}_1 = \eta_1 \rho_1 \nabla \cdot \mathbf{v}_1$$

16.2.2   *The equations of continuity*

The megascopic equations of continuity (Spanos 2002) are

$$\partial_t(\rho_1 \eta S_1) + \rho_1 \eta S_1 \nabla \cdot \mathbf{v}_1 = 0 \tag{16.1}$$

$$\partial_t(\rho_2 \eta S_2) + \rho_2 \eta S_2 \nabla \cdot \mathbf{v}_2 = 0 \tag{16.2}$$

We can solve for the fluid divergences $\nabla \cdot \mathbf{v}_1$ and $\nabla \cdot \mathbf{v}_2$. This yields

$$\nabla \cdot \mathbf{v}_1 = -\frac{1}{\eta}\partial_t \eta - \frac{1}{S_1}\partial_t S_1 - \frac{\partial_t \rho_1}{\rho_1} \tag{16.3}$$

$$\nabla \cdot \mathbf{v}_2 = -\frac{1}{\eta}\partial_t \eta - \frac{1}{S_2}\partial_t S_2 - \frac{\partial_t \rho_2}{\rho_2} \tag{16.4}$$

Since the system of equations is isothermal, the fluid densities depend only on pressure. Then

$$\frac{\partial_t \rho_1}{\rho_1} = \frac{c_{f1}}{K_1}\partial_t P_1 \qquad \frac{\partial_t \rho_2}{\rho_2} = \frac{c_{f2}}{K_2}\partial_t P_2 \tag{16.5}$$

where the bulk modulus of each fluid is K_1 and K_2 respectively, and the compressibility factors c_{f1} and c_{f2} are utilized as bookkeeping parameters to indicate the compressible or incompressible case. They have the values

$$c_{f1}, c_{f2} = \begin{cases} 1 & \text{compressible case} \\ 0 & \text{incompressible case} \end{cases} \tag{16.6}$$

Substituting the pressure equations into the fluid divergence equations yields

$$\nabla \cdot \mathbf{v}_1 = -\frac{1}{\eta}\partial_t \eta - \frac{1}{S_1}\partial_t S_1 - \frac{c_{f1}}{K_1}\partial_t P_1 \tag{16.7}$$

$$\nabla \cdot \mathbf{v}_2 = -\frac{1}{\eta}\partial_t \eta - \frac{1}{S_2}\partial_t S_2 - \frac{c_{f2}}{K_2}\partial_t P_2 \tag{16.8}$$

Then in the incompressible limit, and using $\partial_t S_2 = -\partial_t S_1$, the fluid divergences become

$$\nabla \cdot \mathbf{v}_1 = -\frac{1}{\eta}\partial_t\eta - \frac{1}{S_1}\partial_t S_1 \tag{16.9}$$

$$\nabla \cdot \mathbf{v}_2 = -\frac{1}{\eta}\partial_t\eta + \frac{1}{S_2}\partial_t S_1 \tag{16.10}$$

Since the fluids are being considered incompressible, then it is useful to use the volumetric flow in expressing some of the results. The volumetric flow for each fluid is

$$\mathbf{q}_1 = \eta_1 \mathbf{v}_1 \tag{16.11}$$

$$\mathbf{q}_2 = \eta_2 \mathbf{v}_2 \tag{16.12}$$

and the total volumetric flow is

$$\mathbf{q} = \mathbf{q}_1 + \mathbf{q}_2 \tag{16.13}$$

The volumetric filter velocity $\mathbf{v}_q$ is defined by

$$\mathbf{v}_q = \frac{1}{\eta}\mathbf{q} = S_1 \mathbf{v}_1 + S_2 \mathbf{v}_2 \tag{16.14}$$

Since the divergences of the volumetric flows are

$$\nabla \cdot \mathbf{q}_1 = -S_1 \partial_t\eta - \eta\partial_t S_1 \tag{16.15}$$

$$\nabla \cdot \mathbf{q}_2 = -S_2 \partial_t\eta + \eta\partial_t S_1 \tag{16.16}$$

then

$$\nabla \cdot \mathbf{q} = -\partial_t\eta \tag{16.17}$$

or alternatively

$$\nabla \cdot \mathbf{v}_q = -\frac{1}{\eta}\partial_t\eta \tag{16.18}$$

16.2.3 *The equations of motion*

In the absence of gravity, the megascopic equations of motion for the solid and each fluid (Hickey 1994, Spanos 2002) are

$$\eta_s\rho_s\partial_t^2\mathbf{u}_s = \eta_s K_s\nabla(\nabla \cdot \mathbf{u}_s) + \mu_{Mss}\left(\nabla^2\mathbf{u}_s + \frac{1}{3}\nabla(\nabla \cdot \mathbf{u}_s)\right) - K_s\nabla\eta$$
$$+ Q_{s1}(\mathbf{v}_1 - \mathbf{v}_s) + Q_{s2}(\mathbf{v}_2 - \mathbf{v}_s) + \varrho_{s1}\partial_t(\mathbf{v}_1 - \mathbf{v}_s) + \varrho_{s2}\partial_t(\mathbf{v}_2 - \mathbf{v}_s) \tag{16.19}$$

$$\eta_1\rho_1\partial_t\mathbf{v}_1 = -\eta_1\nabla P_1 + \mu_{M11}\left(\nabla^2\mathbf{v}_1 + \frac{1}{3}\nabla(\nabla \cdot \mathbf{v}_1)\right) + \eta_1\xi_1\nabla(\nabla \cdot \mathbf{v}_1) + \xi_1\nabla\partial_t\eta_1$$
$$+ \mu_{M1s}\left(\nabla^2\mathbf{v}_s + \frac{1}{3}\nabla(\nabla \cdot \mathbf{v}_s)\right) + \mu_{M12}\left(\nabla^2\mathbf{v}_2 + \frac{1}{3}\nabla(\nabla \cdot \mathbf{v}_2)\right) \tag{16.20}$$
$$-\eta^2\frac{\mu_1}{K}R_{11}(\mathbf{v}_1 - \mathbf{v}_s) + \eta^2\frac{\mu_1}{K}R_{12}(\mathbf{v}_2 - \mathbf{v}_s) - \varrho_{11}\partial_t(\mathbf{v}_1 - \mathbf{v}_s) + \varrho_{12}\partial_t(\mathbf{v}_2 - \mathbf{v}_s)$$

$$\eta_2\rho_2\partial_t\mathbf{v}_2 = -\eta_2\nabla P_2 + \mu_{M22}\left(\nabla^2\mathbf{v}_2 + \frac{1}{3}\nabla(\nabla \cdot \mathbf{v}_2)\right) + \eta_2\xi_2\nabla(\nabla \cdot \mathbf{v}_2) + \xi_2\nabla\partial_t\eta_2$$
$$+ \mu_{M2s}\left(\nabla^2\mathbf{v}_s + \frac{1}{3}\nabla(\nabla \cdot \mathbf{v}_s)\right) + \mu_{M21}\left(\nabla^2\mathbf{v}_1 + \frac{1}{3}\nabla(\nabla \cdot \mathbf{v}_1)\right) \tag{16.21}$$
$$+ \eta^2\frac{\mu_2}{K}R_{21}(\mathbf{v}_1 - \mathbf{v}_s) - \eta^2\frac{\mu_2}{K}R_{22}(\mathbf{v}_2 - \mathbf{v}_s) + \varrho_{21}\partial_t(\mathbf{v}_1 - \mathbf{v}_s) - \varrho_{22}\partial_t(\mathbf{v}_2 - \mathbf{v}_s)$$

In these equations, the megascopic shear modulus μ_{Mss} and megascopic viscosities μ_{Mij} are given by

$$\begin{bmatrix} \mu_{Mss} & & \\ \mu_{M1s} & \mu_{M11} & \mu_{M12} \\ \mu_{M2s} & \mu_{M21} & \mu_{M22} \end{bmatrix} = \begin{bmatrix} \eta_s\mu_s(1-(\lambda_{s1}+\lambda_{s2})) & & \\ \eta_s\mu_1\lambda_{s1} & \eta_1\mu_1(1-\lambda_{12}) & \eta_2\mu_1\lambda_{21} \\ \eta_s\mu_2\lambda_{s2} & \eta_1\mu_2\lambda_{12} & \eta_2\mu_2(1-\lambda_{21}) \end{bmatrix}$$

$$(16.22)$$

where the λ_{ij} are shear and viscosity "strength" parameters. The λ_{ij} parameters, the relative Darcy resistance coefficients R_{ij}, and the induced mass effect coefficients ϱ_{ij} are defined by the surface averages that arise in the volume averaging scheme and are presumed to be single valued functions of saturation S_1. Here the induced mass effects will be considered negligible and will no longer appear in the analysis below.

16.2.4   *The porosity and saturation equations*

The porosity equation (de la Cruz & Spanos 1985, Hickey 1994) is

$$\partial_t\eta = \delta_s\nabla\cdot\mathbf{v}_s - \delta_{f1}\nabla\cdot\mathbf{v}_1 - \delta_{f2}\nabla\cdot\mathbf{v}_2 \tag{16.23}$$

and it describes the relative volumetric changes of the solid and two fluids during a dilatation. The choice of the parameters δ_s, δ_{f1} and δ_{f2} selects a specific thermodynamic process of dilatational motion. The porosity equation is needed to complete the system of equations for dilatational waves.

In steady state flow, the megascopic capillary pressure $P_c(S_1)$ is utilized to account for surface tension effects. In such situations one normally lets the megascopic pressure difference, $P_{21} = P_2 - P_1$ be equal to the megascopic capillary pressure, i.e. $P_{21} = P_c$. However for dynamic processes like wave motion, the value of P_{21} can differ substantially from P_c. The dynamic value of P_{21} is now related to temporal changes in saturation. This relationship is expressed by the saturation equation (de la Cruz et al. 1995, Spanos 2002)

$$\partial_t(P_{21}) = \partial_t(P_2 - P_1) = -\beta_1\partial_t S_1 \tag{16.24}$$

where β_1 is assumed to be a single valued function of saturation. Here β_1 is taken to be

$$\beta_1 = \beta_c + \beta_h(S_1) \tag{16.25}$$

where

$$\beta_c = -\frac{dP_c(S_1)}{dS_1} \tag{16.26}$$

and β_h is an empirically determined function of saturation that accounts for dynamic pressure hysteresis. The saturation equation may also be referred to as the dynamic pressure difference equation.

16.3   DILATATIONAL WAVES

16.3.1   *The Helmholtz decomposition*

When one is looking at waves in a multi-component system, conventionally one splits the problem into an examination of dilatational (irrotational) waves and rotational (shear) waves. Now the Helmholtz decomposition of the velocities $\mathbf{v}_A$ $(A=s,1,2)$ is

$$\mathbf{v}_A = -\nabla\varphi_A + \nabla\times\Omega_A \qquad \nabla\cdot\Omega_A = 0 \tag{16.27}$$

where φ_A and Ω_A are respectively the scalar and solenoidal vector potentials. From this decomposition we can identify the divergence of the velocities as the dilatational part of the velocity fields since

$$\nabla \cdot \mathbf{v}_A = -\nabla^2 \varphi_A \tag{16.28}$$

and the rotational parts of the velocities Ω_A are no longer present. Therefore, the divergence of the equations of motion produces the dilatational wave equations. Instead of the scalar potentials φ_s, φ_1 and φ_2, the divergences $\nabla \cdot \mathbf{u}_s$, $\nabla \cdot \mathbf{v}_1$ and $\nabla \cdot \mathbf{v}_2$ will be taken as the primary independent variables.

16.3.2   *The dilatational wave equations*

The dilatational wave equations are obtained via a two step process. First, the divergence operator is applied to the equations of motion to produce the intermediate form of the dilatational wave equations. Secondly, the fluid divergences $\nabla \cdot \mathbf{v}_1$ and $\nabla \cdot \mathbf{v}_2$ as given by Equations (16.9) and (16.10) are substituted into the intermediate equations; this yields the primary form of the dilatational wave equations.

Now, taking the divergence of the equations of motion, one obtains

$$\eta_s \rho_s \partial_t^2 \nabla \cdot \mathbf{u}_s + (Q_{s1} + Q_{s2}) \nabla \cdot \partial_t \mathbf{u}_s - \eta_s \left(K_s + \frac{4}{3} \frac{\mu_{Mss}}{\eta_s} \right) \nabla^2 (\nabla \cdot \mathbf{u}_s)$$
$$\tag{16.29}$$
$$- Q_{s1} \nabla \cdot \mathbf{v}_1 - Q_{s2} \nabla \cdot \mathbf{v}_2 + K_s \nabla^2 \eta = 0$$

$$-\eta S_1 \rho_1 \partial_t (\nabla \cdot \mathbf{v}_1) - \eta S_1 \nabla^2 P_1 + \left(\eta S_1 \xi_1 + \frac{4}{3} \mu_{M11} \right) \nabla^2 (\nabla \cdot \mathbf{v}_1) + S_1 \xi_1 \nabla^2 \partial_t \eta$$

$$+ \eta \xi_1 \nabla^2 \partial_t S_1 - \frac{4}{3} \mu_{M1s} \nabla^2 (\nabla \cdot \mathbf{v}_s) + \frac{4}{3} \mu_{M12} \nabla^2 (\nabla \cdot \mathbf{v}_2) - \eta^2 \frac{\mu_1}{K} R_{11} (\nabla \cdot \mathbf{v}_1) \tag{16.30}$$

$$+ \eta^2 \frac{\mu_1}{K} R_{12} (\nabla \cdot \mathbf{v}_2) + \eta^2 \frac{\mu_1}{K} (R_{11} - R_{12})(\nabla \cdot \mathbf{v}_s) = 0$$

$$-\eta S_2 \rho_2 \partial_t (\nabla \cdot \mathbf{v}_2) - \eta S_2 \nabla^2 P_2 + \left(\eta S_2 \xi_2 + \frac{4}{3} \mu_{M22} \right) \nabla^2 (\nabla \cdot \mathbf{v}_2) + S_2 \xi_2 \nabla \partial_t \eta$$

$$- \eta \xi_2 \nabla \partial_t S_1 - \frac{4}{3} \mu_{M2s} \nabla^2 (\nabla \cdot \mathbf{v}_s) + \frac{4}{3} \mu_{M21} \nabla^2 (\nabla \cdot \mathbf{v}_1) + \eta^2 \frac{\mu_2}{K} R_{21} (\nabla \cdot \mathbf{v}_1) \tag{16.31}$$

$$- \eta^2 \frac{\mu_2}{K} R_{22} (\nabla \cdot \mathbf{v}_2) + \eta^2 \frac{\mu_2}{K} (R_{22} - R_{21})(\nabla \cdot \mathbf{v}_s) = 0$$

Then, substitution of the fluid divergences $\nabla \cdot \mathbf{v}_1$ and $\nabla \cdot \mathbf{v}_2$ into these intermediate equations yields

$$\eta_s \rho_s \partial_t^2 (\nabla \cdot \mathbf{u}_s) + (Q_{s1} + Q_{s2}) \partial_t (\nabla \cdot \mathbf{u}_s) - \eta_s \left(K_s + \frac{4}{3} \frac{\mu_{Mss}}{\eta_s} \right) \nabla^2 (\nabla \cdot \mathbf{u}_s)$$
$$\tag{16.32}$$
$$+ \frac{Q_{s1} + Q_{s2}}{\eta} \partial_t \eta + K_s \nabla^2 \eta + \left(\frac{Q_{s1}}{S_1} - \frac{Q_{s2}}{S_2} \right) \partial_t S_1 = 0$$

$$\eta \frac{\mu_1}{K} (R_{11} - R_{12}) \partial_t (\nabla \cdot \mathbf{u}_s) + \frac{4}{3} \frac{\mu_{M1s}}{\eta} \nabla^2 \partial_t (\nabla \cdot \mathbf{u}_s) + S_1 \rho_1 \frac{1}{\eta} \partial_t^2 \eta + \eta \frac{\mu_1}{K} (R_{11} - R_{12}) \frac{1}{\eta} \partial_t \eta$$

$$- \frac{4}{3} \frac{\mu_{M11} + \mu_{M12}}{\eta} \frac{1}{\eta} \nabla^2 \partial_t \eta + \rho_1 \partial_t^2 S_1 + \eta \frac{\mu_1}{K} \left(\frac{R_{11}}{S_1} + \frac{R_{12}}{S_2} \right) \partial_t S_1 \tag{16.33}$$

$$- \frac{4}{3} \frac{1}{\eta} \left(\frac{\mu_{M11}}{S_1} - \frac{\mu_{M12}}{S_2} \right) \nabla^2 \partial_t S_1 - S_1 \nabla^2 P_1 = 0$$

$$\eta\frac{\mu_2}{K}(R_{22}-R_{21})\partial_t(\nabla\cdot\mathbf{u}_s)+\frac{4}{3}\frac{\mu_{M2s}}{\eta}\nabla^2\partial_t(\nabla\cdot\mathbf{u}_s)+S_2\rho_2\frac{1}{\eta}\partial_t^2\eta+\eta\frac{\mu_2}{K}(R_{22}-R_{21})\frac{1}{\eta}\partial_t\eta$$

$$-\frac{4}{3}\frac{\mu_{M21}+\mu_{M22}}{\eta}\frac{1}{\eta}\nabla^2\partial_t\eta-\rho_2\partial_t^2 S_1-\eta\frac{\mu_2}{K}\left(\frac{R_{21}}{S_1}+\frac{R_{22}}{S_2}\right)\partial_t S_1 \tag{16.34}$$

$$+\frac{4}{3}\frac{1}{\eta}\left(\frac{\mu_{M22}}{S_2}-\frac{\mu_{M21}}{S_1}\right)\nabla^2\partial_t S_1-S_2\nabla^2 P_2=0$$

These equations are the dilatational wave equations. They can be restated in a more compact form by defining some coefficients.

For the solid dilatational wave equation, one first divides the equation by η_s and then defines

$$A_{s\eta}=\frac{1}{2}\frac{Q_{s1}+Q_{s2}}{\eta_s} \tag{16.35}$$

$$K_{Mss}=K_s+\frac{4}{3}\frac{\mu_{Mss}}{\eta_s} \tag{16.36}$$

$$K_{s\eta}=K_s\frac{\eta}{\eta_s} \tag{16.37}$$

$$A_{sS_1}=\frac{1}{2}\frac{1}{\eta_s}\left(\frac{Q_{s1}}{S_1}-\frac{Q_{s2}}{S_2}\right) \tag{16.38}$$

Then the solid's dilatational wave equation becomes

$$\rho_s\partial_t^2(\nabla\cdot\mathbf{u}_s)+2A_{s\eta}\partial_t(\nabla\cdot\mathbf{u}_s)-K_{Mss}\nabla^2(\nabla\cdot\mathbf{u}_s)+2A_{s\eta}\frac{1}{\eta}\partial_t\eta+K_{s\eta}\frac{1}{\eta}\nabla^2\eta+2A_{sS_1}\partial_t S_1=0 \tag{16.39}$$

For fluid 1 one defines

$$A_{1\eta}=\frac{1}{2}\eta\frac{\mu_1}{K}(R_{11}-R_{12}) \tag{16.40}$$

$$B_{1s}=\frac{2}{3}\frac{\mu_{M1s}}{\eta} \tag{16.41}$$

$$B_{1\eta}=\frac{2}{3}\frac{\mu_{M11}+\mu_{M12}}{\eta} \tag{16.42}$$

$$A_{1S_1}=\frac{1}{2}\eta\frac{\mu_1}{K}\left(\frac{R_{11}}{S_1}+\frac{R_{12}}{S_2}\right) \tag{16.43}$$

$$B_{1S_1}=\frac{2}{3}\frac{1}{\eta}\left(\frac{\mu_{M11}}{S_1}-\frac{\mu_{M12}}{S_2}\right) \tag{16.44}$$

Consequently the simplified form of the dilatational wave equation for fluid 1 is

$$2A_{1\eta}\partial_t(\nabla\cdot\mathbf{u}_s)+2B_{1s}\nabla^2\partial_t(\nabla\cdot\mathbf{u}_s)+S_1\rho_1\frac{1}{\eta}\partial_t^2\eta+2A_{1\eta}\frac{1}{\eta}\partial_t\eta-2B_{1\eta}\frac{1}{\eta}\nabla^2\partial_t\eta$$

$$+\rho_1\partial_t^2 S_1+2A_{1S_1}\partial_t S_1-2B_{1S_1}\nabla^2\partial_t S_1-S_1\nabla^2 P_1=0 \tag{16.45}$$

Finally, for fluid 2 one defines

$$A_{2\eta}=\frac{1}{2}\eta\frac{\mu_2}{K}(R_{22}-R_{21}) \tag{16.46}$$

$$B_{2s}=\frac{2}{3}\frac{\mu_{M2s}}{\eta} \tag{16.47}$$

$$B_{2\eta} = \frac{2}{3} \frac{\mu_{M21} + \mu_{M22}}{\eta} \tag{16.48}$$

$$A_{2S_1} = \frac{1}{2} \eta \frac{\mu_2}{K} \left(\frac{R_{21}}{S_1} + \frac{R_{22}}{S_2} \right) \tag{16.49}$$

$$B_{2S_1} = \frac{2}{3} \frac{1}{\eta} \left(\frac{\mu_{M22}}{S_2} - \frac{\mu_{M21}}{S_1} \right) \tag{16.50}$$

Consequently the simplified form of the dilatational wave equation for fluid 2 is

$$2A_{2\eta} \partial_t (\nabla \cdot \mathbf{u}_s) + 2B_{2s} \nabla^2 \partial_t (\nabla \cdot \mathbf{u}_s) + S_2 \rho_2 \frac{1}{\eta} \partial_t^2 \eta + 2A_{2\eta} \frac{1}{\eta} \partial_t \eta - 2B_{2\eta} \frac{1}{\eta} \nabla^2 \partial_t \eta$$

$$- \rho_2 \partial_t^2 S_1 - 2A_{2S_1} \partial_t S_1 + 2B_{2S_1} \nabla^2 \partial_t S_1 - S_2 \nabla^2 P_2 = 0 \tag{16.51}$$

In these equations, terms of the form A_{ij} are attenuation densities with dimensionality [attenuation $(s^{-1}) \times$ density (kg/m^3)]. Terms of the form B_{ij} are bulk attenuation densities with dimensionality [attenuation density $\times$ area (m^2)]. Finally, the coefficients of the form K_{ij} are bulk moduli with units of pressure (Pa).

To obtain the porosity dilatational wave equation, one substitutes the fluid divergences $\nabla \cdot \mathbf{v}_1$ and $\nabla \cdot \mathbf{v}_2$ into the porosity equation which is given by Equation (16.23). This operation yields

$$\partial_t \nabla \cdot \mathbf{u}_s - \frac{\alpha_\eta}{\eta} \partial_t \eta + \alpha_{S_1} \partial_t S_1 = 0 \tag{16.52}$$

where

$$\alpha_\eta = \frac{\eta - \delta_{f1} - \delta_{f2}}{\delta_s} \tag{16.53}$$

$$\alpha_{f_1} = \frac{\delta_{f1}}{\delta_s} \tag{16.54}$$

$$\alpha_{f_2} = \frac{\delta_{f2}}{\delta_s} \tag{16.55}$$

$$\alpha_{S_1} = \frac{\alpha_{f1}}{S_1} - \frac{\alpha_{f2}}{S_2} \tag{16.56}$$

The saturation equation, Equation (16.24), may be rewritten in a more useful form by expanding terms and putting them all on the left hand. Then one obtains

$$\beta_1 \partial_t S_1 + \partial_t P_2 - \partial_t P_1 = 0 \tag{16.57}$$

In summary, five dilatational wave equations have been obtained. These equations govern the wave behaviour of five variables, namely the solid's divergence of displacement $\nabla \cdot \mathbf{u}_s$, the porosity η, the saturation S_1, and the two fluid pressures P_1 and P_2. The analysis of these dilatational wave equations will be facilitated by recasting them into a wave operator format.

16.3.3 *The dilatational wave operator matrix equation*

To establish the wave operator formalism, define a differential wave operator basis W_b by letting

$$W_b = [\partial_t^2, \partial_t, \nabla^2 \partial_t, \nabla^2] \tag{16.58}$$

Let $\psi = \psi(\mathbf{x}, t)$ be a wave variable associated with a wave operator W_ψ defined formally by

$$\begin{aligned} W_\psi &= [w_o, w_1,, w_2, w_3] \cdot W_b \\ &= w_o \partial_t^2 + w_1 \partial_t + w_2 \nabla^2 \partial_t + w_3 \nabla^2 \end{aligned} \tag{16.59}$$

Then the application of the wave operator W_ψ to the variable ψ is

$$
\begin{aligned}
W_\psi \cdot \psi &= [w_o, w_1, w_2, w_3] \cdot W_b \cdot \psi \\
&= w_o \partial_t^2 \psi + w_1 \partial_t \psi + w_2 \nabla^2 \partial_t \psi + w_3 \nabla^2 \psi
\end{aligned}
\tag{16.60}
$$

For a practical example, suppose the variable ψ has associated with it a density ρ, an attenuation density A, a bulk attenuation density B, and a bulk modulus K_b. Then the damped wave equation operator W_d for ψ is

$$
W_d = [\rho, 2A, -2B, -K_b] \cdot W_b
\tag{16.61}
$$

and the damped wave equation is then

$$
W_d \cdot \psi = 0
\tag{16.62}
$$

or

$$
\rho \partial_t^2 \psi + 2A \partial_t \psi - 2B \nabla^2 \partial_t \psi - K_b \nabla^2 \psi = 0
\tag{16.63}
$$

The dilatational wave equations for the solid and the two fluids can now be rewritten using this wave operator formalism. The solid's dilatational wave equation becomes

$$
W_{ss} \nabla \cdot \mathbf{u}_s + W_{s\eta} \frac{\eta}{\eta^o} + 2A_{sS_1} \partial_t S_1 = 0
\tag{16.64}
$$

where the wave operators W_{ss} and $W_{s\eta}$ are defined by

$$
W_{ss} = [\rho_s, 2A_{s\eta}, 0, -K_{Mss}] \cdot W_b
\tag{16.65}
$$

$$
W_{s\eta} = [0, 2A_{s\eta}, 0, K_{s\eta}] \cdot W_b
\tag{16.66}
$$

The dilatational wave equation for fluid 1 becomes

$$
W_{1s} \nabla \cdot \mathbf{u}_s + W_{1\eta} \frac{\eta}{\eta^o} + W_{1S_1} S_1 - S_1 \nabla^2 P_1 = 0
\tag{16.67}
$$

where

$$
W_{1s} = [0, 2A_{1\eta}, 2B_{1s}, 0] \cdot W_b
\tag{16.68}
$$

$$
W_{1\eta} = [S_1 \rho_1, 2A_{1\eta}, -2B_{1\eta}, 0] \cdot W_b
\tag{16.69}
$$

$$
W_{1S_1} = [\rho_1, 2A_{1S_1}, -2B_{1S_1}, 0] \cdot W_b
\tag{16.70}
$$

Finally, the dilatational wave equation for fluid 2 is

$$
W_{2s} \nabla \cdot \mathbf{u}_s + W_{2\eta} \frac{\eta}{\eta^o} - W_{2S_1} S_1 - S_2 \nabla^2 P_2 = 0
\tag{16.71}
$$

where

$$
W_{2s} = [0, 2A_{2\eta}, 2B_{2s}, 0] \cdot W_b
\tag{16.72}
$$

$$
W_{2\eta} = [S_2 \rho_2, 2A_{2\eta}, -2B_{2\eta}, 0] \cdot W_b
\tag{16.73}
$$

$$
W_{2S_1} = [\rho_2, 2A_{2S_1}, -2B_{2S_1}, 0] \cdot W_b
\tag{16.74}
$$

The three rewritten dilatational wave equations, the porosity dilatational wave equation, and the saturation equation may now be assembled into a wave operator matrix equation. The result is

$$
\begin{bmatrix}
W_{ss} & W_{s\eta} & 2A_{sS_1} \partial_t & 0 & 0 \\
W_{1s} & W_{1\eta} & W_{1S_1} & -S_1 \nabla^2 & 0 \\
W_{2s} & W_{2\eta} & -W_{2S_1} & 0 & -S_2 \nabla^2 \\
\partial_t & -\alpha_\eta \partial_t & \alpha_{S_1} \partial_t & 0 & 0 \\
0 & 0 & \beta_1 \partial_t & -\partial_t & \partial_t
\end{bmatrix}
\begin{bmatrix}
\nabla \cdot \mathbf{u}_s \\
\eta/\eta^o \\
S_1 \\
P_1 \\
P_2
\end{bmatrix}
=
\begin{bmatrix}
0 \\
0 \\
0 \\
0 \\
0
\end{bmatrix}
\tag{16.75}
$$

The wave solutions of this matrix equation are coupled porosity and saturation waves. There are two limiting cases one may examine to help one understand the complexity of these waves. The first limiting case is the porosity wave for a porous medium consisting of a solid and a single fluid. The second case is the saturation wave when the solid is completely rigid so that no porosity wave can arise. The trial solutions and dispersion relations for these two cases can provide substantial insight into the trial solution and dispersion relation for the more general case.

16.3.4   *Wave operator trial solutions*

A standard technique to analyse waves governed by a wave equation is to substitute a trial solution, typically a homogeneous plane wave, into the wave equation to obtain a dispersion relation. An analysis of the dispersion relation reveals valuable information about the waves such as attenuation and speed of propagation.

Recall the example of a damped wave equation for a variable ψ, i.e.

$$W_d \cdot \psi = [\rho, 2A, -2B, -K_b] \cdot W_b \cdot \psi = 0 \qquad (16.76)$$

Let a trial solution of this damped wave equation for ψ consist of an unperturbed value ψ^o plus a one dimensional wave perturbation. One writes this as

$$\psi(\mathbf{x}, t) = \psi^o + \delta\psi \qquad \delta\psi = \mathcal{A}_\psi^o\, e^{nx+wt} \qquad (16.77)$$

where the perturbation $\delta\psi$ represents a homogeneous plane wave of amplitude $\mathcal{A}_\psi^o$ moving in the x direction with complex wave number n and complex frequency w.

Applying the wave operator basis W_b to the trial solution yields

$$\begin{aligned} W_b \cdot \psi &= [\partial_t^2, \partial_t, \nabla^2\partial_t, \nabla^2] \cdot \delta\psi \\ &= [w^2, w, wn^2, n^2] \cdot \delta\psi \\ &= \mathcal{P}_b(n, w) \cdot \delta\psi \end{aligned} \qquad (16.78)$$

where the dispersion relation basis $\mathcal{P}_b(n, w)$ is defined as

$$\mathcal{P}_b(n, w) = [w^2, w, wn^2, n^2] \qquad (16.79)$$

Now application of the damped wave operator to the trial solution yields

$$\begin{aligned} W_d \cdot \psi &= [\rho, 2A, -2B, -K_b] \cdot W_b \cdot \delta\psi \\ &= [\rho, 2A, -2B, -K_b] \cdot \mathcal{P}_b(n, w) \cdot \delta\psi \\ &= \mathcal{P}_d \cdot \delta\psi \end{aligned} \qquad (16.80)$$

where the dispersion polynomial $\mathcal{P}_d(n, w)$ corresponding to the wave operator W_d is

$$\begin{aligned} \mathcal{P}_d(n, w) &= [\rho, 2A, -2B, -K_b] \cdot \mathcal{P}_b(n, w) \\ &= [\rho, 2A, -2B, -K_b] \cdot [w^2, w, wn^2, n^2] \\ &= \rho w^2 + 2Aw - 2Bwn^2 - K_b n^2 \end{aligned} \qquad (16.81)$$

So, when the trial solution is substituted into the damped wave equation $W_d \cdot \psi = 0$, one obtains

$$\mathcal{P}(n, w)\mathcal{A}_\psi^o\, e^{nx+wt} = 0 \qquad (16.82)$$

A non-trivial solution $\mathcal{A}_\psi^o \neq 0$ now requires

$$\mathcal{P}(n, w) = 0 \qquad (16.83)$$

and this produces the dispersion relation for the trial solution, namely

$$\rho w^2 + 2Aw - 2Bwn^2 - K_b n^2 = 0 \tag{16.84}$$

Dividing by ρ yields

$$w^2 + 2aw - 2bwn^2 - v_o^2 n^2 = 0 \tag{16.85}$$

where the attenuation a, bulk attenuation b, and undamped wave speed v_o are defined by

$$a = \frac{A}{\rho} \tag{16.86}$$

$$b = \frac{B}{\rho} \tag{16.87}$$

$$v_o^2 = \frac{K_b}{\rho} \tag{16.88}$$

Choosing w as the independent variable, the solution for n^2 is

$$n^2 = \frac{w^2 + 2aw}{v_o^2 + 2bw} \tag{16.89}$$

The frequency domain for this trial solution is selected by letting the complex wave frequency be $w = -i\omega$ where ω is the angular frequency of the wave. The equation for n becomes

$$n^2 = -\frac{\omega^2 + 2a\omega i}{v_o^2 - 2b\omega i} \tag{16.90}$$

One now seeks a solution for n of the form

$$n = -\kappa + ik \tag{16.91}$$

where κ and k are the spatial attenuation and real wave number respectively. Then the trial solution is

$$\psi = \psi^o + \mathcal{A}_\psi^o \, e^{-\kappa x} \, e^{i(kx - \omega t)} \tag{16.92}$$

which is a damped homogeneous plane wave travelling "to the right in the x direction" with the wave speed v given by

$$v = \frac{\omega}{k} \tag{16.93}$$

If $b \neq 0$, then the solution for n yields a wave speed v whose asymptotic behaviour is

$$\lim_{\omega \to \infty} v = 2\sqrt{|b|}\sqrt{\omega} \tag{16.94}$$

that is, the maximum wave speed is frequency dependent. Such behaviour is unphysical since one would expect all wave speeds to be bounded above by the unattenuated wave speed v_o. Now, letting $b = 0$, the solution to

$$n^2 = -\frac{\omega^2 + 2a\omega i}{v_o^2} \tag{16.95}$$

is

$$n = -\kappa + ik \tag{16.96}$$

$$\kappa = \frac{1}{v_o \sqrt{2}} \sqrt{\sqrt{\omega^4 + 4a^2\omega^2} - \omega^2} \tag{16.97}$$

$$k = \frac{1}{v_o \sqrt{2}} \sqrt{\sqrt{\omega^4 + 4a^2\omega^2} + \omega^2} \tag{16.98}$$

and the wave speed is

$$v = \frac{\omega}{k} = v_o \frac{\sqrt{2}\,\omega}{\sqrt{\sqrt{\omega^4 + 4a^2\omega^2} + \omega^2}} \tag{16.99}$$

One observes that $v(\omega) < v_o$ and $\lim_{\omega \to \infty} v(\omega) = v_o$ Also, if the attenuation a were 0, then one would have $v = v_o$.

16.4   POROSITY WAVES

16.4.1   *The porosity wave equation*

Porosity waves were first examined for the case of a solid and a single fluid. The two fluid equations of motion now collapse into a single fluid equation. Consequently our equations of motion become

$$\eta_s \rho_s \partial_t^2 \mathbf{u}_s = \eta_s K_s \nabla(\nabla \cdot \mathbf{u}_s) + \mu_{Mss}\left(\nabla^2 \mathbf{u}_s + \frac{1}{3}\nabla(\nabla \cdot \mathbf{u}_s)\right) - K_s \nabla \eta + \eta^2 \frac{\mu_f}{K}(\mathbf{v}_f - \mathbf{v}_s) \tag{16.100}$$

$$\eta_f \rho_f \partial_t \mathbf{v}_f = -\eta_f \nabla P_f + \mu_{Mff}\left(\nabla^2 \mathbf{v}_f + \frac{1}{3}\nabla(\nabla \cdot \mathbf{v}_f)\right) + \eta \xi_f \nabla(\nabla \cdot \mathbf{v}_f) \tag{16.101}$$

$$+ \xi_f \nabla \partial_t \eta + \mu_{Mfs}\left(\nabla^2 \mathbf{v}_s + \frac{1}{3}\nabla(\nabla \cdot \mathbf{v}_s)\right) - \eta^2 \frac{\mu_f}{K}(\mathbf{v}_f - \mathbf{v}_s)$$

while the porosity equation becomes

$$\partial_t \eta = \delta_s \nabla \cdot \mathbf{v}_s - \delta_f \nabla \cdot \mathbf{v}_f \tag{16.102}$$

Taking the divergence of these equations of motion produces

$$\eta_s \rho_s \partial_t^2 (\nabla \cdot \mathbf{u}_s) + \frac{\eta^2 \mu_f}{K}\partial_t(\nabla \cdot \mathbf{u}_s) - \eta_s\left(K_s + \frac{4}{3}\frac{\mu_{Mss}}{\eta_s}\right)\nabla^2(\nabla \cdot \mathbf{u}_s) \tag{16.103}$$

$$- \eta^2 \frac{\mu_f}{K}\nabla \cdot \mathbf{v}_f + K_s \nabla^2 \eta = 0$$

$$-\eta^2 \frac{\mu_f}{K}\partial_t(\nabla \cdot \mathbf{u}_s) - \frac{4}{3}\mu_{Mfs}\nabla^2 \partial_t(\nabla \cdot \mathbf{u}_s) + \eta_f \rho_f \partial_t \nabla \cdot \mathbf{v}_f + \eta^2 \frac{\mu_f}{K}(\nabla \cdot \mathbf{v}_f) \tag{16.104}$$

$$- \left(\eta \xi_f + \frac{4}{3}\mu_{Mff}\right)\nabla^2(\nabla \cdot \mathbf{v}_f) - \xi_f \nabla^2 \partial_t \eta + \eta_f \nabla^2 P_f = 0$$

Inserting the divergence of the fluid velocity in the incompressible limit

$$\nabla \cdot \mathbf{v}_f = -\frac{1}{\eta}\partial_t \eta \tag{16.105}$$

into these equations and the porosity equation yields

$$\eta_s \rho_s \partial_t^2(\nabla \cdot \mathbf{u}_s) + \frac{\eta^2 \mu_f}{K}\partial_t(\nabla \cdot \mathbf{u}_s) - \eta_s\left(K_s + \frac{4}{3}\frac{\mu_{Mss}}{\eta_s}\right)\nabla^2(\nabla \cdot \mathbf{u}_s) + \eta\frac{\mu_f}{K}\partial_t \eta + K_s \nabla^2 \eta = 0 \tag{16.106}$$

$$\eta^2 \frac{\mu_f}{K} \partial_t (\nabla \cdot \mathbf{u}_s) + \frac{4}{3} \mu_{Mfs} \nabla^2 \partial_t (\nabla \cdot \mathbf{u}_s) + \rho_f \partial_t^2 \eta + \eta \frac{\mu_f}{K} \partial_t \eta - \frac{4}{3} \frac{\mu_{Mff}}{\eta} \nabla^2 \partial_t \eta - \eta \nabla^2 P_f = 0$$

(16.107)

$$\partial_t (\nabla \cdot \mathbf{u}_s) - \frac{\alpha_\eta}{\eta} \partial_t \eta = 0$$

(16.108)

where

$$\alpha_\eta = \frac{\eta - \delta_f}{\delta_s}$$

(16.109)

For dilatational wave motions, Equation (16.108) may be integrated in time and becomes

$$\nabla \cdot \mathbf{u}_s = \frac{\alpha_\eta}{\eta} (\eta - \eta^o)$$

(16.110)

Now, substituting this result into the solid dilatational equations produces a damped wave equation for the porosity

$$\rho_\eta \partial_t^2 \eta + 2 A_\eta \partial_t \eta - K_\eta \nabla^2 \eta = 0$$

(16.111)

where the porosity's density, attenuation density and bulk modulus are

$$\rho_\eta = \alpha_\eta \rho_s$$

(16.112)

$$A_\eta = \frac{1}{2} \frac{(\alpha_\eta + 1) \eta^2}{\eta_s} \frac{\mu_f}{K}$$

(16.113)

$$K_\eta = K_s \left(\alpha_\eta - \frac{\eta}{\eta_s} \right) + \frac{4}{3} \alpha_\eta \frac{\mu_{Mss}}{\eta_s}$$

(16.114)

In wave operator format, this equation is

$$W_\eta \cdot \eta = 0$$

(16.115)

where the porosity wave operator is

$$W_\eta = [\rho_\eta, 2 A_\eta, 0, -K_\eta] \cdot W_b$$

(16.116)

The solutions of this wave equation are porosity waves.

Substitution of $\nabla \cdot \mathbf{u}_s$ as given by Equation (16.110) into the fluid's dilatational wave equation shows that the fluid pressure responds to the porosity wave via

$$\nabla^2 P_f = \frac{1}{\eta} (\rho_f \partial_t^2 \eta + 2 A_{f\eta} \partial_t \eta - 2 B_{f\eta} \nabla^2 \partial_t \eta)$$

(16.117)

where the attenuation density $A_{f\eta}$ and bulk attenuation density $B_{f\eta}$ are given by

$$A_{f\eta} = \eta (\alpha_\eta + 1) \frac{\mu_f}{K}$$

(16.118)

$$B_{f\eta} = \frac{2}{3} (\mu_f - \alpha_\eta \mu_{Mfs})$$

(16.119)

The porosity wave's attenuation a_η and undamped wave speed $v_{o\eta}$ are given by

$$a_\eta = \frac{A_\eta}{\rho_\eta}$$

(16.120)

$$v_{o\eta}^2 = \frac{K_\eta}{\rho_\eta}$$

(16.121)

In the diffusion limit of these equations, the inertial and bulk attenuation density terms are considered negligible and therefore

$$\partial_t \eta = D_\eta \nabla^2 \eta \tag{16.122}$$

$$\nabla^2 P_f = \frac{2 A_{f\eta}}{\eta} \partial_t \eta \tag{16.123}$$

where the porosity diffusion coefficient is

$$D_\eta = \frac{K_\eta}{2 A_\eta} = \frac{v_o^2}{2 a_\eta} = \frac{\eta_s}{\eta} \frac{K_s \left(\alpha_\eta - \frac{\eta}{\eta_s} \right) + \frac{4}{3} \alpha_\eta \frac{\mu_{Mss}}{\eta_s}}{(\alpha_\eta + 1)} \frac{1}{\eta} \frac{K}{\mu_f} \tag{16.124}$$

These results conform to the porosity diffusion results originally obtained by Geilikman et al. (1993).

16.4.2 *The dispersion relation*

Now consider a trial solution of the form

$$\eta = \eta^o + A_\eta^o e^{nx + wt} \tag{16.125}$$

which was previously discussed in Subsection 16.3.4. Substitution of this trial solution into the porosity wave equation $W_\eta \cdot \eta = 0$ produces the dispersion relation

$$\rho_\eta w^2 + 2 A_\eta w - K_\eta n^2 = 0 \tag{16.126}$$

or, dividing by ρ_η,

$$w^2 + 2 a_\eta w - v_{on}^2 n^2 = 0 \tag{16.127}$$

As shown in Subsection 16.3.4, for a given $w = -i\omega$, the solution of the dispersion relation is

$$n = -\kappa + ik \tag{16.128}$$

$$\kappa = \frac{1}{v_{on} \sqrt{2}} \sqrt{\sqrt{\omega^4 + 4 a_\eta^2 \omega^2} - \omega^2} \tag{16.129}$$

$$k = \frac{1}{v_{on} \sqrt{2}} \sqrt{\sqrt{\omega^4 + 4 a_\eta^2 \omega^2} + \omega^2} \tag{16.130}$$

$$v = \frac{\omega}{k} = v_{on} \frac{\sqrt{2}\,\omega}{\sqrt{\sqrt{\omega^4 + 4 a_\eta^2 \omega^2} + \omega^2}} \tag{16.131}$$

16.4.3 *Comparison with pressure diffusion*

The pressure diffusion equation is obtained in its simplest form by taking the solid to be perfectly rigid, i.e. η is constant, and neglecting viscous dissipation and induced mass effects. Then the fluid's equation of motion is governed by the competition between the pressure gradient and the Darcy force; it is given by

$$\eta \rho_f \partial_t \mathbf{v}_f = -\eta \nabla P_f - \eta^2 \frac{\mu_f}{K} \mathbf{v}_f \tag{16.132}$$

The divergence of this equation yields the intermediate wave equation

$$\eta \rho_f \partial_t \nabla \cdot \mathbf{v}_f = -\eta \nabla^2 P_f - \eta^2 \frac{\mu_f}{K} \nabla \cdot \mathbf{v}_f \tag{16.133}$$

For a single fluid, the fluid divergence in the compressional case is

$$\nabla \cdot \mathbf{v}_f = -\frac{1}{K_f} \partial_t P_f \tag{16.134}$$

When this result is substituted into the intermediate wave equation, one obtains a damped wave equation for the pressure

$$\rho_f \partial_t^2 P_f + \eta \frac{\mu_f}{K} \partial_t P_f - K_f \nabla^2 P_f = 0 \tag{16.135}$$

The diffusion limit of this equation is the pressure diffusion equation

$$\partial_t P_f = D_{P_f} \nabla^2 P_f \tag{16.136}$$

where the pressure diffusion coefficient D_{P_f} is

$$D_{P_f} = K_f \frac{1}{\eta} \frac{K}{\mu_f} \tag{16.137}$$

Now, when energy is released in a localized volume in a porous medium, an increase in pressure is created. The pressure can diffuse either by the porosity diffusion process (since an increase in pressure will change the porosity), or by the pressure diffusion process. The porosity diffusion process will be the preferred process if the porosity diffusion coefficient D_η is larger than the pressure diffusion coefficient D_{P_f}, i.e. when

$$\frac{D_\eta}{D_{P_f}} > 1 \tag{16.138}$$

Since the expression for this ratio is

$$\frac{D_\eta}{D_{P_f}} = \frac{\eta_s}{\eta} \frac{K_s(\alpha_\eta - \frac{\eta}{\eta_s}) + \frac{4}{3}\alpha_\eta \frac{\mu_{Mss}}{\eta_s}}{(\alpha_\eta + 1)K_f} \tag{16.139}$$

one can see that an appropriate choice of the parameters can easily lead to a diffusion coefficient ratio that is significantly greater than 1. In such cases the porosity diffusion process is the preferred diffusion mechanism, and the pressure will relax much more quickly than expected by pressure diffusion.

16.5   SATURATION WAVES

16.5.1   *The wave equations*

An analysis of saturation waves was presented by Udey (2009). That analysis will be recapitulated here using the notation and results obtained above.

The solid will be be taken to be completely rigid, so η is a constant and its derivatives are 0. Porosity waves cannot exist in this case. For the fluids, one considers the viscous dissipation and induced mass effects to be negligible. The only forces that remain are pressure gradients and the Darcy forces. In this situation, the only equations of motion are those for the two fluids; they are

$$\eta_1 \rho_1 \partial_t \mathbf{v}_1 = -\eta_1 \nabla P_1 - \eta^2 \frac{\mu_1}{K} R_{11} \mathbf{v}_1 + \eta^2 \frac{\mu_1}{K} R_{12} \mathbf{v}_2 \tag{16.140}$$

$$\eta_2 \rho_2 \partial_t \mathbf{v}_2 = -\eta_2 \nabla P_2 + \eta^2 \frac{\mu_2}{K} R_{21} \mathbf{v}_1 - \eta^2 \frac{\mu_2}{K} R_{22} \mathbf{v}_2 \tag{16.141}$$

Taking the divergence of these Equations (16.140) and (16.141) produces

$$\eta S_1 \rho_1 \frac{\partial \nabla \cdot \mathbf{v}_1}{\partial t} = -\eta S_1 \nabla^2 P_1 - \eta^2 \frac{\mu_1}{K} R_{11} \nabla \cdot \mathbf{v}_1 + \eta^2 \frac{\mu_1}{K} R_{12} \nabla \cdot \mathbf{v}_2 \qquad (16.142)$$

$$\eta S_2 \rho_2 \frac{\partial \nabla \cdot \mathbf{v}_2}{\partial t} = -\eta S_2 \nabla^2 P_2 + \eta^2 \frac{\mu_2}{K} R_{21} \nabla \cdot \mathbf{v}_1 - \eta^2 \frac{\mu_2}{K} R_{22} \nabla \cdot \mathbf{v}_2 \qquad (16.143)$$

For a rigid solid, $\partial_t \eta = 0$, so Equations (16.9) and (16.10) for the fluid divergences become

$$\nabla \cdot \mathbf{v}_1 = -\frac{1}{S_1} \partial_t S_1 \qquad (16.144)$$

$$\nabla \cdot \mathbf{v}_2 = \frac{1}{S_2} \partial_t S_1 \qquad (16.145)$$

Substituting these fluid divergences into Equations (16.142) and (16.143) yields the dilatational wave equations for the two fluids

$$\rho_1 \partial_t^2 S_1 + 2A_{1S_1} \partial_t S_1 - S_1 \nabla^2 P_1 = 0 \qquad (16.146)$$

$$-\rho_2 \partial_t^2 S_1 - 2A_{2S_1} \partial_t S_1 - S_2 \nabla^2 P_2 = 0 \qquad (16.147)$$

where the attenuation densities A_{1S_1} and A_{2S_1} were defined by Equations (16.43) and (16.49) respectively. These two equations and the saturation equation

$$\beta_1 \partial_t S_1 + \partial_t P_2 - \partial_t P_1 = 0 \qquad (16.148)$$

constitute three wave equations in the three unknowns S_1, P_1, and P_2. These equations may be assembled into a wave operator matrix equation, namely

$$\begin{bmatrix} \rho_1 \partial_t^2 + 2A_{1S_1} \partial_t & -S_1 \nabla^2 & 0 \\ -(\rho_2 \partial_t^2 + 2A_{2S_1} \partial_t) & 0 & -S_2 \nabla^2 \\ \beta_1 \partial_t & -\partial_t & \partial_t \end{bmatrix} \begin{bmatrix} S_1 \\ P_1 \\ P_2 \end{bmatrix} = \begin{bmatrix} 0 \\ 0 \\ 0 \end{bmatrix} \qquad (16.149)$$

Solutions of this matrix wave equation are saturation waves; they are dilatational dispersion waves in the porous medium. Also, note that this wave matrix equation is a sub-matrix of the more general matrix wave Equation (16.75) with the bulk attenuation densities being absent.

16.5.2 *The dispersion relation*

Now consider a trial solution

$$[S_1, P_1, P_2] = [S_1^o, P_1^o, P_2^o] + [A_{S_1}^o, A_{P_1}^o, A_{P_2}^o] e^{nx+wt} \qquad (16.150)$$

whose form was previously discussed in Subsection 16.3.4. Substitution of this trial solution into the matrix wave equations yields a dispersion matrix equation given by

$$\begin{bmatrix} \rho_1 w^2 + 2A_{1S_1} w & -S_1 n^2 & 0 \\ -(\rho_2 w^2 + 2A_{2S_1} w) & 0 & -S_2 n^2 \\ \beta_1 w & -w & w \end{bmatrix} \begin{bmatrix} A_{S_1}^o \\ A_{P_1}^o \\ A_{P_2}^o \end{bmatrix} = \begin{bmatrix} 0 \\ 0 \\ 0 \end{bmatrix} \qquad (16.151)$$

We can only have non-trivial solutions if the determinant of the matrix on the left hand side of this equation is 0, i.e.

$$\Delta(n, w) = \begin{vmatrix} \rho_1 w^2 + 2A_{1S_1} w & -S_1 n^2 & 0 \\ -(\rho_2 w^2 + 2A_{2S_1} w) & 0 & -S_2 n^2 \\ \beta_1 w & -w & w \end{vmatrix} = 0 \qquad (16.152)$$

Now evaluation of the determinant yields the dispersion relation

$$wn^2(\rho_{S_1}w^2 + 2A_{S_1}w - K_{S_1}n^2) = 0 \tag{16.153}$$

where the density, attenuation density, and bulk modulus of the saturation wave are

$$\rho_{S_1} = S_1\rho_2 + S_2\rho_1 \tag{16.154}$$

$$A_{S_1} = S_1A_{2S_1} + S_2A_{1S_1} \tag{16.155}$$

$$K_{S_1} = S_1S_2\beta_1 \tag{16.156}$$

The attenuation density may be evaluated utilizing Equations (16.43) and (16.49). Then

$$
\begin{aligned}
A_{S_1} &= S_1A_{2S_1} + S_2A_{1S_1} \\
&= S_1\frac{1}{2}\eta\frac{\mu_2}{K}\left(\frac{R_{21}}{S_1} + \frac{R_{22}}{S_2}\right) + S_2\frac{1}{2}\eta\frac{\mu_1}{K}\left(\frac{R_{11}}{S_1} + \frac{R_{12}}{S_2}\right) \\
&= \frac{1}{2}\eta\frac{\mu_e}{K}
\end{aligned}
\tag{16.157}
$$

where the effective viscosity μ_e is defined as

$$\mu_e = \mu_1 S_2\left(\frac{R_{11}}{S_1} + \frac{R_{12}}{S_2}\right) + \mu_2 S_1\left(\frac{R_{21}}{S_1} + \frac{R_{22}}{S_2}\right) \tag{16.158}$$

Discarding the trivial solutions $w=0$ and $n=0$ in the dispersion relation reduces the dispersion relation to

$$\rho_{S_1}w^2 + 2A_{S_1}w - K_{S_1}n^2 = 0 \tag{16.159}$$

Dividing this equation by the density ρ_{S_1} produces

$$w^2 + 2a_{S_1}w - v_{oS_1}^2n^2 = 0 \tag{16.160}$$

where the attenuation and unattenuated wave speed of the saturation wave are

$$a_{S_1} = \frac{A_{S_1}}{\rho_{S_1}} \tag{16.161}$$

$$v_{oS_1}^2 = \frac{K_{S_1}}{\rho_{S_1}} \tag{16.162}$$

As shown in Subsection 16.3.4, for a given $w = -i\omega$, the solution of this dispersion relation is

$$n = -\kappa + ik \tag{16.163}$$

$$\kappa = \frac{1}{v_{oS_1}\sqrt{2}}\sqrt{\sqrt{\omega^4 + 4a_{S_1}^2\omega^2} - \omega^2} \tag{16.164}$$

$$k = \frac{1}{v_{oS_1}\sqrt{2}}\sqrt{\sqrt{\omega^4 + 4a_{S_1}^2\omega^2} + \omega^2} \tag{16.165}$$

$$v = \frac{\omega}{k} = v_{oS_1}\frac{\sqrt{2}\,\omega}{\sqrt{\sqrt{\omega^4 + 4a_{S_1}^2\omega^2} + \omega^2}} \tag{16.166}$$

16.6 COUPLED POROSITY AND SATURATION WAVES

16.6.1 *The dispersion relation*

The dilatational wave equations for coupled porosity and saturations waves were derived in Subsection 16.3.2 and Subsection 16.3.3 and gathered together into a wave operator matrix equation, namely Equation (16.75). Using the porosity wave and saturation wave solutions from Subsection 16.4.2 and Subsection 16.5.2 as guides, one may now examine the dispersion relation that arises from the application of a trial solution to the full set of wave equations.

Let the trial solution be

$$\left[\nabla \cdot \mathbf{u}_s, \frac{\eta}{\eta^o}, S_1 - S_1^o, P_1 - P_1^o, P_2 - P_2^o \right] = [\mathcal{A}_s^o, \mathcal{A}_\eta^o, \mathcal{A}_{S_1}^o, \mathcal{A}_{P_1}^o, \mathcal{A}_{P_2}^o] e^{nx+wt} \tag{16.167}$$

whose form was analysed in Subsection 16.3.4. Substituting this trial solution into Equation (16.75) and utilizing the dispersion polynomial definition from Equation (16.80), one obtains the dispersion matrix equation

$$\begin{bmatrix} \mathcal{P}_{ss} & \mathcal{P}_{s\eta} & 2A_{sS_1}w & 0 & 0 \\ \mathcal{P}_{1s} & \mathcal{P}_{1\eta} & \mathcal{P}_{1S_1} & -S_1n^2 & 0 \\ \mathcal{P}_{2s} & \mathcal{P}_{2\eta} & -\mathcal{P}_{2S_1} & 0 & -S_2n^2 \\ w & -\alpha_\eta w & \alpha_{S_1}w & 0 & 0 \\ 0 & 0 & \beta_1 w & -w & w \end{bmatrix} \begin{bmatrix} \mathcal{A}_s^o \\ \mathcal{A}_\eta^o \\ \mathcal{A}_{S_1}^o \\ \mathcal{A}_{P_1}^o \\ \mathcal{A}_{P_2}^o \end{bmatrix} = \begin{bmatrix} 0 \\ 0 \\ 0 \\ 0 \\ 0 \end{bmatrix} \tag{16.168}$$

where

$$\mathcal{P}_{ss} = \rho_s w^2 + 2A_{s\eta}w - K_{Mss}n^2 \tag{16.169}$$

$$\mathcal{P}_{s\eta} = 2A_{s\eta}w + K_{s\eta}n^2 \tag{16.170}$$

$$\mathcal{P}_{1s} = 2A_{1\eta}w + 2B_{1s}wn^2 \tag{16.171}$$

$$\mathcal{P}_{1\eta} = S_1\rho_1 w^2 + 2A_{1\eta}w - 2B_{1\eta}wn^2 \tag{16.172}$$

$$\mathcal{P}_{1S_1} = \rho_1 w^2 + 2A_{1S_1}w - 2B_{1S_1}wn^2 \tag{16.173}$$

$$\mathcal{P}_{2s} = 2A_{2\eta}w + 2B_{2s}wn^2 \tag{16.174}$$

$$\mathcal{P}_{2\eta} = S_2\rho_2 w^2 + 2A_{2\eta}w - 2B_{2\eta}wn^2 \tag{16.175}$$

$$\mathcal{P}_{2S_1} = \rho_2 w^2 + 2A_{2S_1}w - 2B_{2S_1}wn^2 \tag{16.176}$$

For non-trivial solutions, the determinant of the matrix in Equation (16.168) must be 0. This condition is

$$\Delta(n, w) = \begin{Vmatrix} \mathcal{P}_{ss} & \mathcal{P}_{s\eta} & 2A_{sS_1}w & 0 & 0 \\ \mathcal{P}_{1s} & \mathcal{P}_{1\eta} & \mathcal{P}_{1S_1} & -S_1n^2 & 0 \\ \mathcal{P}_{2s} & \mathcal{P}_{2\eta} & -\mathcal{P}_{2S_1} & 0 & -S_2n^2 \\ w & -\alpha_\eta w & \alpha_{S_1}w & 0 & 0 \\ 0 & 0 & \beta_1 w & -w & w \end{Vmatrix} = 0 \tag{16.177}$$

The computation of the determinant yields the structure of of $\Delta(n, w)$. One finds that

$$\Delta(n, w) = n^2 w^2 \mathcal{D}(n, w) \tag{16.178}$$

where

$$\mathcal{D}(n, w) = (D_{42}w^2 + D_{41}w + D_{40})n^4 + (D_{23}w^3 + D_{22}w^2 + D_{21}w)n^2$$
$$+ (D_{04}w^4 + D_{03}w^3 + D_{02}w^2) \tag{16.179}$$

and the coefficients D_{ij} are

$$D_{42} = 0 \tag{16.180}$$

$$
\begin{aligned}
D_{41} = {}& 2(0)(S_1 S_2 \beta_1) + 2(S_1 B_{2S_1} + S_2 B_{1S_1})(\alpha_\eta K_{Mss} - K_{s\eta}) \\
& - 2\alpha_{S_1}(K_{Mss}(S_1 B_{2\eta} - S_2 B_{1\eta}) - K_{s\eta}(S_1 B_{2s} - S_2 B_{1s})))
\end{aligned}
\tag{16.181}
$$

$$D_{40} = (\alpha_\eta K_{Mss} - K_{s\eta})(S_1 S_2 \beta_1) \tag{16.182}$$

$$D_{23} = -2(\alpha_\eta \rho_s)(S_1 B_{2S_1} + S_2 B_{1S_1}) - 2(S_1 \rho_2 + S_2 \rho_1)(0) + 2\alpha_{S_1} \rho_s(S_1 B_{2\eta} - S_2 B_{1\eta}) \tag{16.183}$$

$$
\begin{aligned}
D_{22} = {}& -(\alpha_\eta \rho_s)(S_1 S_2 \beta_1) - (S_1 \rho_2 + S_2 \rho_1)(\alpha_\eta K_{Mss} - K_{s\eta}) \\
& - 4((\alpha_\eta + 1) A_{s\eta})(S_1 B_{2S_1} + S_2 B_{1S_1}) - 4(S_1 A_{2S_1} + S_2 A_{1S_1})(0) \\
& - \alpha_{S_1}(K_{Mss} S_1 S_2(\rho_1 - \rho_2) - 4 A_{s\eta}((S_1 B_{2s} - S_2 B_{1s}) + (S_1 B_{2\eta} - S_2 B_{1\eta}))) \\
& + 4 A_{sS_1}(\alpha_\eta(S_1 B_{2s} - S_2 B_{1s}) - (S_1 B_{2\eta} - S_2 B_{1\eta}))
\end{aligned}
\tag{16.184}
$$

$$
\begin{aligned}
D_{21} = {}& -2((\alpha_\eta + 1) A_{s\eta})(S_1 S_2 \beta_1) - 2(S_1 A_{2S_1} + S_2 A_{1S_1})(\alpha_\eta K_{Mss} - K_{s\eta}) \\
& + 2\alpha_{S_1}(S_1 A_{2\eta} - S_2 A_{1\eta})(K_{Mss} + K_{s\eta})
\end{aligned}
\tag{16.185}
$$

$$D_{04} = (\alpha_\eta \rho_s)(S_1 \rho_2 + S_2 \rho_1) + \alpha_{S_1} \rho_s S_1 S_2(\rho_1 - \rho_2) \tag{16.186}$$

$$
\begin{aligned}
D_{03} = {}& 2(\alpha_\eta \rho_s)(S_1 A_{2S_1} + S_2 A_{1S_1}) + 2(S_1 \rho_2 + S_2 \rho_1)((\alpha_\eta + 1) A_{s\eta}) \\
& + 2\alpha_{S_1}(S_1 S_2 A_{s\eta}(\rho_1 - \rho_2) - \rho_s(S_1 A_{2\eta} - S_2 A_{1\eta})) - 2 A_{sS_1} S_1 S_2(\rho_1 - \rho_2)
\end{aligned}
\tag{16.187}
$$

$$D_{02} = 4((\alpha_\eta + 1) A_{s\eta})(S_1 A_{2S_1} + S_2 A_{1S_1}) + 4(\alpha_\eta + 1) A_{sS_1}(S_1 A_{2\eta} - S_2 A_{1\eta}) \tag{16.188}$$

These coefficients are presented in a format to elucidate their structure. In particular, the term (0) in Equation (16.181) is used to indicate that the term $S_1 S_2 \beta_1$ is not in that equation. Similarly, the term (0) is used in Equation (16.183) to indicate the absence of the term $S_1 \rho_2 + S_2 \rho_1$ in that equation; and the term (0) in Equation (16.184) indicates the absence of the term $S_1 A_{2S_1} + S_2 A_{1S_1}$.

Now the dispersion relation $\Delta(n, w) = n^2 w^2 \mathcal{D}(n, w) = 0$ contains the two trivial solutions $n = 0$ and $w = 0$. Consequently, the non-trivial solutions must arise from the condition

$$\mathcal{D}(n, w) = 0 \tag{16.189}$$

For a given value of w, this condition is simple to solve for n since this condition is a quadratic equation in n^2, i.e.

$$a(w)(n^2)^2 + b(w)(n^2) + c(w) = 0 \tag{16.190}$$

with complex coefficients

$$
\begin{aligned}
a(w) &= D_{42} w^2 + D_{41} w + D_{40} \\
b(w) &= D_{23} w^3 + D_{22} w^2 + D_{21} w \\
c(w) &= D_{04} w^4 + D_{03} w^3 + D_{02} w^2
\end{aligned}
\tag{16.191}
$$

Thus there are two solutions for the coupled porosity and saturation wave.

The numerical solution of Equation (16.190) is simple once all of the physical and model parameters have been specified. One may undertake a numerical study to explore the solution space by varying the model parameters. Unfortunately it is all too easy to pick model parameters that produce unphysical results; typically the unphysical result is that the wave gains energy as it propagates. However, one can improve the exploration of the solution space by varying the model parameters about a known physical solution.

16.6.2 *Factorization of the dispersion relation*

Since the limiting cases of the wave equations are a pure (uncoupled) porosity wave and a pure saturation wave, one may hypothesize that the dispersion relation may be expressed as the product of a porosity wave solution and a saturation wave solution, i.e.

$$\mathcal{D}(n, w) = (\rho_\eta w^2 + 2A_\eta w - 2B_\eta n^2 w - K_\eta n^2) \times (\rho_{S_1} w^2 + 2A_{S_1} w - 2B_{S_1} n^2 w - K_{S_1} n^2) \quad (16.192)$$

subject to some constraints that are to be determined.

Expanding the right hand side of Equation (16.192) yields

$$\mathcal{D}(n, w) = (C_{42} w^2 + C_{41} w + C_{40}) n^4 + (C_{23} w^3 + C_{22} w^2 + C_{21} w) n^2 \\ + (C_{04} w^4 + C_{03} w^3 + C_{02} w^2) \quad (16.193)$$

where

$$C_{42} = 4B_\eta B_{S_1} \quad (16.194)$$

$$C_{41} = 2B_\eta K_{S_1} + 2B_{S_1} K_\eta \quad (16.195)$$

$$C_{40} = K_\eta K_{S_1} \quad (16.196)$$

$$C_{23} = -2\rho_\eta B_{S_1} - 2\rho_{S_1} B_\eta \quad (16.197)$$

$$C_{22} = -\rho_\eta K_{S_1} - \rho_{S_1} K_\eta - 4A_\eta B_{S_1} - 4A_{S_1} B_\eta \quad (16.198)$$

$$C_{21} = -2A_\eta K_{S_1} - 2A_{S_1} K_\eta \quad (16.199)$$

$$C_{04} = \rho_\eta \rho_{S_1} \quad (16.200)$$

$$C_{03} = 2\rho_\eta A_{S_1} + 2\rho_{S_1} A_\eta \quad (16.201)$$

$$C_{02} = 4A_\eta A_{S_1} \quad (16.202)$$

Thus a comparison of Equation (16.193) with Equation (16.179) term by term will determine the values of the terms ρ_η, A_η, etc. appearing in Equation (16.192) and any required constraints.

The comparison $C_{40} = D_{40}$ immediately yields

$$K_\eta = \alpha_\eta K_{Mss} - K_{s\eta} \qquad K_{S_1} = S_1 S_2 \beta_1 \quad (16.203)$$

The expressions for K_{Mss} and $K_{s\eta}$ from Equations (16.36) and (16.37) may be utilized to re-express K_η. One obtains

$$K_\eta = K_s \left(\alpha_\eta - \frac{\eta}{\eta_s} \right) + \frac{4}{3} \alpha_\eta \frac{\mu_{Mss}}{\eta_s} \quad (16.204)$$

which agrees with Equation (16.114).

Next, the comparison $C_{42} = D_{42}$ implies that

$$B_\eta = 0 \qquad \text{or} \qquad B_{S_1} = 0 \quad (16.205)$$

Now in the comparison $C_{41} = D_{41}$ one immediately deduces that

$$B_\eta = 0 \quad (16.206)$$

since the term $S_1 S_2 \beta_1$ is absent in the coefficient D_{41}. The remainder of this comparison yields

$$B_{S_1} = S_1 B_{2S_1} + S_2 B_{1S_1} \quad (16.207)$$

and the first constraint is

$$\alpha_{S_1} = 0 \quad (16.208)$$

The comparison $C_{04} = D_{04}$ produces

$$\rho_\eta = \alpha_\eta \rho_s \qquad \rho_{S_1} = S_1 \rho_2 + S_2 \rho_1 \quad (16.209)$$

Subsequently, the comparison $C_{23} = D_{23}$ is trivially satisfied. From the comparison $C_{21} = D_{21}$ one obtains the attenuation densities

$$A_\eta = (\alpha_\eta + 1)A_{s\eta} \qquad A_{S_1} = S_1 A_{2S_1} + S_2 A_{1S_1} \qquad (16.210)$$

Note that the expression here for A_{S1} is same as Equation (16.155); and therefore from Equation (16.157) one has

$$A_{S_1} = \frac{1}{2}\eta\frac{\mu_e}{K} \qquad (16.211)$$

Now substitution of the attenuation densities in Equation (16.210) into the comparison $C_{02} = D_{02}$ leads to the second constraint

$$A_{sS_1} = 0 \qquad (16.212)$$

Finally, with the two constraints and all of the terms having been defined, the remaining comparisons $C_{03} = D_{03}$ and $C_{22} = D_{22}$ are trivially satisfied.

The first constraint is

$$0 = \alpha_{S_1} = \frac{\alpha_{f1}}{S_1} - \frac{\alpha_{f2}}{S_2} = \frac{\delta_{f1}}{S_1\delta_s} - \frac{\delta_{f2}}{S_2\delta_s} \qquad (16.213)$$

Define $\delta_f(S_1)$ by letting

$$\frac{\delta_{f2}}{S_2} = \delta_f \qquad (16.214)$$

Now treating δ_f as the independent function of saturation, then δ_{f2} depends on δ_f by

$$\delta_{f2} = S_2\delta_f \qquad (16.215)$$

and the constraint, Equation (16.213), gives us

$$\delta_{f1} = S_1\delta_f \qquad (16.216)$$

Consequently the porosity equation becomes

$$\begin{aligned}
\partial_t\eta &= \delta_s\nabla\cdot\mathbf{v}_s - \delta_{f1}\nabla\cdot\mathbf{v}_1 - \delta_{f2}\nabla\cdot\mathbf{v}_2 \\
&= \delta_s\nabla\cdot\mathbf{v}_s - S_1\delta_f\nabla\cdot\mathbf{v}_1 - S_2\delta_f\nabla\cdot\mathbf{v}_2 \\
&= \delta_s\nabla\cdot\mathbf{v}_s - \delta_f\nabla\cdot\mathbf{v}_q
\end{aligned} \qquad (16.217)$$

and the porosity dilatational wave equation becomes

$$\partial_t\nabla\cdot\mathbf{u}_s - \frac{\alpha_\eta}{\eta}\partial_t\eta = 0 \qquad (16.218)$$

where

$$\alpha_\eta = \frac{\eta - \delta_{f1} - \delta_{f2}}{\delta_s} = \frac{\eta - S_1\delta_f - S_2\delta_f}{\delta_s} = \frac{\eta - \delta_f}{\delta_s} \qquad (16.219)$$

The second constraint is

$$0 = A_{sS_1} = \frac{1}{2}\frac{1}{\eta_s}\left(\frac{Q_{s1}}{S_1} - \frac{Q_{s2}}{S_2}\right) \qquad (16.220)$$

An effective viscosity function $\mu_f = \mu_f(S_1)$ may be defined for the two fluids viewed as a single fluid by letting

$$\mu_f = \frac{K}{\eta^2}\frac{Q_{s2}}{S_2} \qquad (16.221)$$

Now viewing μ_f as the independent function of saturation, then Q_{s2} depends on μ_f as

$$Q_{s2} = S_2 \eta^2 \frac{\mu_f}{K} \tag{16.222}$$

and the constraint now implies

$$Q_{s1} = S_1 \eta^2 \frac{\mu_f}{K} \tag{16.223}$$

Consequently, in the solid's equation of motion, the Darcy force is

$$Q_{s1}(\mathbf{v}_1 - \mathbf{v}_s) + Q_{s2}(\mathbf{v}_2 - \mathbf{v}_s) = \eta^2 \frac{\mu_f}{K}(\mathbf{v}_q - \mathbf{v}_s) \tag{16.224}$$

Furthermore, the expression for $A_{s\eta}$ in Equation (16.35) becomes

$$A_{s\eta} = \frac{1}{2}\frac{Q_{s1} + Q_{s2}}{\eta_s} = \frac{1}{2}\frac{\eta^2}{\eta_s}\frac{\mu_f}{K} \tag{16.225}$$

and subsequently the expression for A_η in Equation (16.210) becomes

$$A_\eta = \frac{1}{2}\frac{(\alpha_\eta + 1)\eta^2}{\eta_s}\frac{\mu_f}{K} \tag{16.226}$$

This last result agrees with Equation (16.113) except that now μ_f is a function of saturation instead of a constant.

In summary, the two constraints $\alpha_{S_1} = 0$ and $A_{sS_1} = 0$ permit the solution of the dispersion relation $\mathcal{D}(n, w) = 0$ to be expressed as the product of the porosity wave dispersion polynomial and the saturation wave dispersion polynomial. Since $B_\eta = 0$, the porosity wave dispersion relation is

$$\rho_\eta w^2 + 2A_\eta w - K_\eta n^2 = 0 \tag{16.227}$$

where

$$\rho_\eta = \alpha_\eta \rho_s \tag{16.228}$$

$$A_\eta = \frac{1}{2}\frac{(\alpha_\eta + 1)\eta^2}{\eta_s}\frac{\mu_f}{K} \tag{16.229}$$

$$K_\eta = K_s(\alpha_\eta - \frac{\eta}{\eta_s}) + \frac{4}{3}\alpha_\eta \frac{\mu_{Mss}}{\eta_s} \tag{16.230}$$

The saturation wave dispersion relation is

$$\rho_{S_1} w^2 + 2A_{S_1} w - 2B_{S_1} n^2 w - K_{S_1} n^2 = 0 \tag{16.231}$$

where the coefficients are given by

$$\rho_{S_1} = S_1 \rho_2 + S_2 \rho_1 \tag{16.232}$$

$$A_{S_1} = \frac{1}{2}\eta\frac{\mu_e}{K} \tag{16.233}$$

$$B_{S_1} = S_1 B_{2S_1} + S_2 B_{1S_1} \tag{16.234}$$

$$K_{S_1} = S_1 S_2 \beta_1 \tag{16.235}$$

However, in Subsection 16.3.4 it was noted that the bulk attenuation density

$$b_{S_1} = \frac{B_{S_1}}{\rho_{S_1}} \tag{16.236}$$

must be zero to avoid the unphysical frequency dependence, Equation (16.94), for the wave speed. Therefore one must impose the physical constraint $B_{S_1} = 0$.

From Equations (16.22), (16.44) and (16.50) one deduces that

$$B_{1S_1} = \frac{2}{3} \left(\frac{\mu_{M11}}{S_1} - \frac{\mu_{M12}}{S_2} \right) = \frac{2}{3} \eta \mu_1 (1 - (\lambda_{12} + \lambda_{21})) \qquad (16.237)$$

$$B_{2S_1} = \frac{2}{3} \left(\frac{\mu_{M22}}{S_2} - \frac{\mu_{M21}}{S_1} \right) = \frac{2}{3} \eta \mu_2 (1 - (\lambda_{12} + \lambda_{21})) \qquad (16.238)$$

Then

$$B_{S_1} = S_1 B_{2S_1} + S_2 B_{1S_1} = \frac{2}{3} \eta (S_2 \mu_1 + S_1 \mu_2)(1 - (\lambda_{12} + \lambda_{21})) \qquad (16.239)$$

The constraint $B_{S_1} = 0$ now implies that

$$\lambda_{12} + \lambda_{21} = 1 \qquad (16.240)$$

This condition can be easily met if one lets

$$\lambda_{12} = S_2 \qquad \lambda_{21} = S_1 \qquad (16.241)$$

so that

$$\begin{bmatrix} \mu_{M11} & \mu_{M12} \\ \mu_{M21} & \mu_{M22} \end{bmatrix} = \begin{bmatrix} \eta S_1^2 \mu_1 & \eta S_1 S_2 \mu_1 \\ \eta S_1 S_2 \mu_2 & \eta S_2^2 \mu_2 \end{bmatrix} \qquad (16.242)$$

The physical solution obtained here for the coupled porosity and saturation waves represents a situation of minimal coupling between the porosity and saturation waves. In order to examine situations where the coupling is stronger, one may start with the solution obtained here, then relax the conditions $\alpha_{S_1} = 0$ and $A_{sS_1} = 0$ by a slight modification of the model parameters, and then numerically solve the full dispersion relation. This approach is still the subject of an ongoing research program.

16.7   A NUMERICAL ILLUSTRATION

For a numerical illustration consider a porous medium consisting of a Berea sandstone, water and a (fictitious) light oil. The physical properties of these materials are presented in Table 16.1 along with the model parameters required for the numerical illustration. In addition to these parameters, one must also specify the saturation dependence of the total fluid viscosity μ_f, the relative Darcy resistances R_{ij} and the megascopic capillary pressure P_c. These functions will be presented as needed.

16.7.1   *The porosity wave*

To illustrate the porosity wave, the total fluid viscosity is chosen to be a parallel combination of the individual fluids weighted by saturation, namely

$$\frac{1}{\mu_f} = \frac{S_1}{\mu_1} + \frac{S_2}{\mu_2} \qquad (16.243)$$

This function along with the parameters in Table 16.1 completely specifies the porosity wave attenuation given by

$$a_\eta = \frac{1}{2} \frac{\eta \mu_f}{\rho_\eta K} \qquad (16.244)$$

A graph of this function is shown in Figure 16.1.

Table 16.1   Material and model parameters.

Quantity	Value	Units	Comment
ρ_s	2650.0	kg/m^3	
K_s	3.3×10^{10}	Pa	
μ_s	2.3×10^{10}	Pa	
ρ_1	1000.0	kg/m^3	
K_1	2.9×10^9	Pa	
μ_1	0.001	Pa s	1 cp
ξ_1	0.002	Pa s	
ρ_2	900.0	kg/m^3	
K_2	2.9×10^9	Pa	
μ_2	0.002	Pa s	
ξ_2	0.004	Pa s	
η	0.3		
S_{1r}	0.2		
S_{2r}	0.05		
K	9.87×10^{-13}	m^2	1 Darcy
α_η	0.85		
λ_s	0.9		
P_e	6894.8	Pa	1 psi
δ_e	0.01		

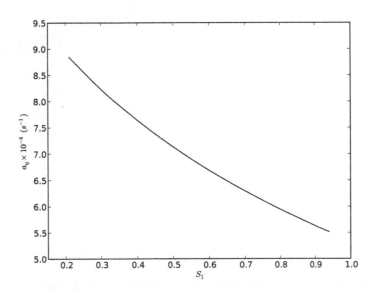

Figure 16.1   Porosity wave attenuation (a_η) versus saturation (S_1).

The undamped porosity wave speed $v_{o\eta}$ depends upon the value of μ_{Mss} which in turn depends upon the values of λ_{s1} and λ_{s2}. If one lets

$$\lambda_{s1} = S_1 \lambda_s \qquad \lambda_{s2} = S_2 \lambda_s \tag{16.245}$$

then

$$\begin{aligned} \mu_{Mss} &= \eta_s \mu_s (1 - \lambda_{s1} - \lambda_{s2}) \\ &= \eta_s \mu_s (1 - \lambda_s) \end{aligned} \tag{16.246}$$

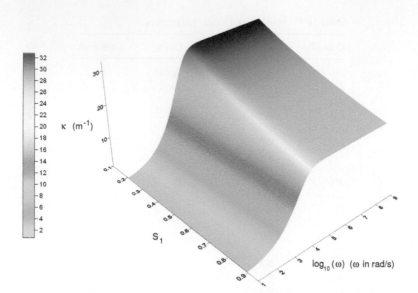

Figure 16.2   Porosity wave spatial attenuation (κ) versus saturation (S_1) and angular frequency ω.

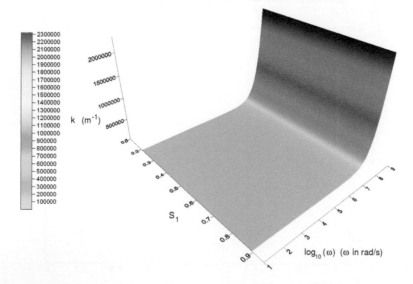

Figure 16.3   Porosity wave, wave number (k) versus saturation (S_1) and angular frequency ω.

By specifying λ_s as a constant, independent of saturation, then the undamped porosity wave speed is independent of saturation. Applying the parameter values given in Table 16.1 yields

$$v_{o\eta} = 2707.64 \, \text{m/s} \qquad (16.247)$$

Now that a_η and $v_{o\eta}$ have been specified, from Equations (16.129) to (16.131) one can determine the spatial attenuation κ, the wave number k, and the wave speed v of the trial solution as functions of saturation S_1 and angular frequency ω. A graph of κ is given in Figure 16.2, a graph of k is given in Figure 16.3, and finally a graph of v is given in Figure 16.4.

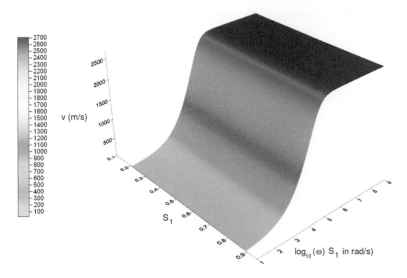

Figure 16.4   Porosity wave, wave speed (v) versus saturation (S_1) and angular frequency ω.

Now consider the comparison of porosity diffusion to the pressure diffusion process. Applying the parameters in Table 16.1 to the diffusion coefficient ratio, Equation (16.139), yields

$$\frac{D_\eta}{D_{P_f}} = 9.47 \qquad (16.248)$$

In this case the porosity diffusion process is clearly the preferred mechanism for the relaxation of pressure to an ambient level.

16.7.2   *The saturation wave*

For a numerical illustration of the saturation wave, one must specify the saturation dependence of the attenuation a_{S_1} and the unattenuated wave speed v_{oS_1}. Then the wave's spatial attenuation κ, wave number k, and wave speed v can be obtained from the dispersion relation solution given by Equations (16.164) to (16.166). Graphs of these functions may then be presented.

Now the expression for a_{S_1} is

$$a_{S_1} = \frac{A_{S_1}}{\rho_{S_1}} = \frac{1}{2} \frac{\eta}{S_1 \rho_2 + S_2 \rho_1} \frac{\mu_e}{K} \qquad (16.249)$$

where μ_e is given by Equation (16.158). So a_{S_1} depends on the relative Darcy resistance functions R_{ij}. To obtain these functions, consider the equations of motion for the two fluids in the case of steady state flow with the viscous terms considered negligible. In this case the equations of motion become

$$0 = -\eta_1 \nabla P_1 - \eta^2 \frac{\mu_1}{K} R_{11} \mathbf{v}_1 + \eta^2 \frac{\mu_1}{K} R_{12} \mathbf{v}_2 \qquad (16.250)$$

$$0 = -\eta_2 \nabla P_2 + \eta^2 \frac{\mu_2}{K} R_{21} \mathbf{v}_1 - \eta^2 \frac{\mu_2}{K} R_{22} \mathbf{v}_2 \qquad (16.251)$$

Solving this system of equations for the the volumetric flows rates $\mathbf{q}_1 = \eta_1 \mathbf{v}_1$ and $\mathbf{q}_2 = \eta_2 \mathbf{v}_2$ yields

$$\mathbf{q}_1 = -\frac{K}{\mu_1} S_1^2 J_{11} \nabla P_1 - \frac{K}{\mu_2} S_1 S_2 J_{12} \nabla P_2 \qquad (16.252)$$

$$\mathbf{q}_2 = -\frac{K}{\mu_1} S_1 S_2 J_{21} \nabla P_1 - \frac{K}{\mu_2} S_2^2 J_{22} \nabla P_2 \qquad (16.253)$$

where the relative Darcy conductivities J_{ij} are defined by

$$\begin{bmatrix} J_{11} & J_{12} \\ J_{21} & J_{22} \end{bmatrix} \begin{bmatrix} R_{11} & -R_{12} \\ -R_{21} & R_{22} \end{bmatrix} = \begin{bmatrix} 1 & 0 \\ 0 & 1 \end{bmatrix} \tag{16.254}$$

So if the J_{ij} are specified, then the functions R_{ij} can be computed using Equation (16.254). The solution for the volumetric flow rates $\mathbf{q}_1$ and $\mathbf{q}_2$ may be rewritten as

$$\mathbf{q}_1 = -\frac{K}{\mu_1} L_{11} \nabla P_1 - \frac{K}{\mu_1} L_{1c} \nabla P_{21} \tag{16.255}$$

$$\mathbf{q}_2 = -\frac{K}{\mu_2} L_{22} \nabla P_2 + \frac{K}{\mu_2} L_{2c} \nabla P_{21} \tag{16.256}$$

where the relative mobilities L_{ij} are given by

$$\begin{bmatrix} L_{11} & L_{1c} \\ L_{22} & L_{2c} \end{bmatrix} = \begin{bmatrix} S_1^2 J_{11} + \dfrac{\mu_1}{\mu_2} S_1 S_2 J_{12} & \dfrac{\mu_1}{\mu_2} S_1 S_2 J_{12} \\[2mm] \dfrac{\mu_2}{\mu_1} S_1 S_2 J_{21} + S_2^2 J_{22} & \dfrac{\mu_2}{\mu_1} S_1 S_2 J_{21} \end{bmatrix} \tag{16.257}$$

As discussed by Udey (2009), one should specify the J_{ij} functions to produce functions L_{11} and L_{22} that are similar to standard relative permeability curves K_{r1} and K_{r2}, i.e.

$$K_{r1} \approx L_{11} \qquad K_{r2} \approx L_{22} \tag{16.258}$$

Furthermore the cross terms L_{1c} and L_{2c} should be negligible at low and high saturations but significant at intermediate saturation (van Genabeek & Rothman 1996).

With the effective saturation S_e defined by

$$S_e(S_1) = \frac{S_1 - S_{1r}}{1 - S_{2r} - S_{1r}} \tag{16.259}$$

the J_{ij} functions are chosen to be (Udey 2009)

$$J_{11} = \frac{S_e}{(1 - S_{2r})^2} \tag{16.260}$$

$$J_{12} = S_e^2 (1 - S_e) \tag{16.261}$$

$$J_{21} = S_e (1 - S_e)^2 \tag{16.262}$$

$$J_{22} = \frac{1 - S_e}{(1 - S_{1r})^2} \tag{16.263}$$

A graph of these functions is shown in Figure 16.5. The relative mobility curves L_{ij} may now be found via Equation (16.257). A graph of the relative mobility curves is shown in Figure 16.6. Note that the curves for L_{11} and L_{22} are similar to standard relative permeability curves (Bear 1972, Spanos et al. 1988).

With the relative Darcy resistances R_{ij} specified by the relative Darcy conductivities J_{ij} via Equation (16.254), one may compute the effective viscosity μ_e as a function of saturation. In turn, this specifies the saturation wave attenuation a_{S_1}. A graph of this attenuation is shown in Figure 16.7.

The unattenuated wave speed v_{oS_1} may be computed from the expression

$$v_{oS_1}^2 = \frac{K_{S_1}}{\rho_{S_1}} = \frac{S_1 S_2 \beta_1}{S_1 \rho_2 + S_2 \rho_1} \tag{16.264}$$

where β_1 is given by

$$\beta_1 = \beta_c + \beta_h(S_1) \tag{16.265}$$

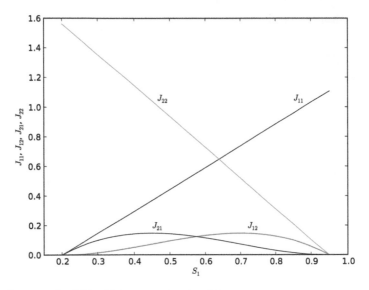

Figure 16.5 Darcy conductivities (J_{ij}) versus saturation (S_1).

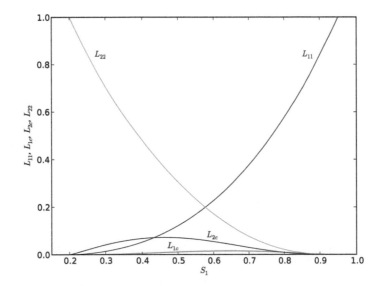

Figure 16.6 Relative mobilities (L_{ij}) versus saturation (S_1).

The megascopic capillary pressure function P_c will be taken to be the Brooks-Corey capillary function with parameter $\lambda = 2$ (Brooks & Corey 1964), namely

$$P_c = P_e \left(\frac{S_e + \delta_e}{1 + \delta_e} \right)^{-\frac{1}{2}} \tag{16.266}$$

Consequently, the function β_c is

$$\beta_c = -\frac{dP_c}{dS_1} = \frac{P_e}{2(1 - S_{2r} - S_{1r})} \left(\frac{S_e + \delta_e}{1 + \delta_e} \right)^{-\frac{3}{2}} \tag{16.267}$$

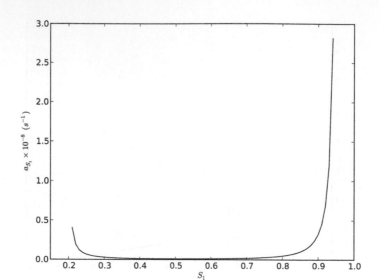

Figure 16.7   Saturation wave attenuation (a_{S_1}) versus saturation (S_1).

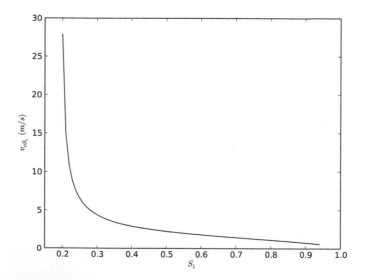

Figure 16.8   Saturation wave unattenuated wave speed (v_{oS_1}) versus saturation (S_1).

For our numerical illustration, capillary hysteresis effects will be considered negligible and one sets $\beta_h = 0$. Therefore $\beta_1 = \beta_c$ and one now has enough information to compute v_{oS_1}. A graph of the unattenuated wave speed is shown in Figure 16.8.

Since a_{S_1} and v_{oS_1} have been specified and graphed in Figures 16.7 and 16.8, one may now compute the curves for the spatial attenuation, wave number, and wave speed. The graph of the spatial attenuation κ is shown in Figure 16.9, the graph of the wave number k is shown in Figure 16.10, and finally the graph of the wave speed v is shown in Figure 16.11.

In Figure 16.11, a cross section of the wave speed v for a given angular frequency ω has the same general shape, or saturation dependence, as the unattenuated wave speed v_{oS_1} as shown in Figure 16.8. Then at all frequencies a saturation wave at low saturation travels faster than a wave at high saturation values. This demonstrates the dispersive nature of the saturation wave.

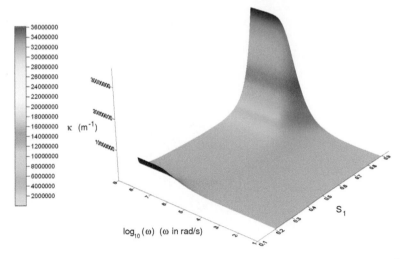

Figure 16.9   Saturation wave spatial attenuation (κ) versus saturation (S_1) and angular frequency ω.

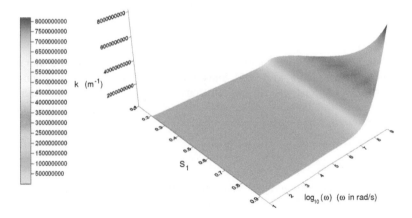

Figure 16.10   Saturation wave, wave number (k) versus saturation (S_1) and angular frequency ω.

16.8   CONCLUSION

The dilatational wave equations were derived for a porous medium consisting of a solid and two incompressible and immiscible fluids. The matrix wave operator representation of these equations is given by Equation (16.75). The solutions of this matrix equation are coupled porosity and saturation waves.

A homogeneous plane wave trial solution to the wave equations was utilized to generate a dispersion relation $\mathcal{D}(n, w) = 0$ whose polynomial structure is delineated in Equations (16.179) to (16.188). There are two solutions to this dispersion relation. In general the waves are strongly coupled and the dispersion relation must be solved numerically to examine the properties of the solutions. Solving the dispersion relation numerically is the subject of an ongoing research program.

Subject to two constraints, Equations (16.208) and (16.212), the dispersion relation was factored into a porosity wave solution and a saturation wave solution. This factored solution has minimal coupling between the waves. When the angular frequency is chosen as the independent

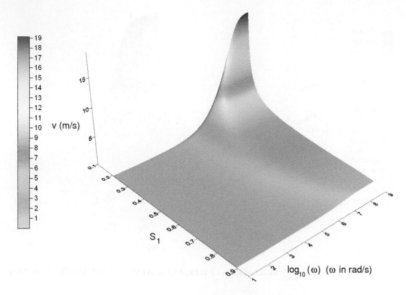

Figure 16.11   Saturation wave, wave Speed (v) versus saturation (S_1) and angular frequency ω.

variable, the solution of the dispersion relation for each of the waves could be found and has the form shown in Equations (16.96) to (16.99). Subsequently, graphs of the spatial attenuation, wave number, and wave speed for each wave as functions of angular frequency and saturation were presented in the numerical illustration. The graphs demonstrate that the solutions for the porosity wave and saturation wave are physically consistent.

REFERENCES

Bear, J. (1972) *Dynamics of Fluids in Porous Media*. New York, Dover Publications Inc.

Brooks, R.H. & Corey, A.T. (1964) *Hydraulic Properties of Porous Media*. Fort Collins, Colorado State University.

de la Cruz, V. & Spanos, T.J.T. (1985) Seismic wave propagation in a porous medium. *Geophysics*, 50 (10), 1556–1565.

de la Cruz, V., Spanos, T.J.T. & Yang, D. (1995) Macroscopic capillary pressure. *Transport in Porous Media*, 19, 67–77.

Dusseault, M.B., Shand, D., Meling, T., Spanos, T. & Davidson, B. (2002) Field applications of pressure pulsing in porous media. *Poromechanics II, Proceedings of the Second Biot Conference on Poromechanics, Grenoble, France, 26–28 August 2002*. Lisse/Abingdon/Exton (PA)/Tokyo, A.A Balkema

Geilikman, M.B., Spanos, T.J.T. & Nyland, E. (1993) Porosity Diffusion in fluid-saturated media. *Tectonophysics*, 217, 111–115.

van Genabeek, O. & Rothman, D.H. (1996) Macroscopic manifestations of microscopic flows through porous media: Phenomenology from simulation. *Annual Review of Earth and Planetary Sciences*, 24, 63–87.

Groenenboom, J., Wong, S., Meling, T., Zschuppe, R. & Davidson, B. (2003) Pulsed water injection during waterflooding. *Proceedings SPE International Improved Oil Recovery Conference in Asia Pacific, 20–21 October 2003, Kuala Lumpur, Malaysia*.

Hickey, C. (1994) *Mechanics of Porous Media*. Ph.D. Dissertation. Edmonton, University of Alberta.

Slattery, J.C. (1967) Flow of viscoelastic fluids through porous media. *AIChE Journal*, 13, 1066–1071.

Spanos, T.J.T., de la Cruz, V. & Hube, J. (1988) An analysis of the theoretical foundations of relative permeability curves. *AOSTRA Journal of Research*, 4 (3), 181–192.

Spanos, T.J.T. (2002) The thermophysics of porous media. *Monographs and Surveys in Pure and Applied Mathematics*, 126. Boca Raton, Chapman & Hall/CRC.

Spanos, T., Shand, D., Davidson, B., Dusseault, M. & Samaroo, M. (2003) Pressure pulsing at the reservoir scale: A new IOR approach. *Journal of Canadian Petroleum Technology*, 42 (2), 16–28

Whitaker, S. (1967) Diffusion and dispersion in porous media. *AIChE Journal* 13, 420–427.

Udey, N. (2009) Dispersion waves of two fluids in a porous medium. *Transport in Porous Media*, 79, 107–115.

Wang, J., Dusseault, M.B., Spanos, T.J.T. & Davidson, B. (1998) Fluid enhancement under liquid pressure pulsing at low frequency. *Proceedings 7th UNITAR International Conference on Heavy Crude and Tar Sands.* 27–28 October 1998, Beijing, China.

Zschuppe, R.P. (2001) *Pulse Flow Enhancement in Two-Phase Media.* M.Sc. Dissertation. Waterloo, University of Waterloo.

Subject index

335

Multiphysics Modeling

Series Editors: Jochen Bundschuh & Mario César Suárez Arriaga

ISSN:1877-0274

Publisher: CRC/Balkema, Taylor & Francis

1. Numerical Modeling of Coupled Phenomena in Science and Engineering:
   Practical Use and Examples
   Editors: M.C. Suárez Arriaga, J. Bundschuh & F.J. Domínguez-Mota
   2009
   ISBN: 978-0-415-47628-72.

2. Introduction to the Numerical Modeling of Groundwater and Geothermal Systems:
   Fundamentals of Mass, Energy and Solute Transport in Poroelastic Rocks
   J. Bundschuh & M.C. Suárez Arriaga
   2010
   ISBN: 978-0-415-40167-83.

3. Drilling and Completion in Petroleum Engineering: Theory and Numerical Applications
   Editors: Xinpu Shen, Mao Bai & William Standifird
   2011
   ISBN: 978-0-415-66527-8

4. Computational Modeling of Shallow Geothermal Systems
   Rafid Al-Khoury
   2011
   ISBN: 978-0-415-59627-5

5. Geochemical Modeling of Groundwater, Vadose and Geothermal Systems
   Editors: J. Bundschuh & M. Zilberbrand
   2011
   ISBN: 978-0-415-668101-1

Printed and bound by CPI Group (UK) Ltd, Croydon, CR0 4YY

18/10/2024

01776253-0005